Partial Differential Equations and Applications

Partial Differential Equations and Applications

A Bridge for Students and Researchers in Applied Sciences

Hong-Ming Yin
Department of Mathematics and Statistics
Washington State University
Pullman, WA, United States

Academic Press is an imprint of Elsevier
125 London Wall, London EC2Y 5AS, United Kingdom
525 B Street, Suite 1650, San Diego, CA 92101, United States
50 Hampshire Street, 5th Floor, Cambridge, MA 02139, United States
The Boulevard, Langford Lane, Kidlington, Oxford OX5 1GB, United Kingdom

Notices

Knowledge and best practice in this field are constantly changing. As new research and experience broaden our understanding, changes in research methods, professional practices, or medical treatment may become necessary.

Practitioners and researchers must always rely on their own experience and knowledge in evaluating and using any information, methods, compounds, or experiments described herein. In using such information or methods they should be mindful of their own safety and the safety of others, including parties for whom they have a professional responsibility.

To the fullest extent of the law, neither the Publisher nor the authors, contributors, or editors, assume any liability for any injury and/or damage to persons or property as a matter of products liability, negligence or otherwise, or from any use or operation of any methods, products, instructions, or ideas contained in the material herein.

ISBN: 978-0-443-18705-6

For information on all Academic Press publications
visit our website at https://www.elsevier.com/books-and-journals

Publisher: Katey Birtcher
Editorial Project Manager: Aleksandra Packowska
Publishing Services Manager: Shereen Jameel
Production Project Manager: Vishnu T. Jiji
Cover Designer: Greg Harris

Typeset by VTeX

Dedicated to my wife
Huimin Lin
for her love, patience, and understanding

Contents

Electronic instructor access available at https://educate.elsevier.com/9780443187056

Preface

Mathematical modeling and analysis is the core of applied mathematics. Ordinary Differential Equations (ODE) and Partial Differential Equations (PDE) are the main tools used to establish various continuous models in biological, ecological and health sciences, various engineering fields, and natural as well as physical sciences. Some continuous and stochastic models in modern economics, finance, and social sciences are also built up by employing ordinary differential and partial differential equations. Moreover, when a set of data is large, a continuous mathematical model is more appropriate than a discrete model. On the other hand, the theory of partial differential equations has its own intrinsic interest and it connects many research fields in mathematical sciences such as differential geometry, complex analysis, harmonic analysis, functional analysis, and many more. It is an emerging need for students and researchers to have some basic knowledge in partial differential equations and some advanced tools to deal with challenging problems arising from different fields.

This is the main motivation for the current textbook. During the past three decades I have taught an advanced course in applied mathematics and courses in partial differential equations at different levels. In addition to senior and graduate students in the mathematical sciences in my class, there was a large number of graduate students from the college of engineering as well as other departments with different majors who took these courses without an advanced mathematical background. For a long time I could not find a suitable textbook that could be used to satisfy the needs of both groups of students. I had to select different topics to motivate students to study the theory of partial differential equations and applications.

This textbook serves as a bridge for this group of students and researchers from different disciplines and backgrounds. Students do not need to have advanced mathematical knowledge (for example, of concepts such as Lebesgue measure theory and functional analysis) to understand the materials in this book. The book begins with elementary materials in partial differential equations, which covers the basic heat equation, wave equation, and the Laplace equation (Chapter 4 to Chapter 6 and partial materials in Chapters 7, 8, and 9). Then it moves onto the materials in research frontiers (several sections in Chapter 7 to Chapter 9). One of the distinguishing features of the book is that I emphasize the modeling process for every physical problem discussed in the book. I give a very detailed step-by-step process to build up each mathematical model, based on physical and experimental laws. Students and researchers can extend these ideas and tools to establish more complicated models in their own research fields.

I hope that students in mathematical-science majors can use the theory and methods in this book as a starting point to study partial differential equations at a more advanced theoretical level. On the other hand, for students and researchers in different fields, they can use the theory and techniques to investigate research topics in their own fields.

This textbook is based on the lectures I used for different courses at the University of Toronto, the University of Notre Dame, and Washington State University. The central theme of the text is to introduce basic concepts and techniques in partial differential equations for students without much advanced knowledge other than advanced calculus and elementary differential equations. An instructor can cover most materials in a one-semester course, with the option to cover more advanced materials in later courses as necessary.

Here is a brief of outline of the book:

In Chapter 1 we introduce basic concepts about partial differential equations. Many of these concepts are similar to those of ordinary differential equations. We then discuss the central questions of well-posedness and ill-posedness for a PDE problem. These questions are much more complicated than those encountered in ordinary differential equations. We present a number of counterexamples to illustrate why these questions are important and difficult to answer for partial differential equations. To motivate students' interest in the subject, we present many important examples arising in pure and applied fields, including fundamental systems in fluid mechanics, electromagnetic fields, material sciences, and biological, ecological, and health sciences. One section in this chapter is devoted to the classification of second-order PDEs, which explains why we only focus on three classes of PDEs.

In Chapter 2 we introduce Banach spaces and Hilbert spaces, which serves as either an introduction or a review for students with different levels of mathematical background. In this chapter we focus only on classical function spaces, describing other spaces (such as Sobolev spaces) in the Appendix. The purpose of these advanced mathematical concepts is to extend results from finite-dimensional vector spaces in linear algebra into infinite-dimensional function spaces. These ideas form the foundation for various methods and techniques we develop later in the book. One section of this chapter is devoted to classical Fourier series, which we use to find series solutions for various equations. In the rest of the chapter we introduce several methods that will be used to deal with solvability for a PDE problem. This includes a powerful contraction mapping principle. As an application, we show how the principle can be used to deal with systems of ODEs and integral equations.

In Chapter 3 we first study the eigenvalue problems in one dimension (the Sturm–Liouville theory). We then extend the idea into higher dimensions. In particular, we investigate under what conditions a set of functions can form a basis for a function space and how a function can be expressed in terms of basis functions. This generalizes the classical Fourier series in one dimension into higher dimensions.

Chapter 4 to Chapter 6 deal with three types of basic equations: the heat equation, the wave equation, and the Laplace equation. For each chapter, we begin with a detailed derivation of the model equation as well as initial and boundary conditions. Then, we find the series solution from one-space dimension to higher-space dimensions. We present many examples to illustrate basic methods and ideas for those PDE problems. The theoretical question about the well-posedness of the problem is investigated in the following sections. We also study the long-term behavior of solutions for the time-dependent problems. We deduce the precise decay rates for the solutions of the heat equation and the wave equation. The maximum principle is a powerful tool for investigating nonlinear equations usually not covered in a PDE course for undergraduate students; instead, we use an elementary approach to show the principle and then employ the principle to demonstrate several applications.

In Chapter 7 we introduce the Fourier-transform method. The method is a powerful way to deal with PDE problems in infinite domains. As an application, we use the method to find solutions for the Cauchy problem associated with each class of equations. We present a number of examples throughout this chapter. One of the applications is to derive the explicit formula for the price of a European call option. We also include a more advanced section designed for advanced graduate students in the mathematical sciences.

In Chapter 8 we introduce the fundamental solutions for the Laplace and the heat equations. We use an elementary methodology to derive the explicit solution representation for these equations in the whole space. We then use the fundamental solutions to construct Green's functions for both types of equations. Green's functions are important for many applications such as boundary-element methods in numerical analysis. The rest of the materials about finite-time blowup and extinction are more advanced, and are optional for instructors. These materials are usually covered in advanced PDE courses. We present some of the results in this chapter to show students complicated dynamics when dealing with nonlinear equations.

Chapter 9 deals with a system of first-order partial differential equations. Again, we use the transportation model and Maxwell's system as part of the motivation. A new method called the characteristic method is introduced to find a solution for the system of PDEs. We establish global existence and uniqueness based on the characteristic method.

Acknowledgments

This textbook is based on my lecture notes from the past several decades. I would like to thank my colleagues and students for their comments and suggestions—in particular, Mr. Guanjun Pan. I also thank my former advisors Professor Lishang Jiang of Peking University, the late Professor John R. Cannon of Washington State University, and Professor John Chadam of the University of Pittsburgh for their constant support and encouragement in my career development. My deep gratitude also goes to my research collaborators, including Professor Tang Q. Bao of the University of Graz, Professor Xinfu Chen at the University of Pittsburgh, Professor W. Fitzgibbon at the University of Houston, Professor Bei Hu at the University of Notre Dame, Professor Jin Liang at Tongji University, Professor Yanping Lin at the Polytechnic University of Hong Kong, Professor Jeffrey Morgan at the University of Houston, Professor William Rundell at Texas A& M University, Professor Ralph Showalter at Oregon State University, Professor Lihe Wang at the University of Iowa, Professor W. Wei at Guizhou University, and Professor Jun Zou at the Chinese University of Hong Kong.

Hong-Ming Yin
Pullman, United States

CHAPTER

1 Basics of partial differential equations

1.1 Introduction

Mathematical modeling and analysis plays a central role in scientific research and industrial applications. The basic approach in deriving a mathematical model is based on data-driven simulations, statistical patterns, and experimental and physical laws. Mathematical models can be broadly divided into discrete type and continuous type. When the data set for a problem is large, a continuous model for the problem is more appropriately used than a discrete version. These continuous models are typically governed by differential equations and systems. This is particularly true in biological and ecological systems, population dynamics, health and medical sciences, fluid mechanics, heat conduction, wave propagation, electromagnetism, and other physical sciences. More recently, continuous models are also used in modern economics, finance, and social sciences. We will see many examples throughout this book.

1.1.1 Basic conceptions of partial differential equations

An equation involving an unknown function of one variable and its derivative is called an ordinary differential equation (ODE). A partial differential equation (PDE) is an equation involving an unknown function of several variables and its partial derivatives. A system of ODEs or PDEs consists of more than one ODE or PDE. The highest order of a partial derivative in a PDE is called the order of the equation.

When a mathematical model governed by an ODE or PDE is established, one needs to find an unknown function (called the solution) and its properties. This leads to a general question of how to find a solution for an ODE or PDE problem and to analyze the solution of the problem. This book is devoted to this central question. We will introduce various methods and techniques of how to solve a PDE problem and to analyze the solution.

Suppose $u(x,t)$ is a function with two variables x and t. We denote by u_x and u_t, respectively, the partial derivatives with respect to x and t. Suppose a physical model can be described by an unknown function $u(x,t)$ that satisfies the following equation:

$$u_t(x,t) - u_x(x,t) = x^2 + t^2, \qquad 0 < x < 1, t \geq 0. \qquad (1.1.1)$$

Naturally, the goal is to find $u(x,t)$ that satisfies Eq. (1.1.1).

Partial Differential Equations and Applications. https://doi.org/10.1016/B978-0-44-318705-6.00007-0

Let us see some other classical examples:

$$u_t - k^2 u_{xx} + 3u = \sin(x+t), \qquad 0 < x < 1, t > 0, \tag{1.1.2}$$

$$u_{tt} - c^2 u_{xx} + u^3 = \cos t, \qquad x \in R^1, t > 0, \tag{1.1.3}$$

$$u_t + u u_x = t^2 e^x, \qquad x \in R^1, t > 0, \tag{1.1.4}$$

where k and c are positive constants.

By definition, Eq. (1.1.1) and Eq. (1.1.4) are first-order PDEs, while Eq. (1.1.2) and Eq. (1.1.3) are second-order PDEs. A solution of a PDE is a function $u(x,t)$ that satisfies the PDE in the classical sense. Sometimes, a solution $u(x,t)$ may not exist in the classical sense, such as a shock-wave solution. To understand such a solution, one needs to extend the concept of classical derivative, called a "weak derivative". The study for a generalized solution (weak solution) is one of the major topics in advanced PDE courses.

1.1.2 Linear and nonlinear, homogeneous, and nonhomogeneous PDEs

Similar to an ordinary differential equation (ODE), it is often convenient to use an operator notation to express a PDE. For example, we denote by L the following partial differential operator

$$L = \frac{\partial}{\partial t} - \frac{\partial}{\partial x},$$

then Eq. (1.1.1) can be written as

$$L[u] = x^2 + t^2, \qquad (x,t) \in (0,1) \times [0,\infty).$$

Eq. (1.1.2) can be expressed as

$$L[u] = \sin(x+t), \qquad 0 < x < 1, t > 0,$$

where

$$L = \frac{\partial}{\partial t} - k^2 \frac{\partial^2}{\partial x^2} + 3.$$

Definition 1.1.1. A PDE, $L[u] = f$, is said to be linear if for any smooth functions u_1, u_2 and any constants k_1, k_2, the following identity holds:

$$L[k_1 u_1 + k_2 u_2] = k_1 L[u_1] + k_2 L[u_2].$$

Otherwise, it is said to be a nonlinear PDE.

For example, Eq. (1.1.1) and Eq. (1.1.2) are linear equations, while Eq. (1.1.3) and Eq. (1.1.4) are nonlinear. A PDE can be written as

$$L[u] = f, \tag{1.1.5}$$

where f is a known function.

When $f = 0$, the PDE (1.1.5) is called homogeneous, otherwise it is called nonhomogeneous. The following proposition is the same as for an ordinary differential equation.

Superposition Property 1.1.1. *(a) Suppose* $\{u_k\}_{k=1}^m$ *are solutions of a linear equation*

$$L[u] = 0,$$

then any linear combination

$$u = \sum_{k=1}^{m} c_k u_k$$

is also a solution of the linear equation $L[u] = 0$*, where* $c_k, k = 1, 2, \cdots, m$*, are constants.*

(b) Suppose $\{u_k\}_{k=1}^m$ *is a basis of solutions for the homogeneous equation*

$$L[u] = 0,$$

and u_p *is a particular solution of the nonhomogeneous equation*

$$L[u] = f,$$

then

$$u = \sum_{k=1}^{m} c_k u_k + u_p$$

is the general solution of the equation $L[u] = f$*.*

Under suitable conditions, the superposition principle holds when $m = \infty$ in some sense. We will discuss this case in more detail in Chapter 2.

Throughout this book, we mainly focus on linear equations and systems. Some nonlinear PDEs are also included to illustrate the complicated dynamics of the system in comparison with the linear problem. When a PDE problem is nonlinear, the structure of the solution is much more complicated. The solution for a nonlinear PDE may produce new unexpected patterns or chaotic phenomena such as shock waves or blowup in finite time. Analysis for a nonlinear PDE often needs more advanced mathematical tools and theories, which is one of the main research topics in modern mathematical sciences.

1.2 Fundamental PDE questions: Well-posedness and ill-posedness

For a physical problem, in addition to a governing PDE or system, initial conditions, boundary conditions or initial and boundary conditions may need to be imposed along with the equation or system. Otherwise, one may find an infinite number of solutions. Therefore a complete PDE model for a physical problem may need to impose appropriate initial, boundary, or initial-boundary conditions. These conditions are often derived from a physical reality, such as Newton's cooling law in heat conduction. We will discuss a number of models arising from physical sciences and engineering on how an initial or initial-boundary conditions are imposed in later chapters.

As we will see many examples in Section 1.3, a partial differential equation often models certain physical phenomenon. The solution provides the essential information to understand the complicated dynamics, to explain experimental observations, and to explore new insights of physical phenomena. However, one needs to ask whether or not the model is appropriate or can be justified. How much tolerance (sensitivity) does a PDE model allow? These are fundamental questions before answering other questions. This leads to the so-called well-posedness or ill-posedness question for a PDE problem.

1.2.1 What is a well-posed or ill-posed PDE problem?

The first question one would ask is whether or not a PDE problem has a solution (*Existence Question or Solvability*). If a PDE model does not have a solution, then the model must be wrong or nonphysical. From the mathematical point of view, the solvability of a differential equation is interesting and has its own intrinsic value. One of Hilbert's questions raised in the world congress of Mathematicians in the 19th century is concerned with this problem. The Cauchy–Kovalevskaya theorem ([7]) and Levy's counterexample (see [22]) were motivated by Hilbert's question. There is a large branch of mathematical analysis devoted to the question (see [13] for example).

The second question is whether or not a PDE problem has one or many solutions (*Uniqueness Question*). When a PDE problem has more than one solution, then numerical computation often becomes unstable. A bifurcation may occur for the solution of an evolution system as time evolves.

The third question is whether or not the solution to a PDE problem has a small change when known data have small changes. This question is called the *Continuous Dependence*. This is important due to the fact that most data obtained from practical measurements or experiments have some errors. If the solution of a PDE problem has a drastic change when the known data have small changes, then one could not be sure if the model is suitable to describe the physical phenomenon. We will see that there are many PDE models that are not well-posed.

We summarize the above discussion to present the following definition.

Definition 1.2.1. A PDE problem is said to be well-posed, if the problem has a unique solution and the solution is continuously dependent upon the known data. Otherwise, the problem is said to be not well-posed or ill-posed.

One of the challenging questions is how to use the theory of PDEs to model a physical problem properly (modeling process). When a PDE model is established, it may be well-posed or ill-posed. This is one of the major research topics in applied mathematics. In this book we will discuss numerous examples to illustrate how a PDE model is built up and how these well-posedness questions are answered.

In addition to the question of well-posedness for a PDE problem, there are many other important questions one may want to answer. Here is a list of some sample questions:

(Q1) What is the behavior of the solution for a time-dependent problem as time evolves?
(Q2) What qualitative properties does the solution have?
(Q3) What is the numerical solution (numerical algorithms)?
(Q4) How can one find numerical solutions efficiently (convergence and convergence rate)?
(Q5) How can one use the solution to explain the physical phenomenon (data interpretation)?

One can ask many more questions such as free-boundary problems and inverse problems that are important in applications (see [4]). In the following subsection we give a few examples to show why the well-posedness for a PDE problem is important.

1.2.2 Some examples about the well-posedness question

In this subsection we give a few examples to illustrate the importance of well-posedness for a PDE problem.

Example 1.2.1. Consider the following PDE problem (a boundary value problem): Find $u(x, y)$ such that

$$u_{xx} + u_{yy} = 1, \qquad (x, y) \in B_1(0) := \{(x, y) \in R^2 : x^2 + y^2 < 1\}, \tag{1.2.1}$$

$$\nabla_\nu u(x, y) = 0, \qquad (x, y) \in \partial B_1(0) = \{(x, y) \in R^2 : x^2 + y^2 = 1\}, \tag{1.2.2}$$

where ν represents the outward unit normal on $\partial B_1(0)$ and $\nabla_\nu u := \nabla u \cdot \nu$ represents the normal derivative of u on $\partial B_1(0)$.

We claim that the PDE problem has no solution $u(x, y)$ satisfying Eq. (1.2.1)–(1.2.2). Indeed, if a solution $u(x, y)$ exists, integrating Eq. (1.2.1) over $B_1(0)$, we see

$$\int\int_{B_1(0)} [u_{xx} + u_{yy}]dxdy = \int\int_{B_1(0)} dxdy = \pi.$$

On the other hand, Gauss's Divergence theorem and the boundary condition (1.2.2) imply

$$\int\int_{B_1(0)}[u_{xx}+u_{yy}]dxdy=\int_{\partial B_1(0)}(\nabla_\nu u)ds=0,$$

i.e., a contradiction.

If we change the constant 1 in the right-hand side of Eq. (1.2.1) to 0, then there exists an infinite number of solutions for (1.2.1)–(1.2.2). In fact, any constant is a solution to (1.2.1)–(1.2.2). A more challenging question is whether or not there are other nonconstant solutions to Eq. (1.2.1)–(1.2.2). The short answer is negative, which will be investigated in Chapter 6. We will prove a general result that gives a completed answer for the Laplace equation.

Example 1.2.2. Consider the following initial-value problem: Find $u(x, y)$ such that

$$u_{xx}+u_{yy}=0, \qquad -\infty<x<\infty,\ y>0, \tag{1.2.3}$$

$$u(x,0)=0,\ u_y(x,0)=0, \qquad -\infty<x<\infty. \tag{1.2.4}$$

It can be shown (see Chapter 7) that $u(x, y) = 0$, $(x, y) \in R \times R^+$ is the unique solution. However, if we make a small perturbation for the initial data, say,

$$u_y(x,0)=\frac{sin(nx)}{n},$$

which converges to 0 uniformly on R^1 as $n \to \infty$, then a unique solution of Eq. (1.2.3)–(1.2.4) can be found as follows:

$$u(x,y)=\frac{sin(nx)(e^{ny}-e^{-ny})}{2n^2},$$

which is unbounded in $R^1 \times R^+$. This implies that the PDE problem (1.2.3)–(1.2.4) does not continuously depend on the initial data.

Example 1.2.3. Consider the following initial-value problem: Find $u(x, t)$ such that

$$u_t-u_{xx}=0, \qquad -\infty<x<\infty, t>0, \tag{1.2.5}$$

$$u(x,0)=0, \qquad -\infty<x<\infty. \tag{1.2.6}$$

It is clear that $u(x, t) = 0$ is a solution. It is natural to ask if the problem has a nontrivial solution.

Define

$$\phi(t)=\begin{cases} e^{-\frac{1}{t^2}}, & \text{if } t\neq 0,\\ 0, & \text{if } t=0.\end{cases}$$

Then, $\phi(t) \in C^\infty(R^1)$. Define

$$u(x,t) = \begin{cases} \sum_{n=0}^{\infty} \frac{d^n}{dt^n}\phi(t)\frac{x^{2n}}{(2n)!}, & \text{if } x \in R^1, t > 0 \\ 0, & \text{if } t = 0. \end{cases}$$

It is easy to verify that

$$u_t - u_{xx} = 0, \qquad x \in R^1, t > 0.$$

It follows that the solution for Eq. (1.2.5)–(1.2.6) is not unique.

Next, we give a few examples to show some physical phenomena that are very different from a linear PDE.

Example 1.2.4. Consider the following boundary value problem:

$$\begin{aligned} u_{xx} + u_{yy} &= u(1-u), && (x,y) \in B_1(0) = \{(x,y) \in R^2 : x^2 + y^2 < 1\}, \\ \nabla_\nu u(x,y) &= 0, && (x,y) \in \partial B_1(0) = \{(x,y) \in R^2 : x^2 + y^2 = 1\}. \end{aligned}$$

It is clear that $u_1(x,y) = 0$ and $u_2(x,y) = 1$ are two solutions. Does the problem have a nonconstant solution? The answer will be given in Chapter 6 (see Example 6.5.1).

Here is another interesting example.

Example 1.2.5. Consider the nonlinear problem:

$$\begin{aligned} -\Delta u &= u^p (\text{or } \delta e^u), && x \in \Omega, \\ u(x) &= 0, && x \in \partial\Omega, \end{aligned}$$

where $p > 1$ and Ω is a domain in R^n.

Clearly, $u(x) = 0$ is a solution to the above nonlinear problem. Does the problem have a nontrivial solution? To answer this question, one needs much deeper analysis. It turns out that the question depends on p, the dimension n, and the geometry of Ω.

If Ω is bounded and

$$1 < p < \infty, \text{ for } n = 1, 2; \; 1 < p < \frac{n+2}{n-2}, \text{ for } n \geq 3,$$

then the nonlinear problem has a nontrivial solution. On the other hand, when

$$p > \frac{n+2}{n-2}, \qquad n \geq 3$$

and Ω is a star-shaped domain, then the nonlinear problem has only the trivial solution (see [8]).

For $p = \frac{n+2}{n-2}$, the existence of a nontrivial solution depends on the geometry of Ω (called the Yamabe problem in differential geometry, see [33]).

If one replaces the nonlinear term u^p by δe^u with some $\delta > 0$, then there are more interesting cases for the existence of multiple solutions. It depends on the dimension and size of the parameter δ ([3]).

Example 1.2.6. (Finite-time blowup) Consider the following nonlinear heat equation: Let $p > 0$. Find $u(x,t)$ such that

$$u_t - \Delta u = u^p, \qquad x \in R^n, t > 0, \tag{1.2.7}$$

$$u(x,0) = f(x) \geq 0 \qquad x \in R^n. \tag{1.2.8}$$

It will be shown (see Chapter 8) that if $1 < p < 1 + \frac{2}{n}$ there exists a constant $T^* > 0$ such that $u(x,t)$ becomes unbounded as $t \to T^*-$ for any $f(x) \geq 0$ with

$$\int_{R^n} f(x)dx > 0.$$

On the other hand, if $p \geq 1 + \frac{2}{n}$, the unique solution $u(x,t)$ exists globally if $f(x)$ is sufficiently small and blows up in finite time if $f(x)$ is suitably large.

Example 1.2.7. Consider the following nonlinear wave equation: Let p be a positive integer. Find $u(x,t)$ such that

$$u_{tt} - \Delta u = -u^p, \qquad x \in R^n, t > 0, \tag{1.2.9}$$

$$u(x,0) = f(x), u_t(x,0) = g(x) \qquad x \in R^n. \tag{1.2.10}$$

It will be seen in Chapter 8 that the global existence or finite-time blowup to the problem (1.2.9)–(1.2.10) depends on the dimension n and p (see [30]).

Example 1.2.8. (Shock-wave phenomenon) Consider the following initial-value problem: Find $u(x,t)$ such that

$$u_t + uu_x = 0, \qquad -\infty < x < \infty, t > 0, \tag{1.2.11}$$

$$u(x,0) = f(x), \qquad -\infty < x < \infty, \tag{1.2.12}$$

where

$$f(x) = \begin{cases} 1, & \text{if } x < 0 \\ 0, & \text{if } x \geq 0. \end{cases}$$

It will be shown in Chapter 9 that $u(x,t) = 1$ is a solution for any $x < 0$ near $t = 0$, while $u(x,t) = 0$ for any $x > 0$ with sufficiently small t. However, near $(0,0)$ there is no solution in the classical sense. Nevertheless, there is a solution in the weak sense (in Sobolev space). The most interesting observation is that the solution forms a shock wave along the line $x = \frac{t}{2}$. We will discuss this class of equations in Chapter 9. Example 1.2.8 shows why a nonlinear problem is much harder to analyze for its solution and why one needs to extend the definition of a solution in the weak sense.

For a general partial differential equation, it is hard to answer the well-posedness question. This is even more difficult when a PDE problem is nonlinear. However, there is a very interesting result such as the Cauchy–Kovalevskaya theorem when the coefficients of a PDE are analytic. The interested reader may find this theorem in an advanced PDE book such as [13].

1.3 Some important equations and systems

In this section we introduce some important equations and systems in applied sciences. These equations and systems are derived from various physical models in natural and physical sciences as well as modern economics, finance, and social sciences. In subsequent chapters, we will study some examples in detail on how to use basic physical and experimental laws to derive various equations and systems and to find explicit solutions for these model problems.

1.3.1 Some model equations in the physical sciences

Example 1.3.1. The heat equation and chemical reaction–diffusion equations.

This example models the heat conduction in a medium occupying a region $\Omega \subset R^n$. Let $u(x,t)$ represent the temperature at position $x \in \Omega$ and time $t \geq 0$. Then, $u(x,t)$ satisfies the following heat equation:

$$u_t - k^2 \Delta u = 0, \qquad x \in \Omega, t > 0, \tag{1.3.1}$$

where

$$\Delta := \sum_{i=1}^{n} \frac{\partial^2}{\partial x_i}, \qquad \text{called the Laplace operator,}$$

where $k > 0$ is the thermal conductivity. A detailed derivation of the physical model will be given in Chapter 4.

Eq. (1.3.1) and its generalization are also used for a very different class of physical models arising from chemical reaction–diffusion processes. Suppose there are m types of chemical substances with concentration $u_i(x,t)$ of the ith chemical substance and $f_i(x,t,u_1,\cdots,u_m)$ represents the output of chemical reaction for the ith component with other chemical substances. Then, u_i satisfies

$$u_{it} - a_i^2 \Delta u_i = f_i(x,t,u_1,\cdots,u_m), \qquad i = 1,\cdots,m, \tag{1.3.2}$$

where $a_i > 0$, $i = 1,\cdots,m$ represents the diffusion coefficient.

There are many other types of physical problems that can be modeled by this type of equation and system. This type of PDE is called parabolic. The precise definition will be given in Section 1.4.

Example 1.3.2. Equation of wave propagation.

Let $u(x,t)$ represent the strength of a sound (pressure) at position $x = (x_1, x_2, x_3) \in \Omega \subset R^3$ and time t. When the sound propagates in a region, $u(x,t)$ satisfies the following equation:

$$u_{tt} - c^2 \Delta u = 0, \qquad x \in \Omega, t > 0, \tag{1.3.3}$$

where c represents the propagation speed of the sound.

A similar equation that describes the motion of a soliton in one-space dimension can be expressed in the following form

$$\psi_{tt} - c^2\psi_{xx} + \sin\psi = 0.$$

This is called the *sine-Gordon* equation in physics.

This class of PDEs is called hyperbolic.

Example 1.3.3. The Laplace equation and Cauchy–Riemann equation.

For a static electric field, the electric field can be expressed by a gradient of potential function $u(x)$:

$$\mathbf{E} = -\nabla u.$$

Gauss's law yields

$$div\mathbf{E} = -\nabla \cdot (\nabla u) = -\Delta u = \rho(x), \tag{1.3.4}$$

where ρ represents the charge density.

Eq. (1.3.4) is called a Poisson equation. When $\rho(x) = 0$, it is called the Laplace equation and a solution of the Laplace equation is called a *harmonic function*.

Another motivation for the Laplace equation (1.3.4) comes from the minimization of kinetic energy. Let $v(x)$ represent a physical quantity such as momentum or pressure in a region $\Omega \subset R^n$. The kinetic energy can be expressed by

$$E(v) := \frac{1}{2}\int_\Omega |\nabla v|^2 dx.$$

A fundamental problem in physics is to find a function that minimizes the energy functional $E(v)$.

It will be seen in Chapter 6 that the minimum of the energy functional $E(v)$ is the solution of the Laplace equation:

$$\Delta u = 0. \tag{1.3.5}$$

In complex analysis, a fundamental problem is to determine whether or not a complex function $f(z)$ is analytic in a domain. It turns out that this problem is equivalent to the following Cauchy–Riemann equation:

$$\frac{\partial f(z)}{\partial \bar{z}} = 0,$$

where $\bar{z} = x - iy$ is the complex conjugate of a complex variable $z = x + iy$.

If we separate $f(z)$ into the real and imaginary parts:

$$f(z) = Ref(z) + i(Imf(z)) := u(x, y) + iv(x, y),$$

then,

$$\frac{\partial f(z)}{\partial \bar{z}} = (\frac{\partial}{\partial x} - i\frac{\partial}{\partial y})(u(x, y) + iv(x, y)) = 0,$$

which is equivalent to the Laplace equation:

$$\Delta u = u_{xx} + u_{yy} = 0; \qquad \Delta v = v_{xx} + v_{yy} = 0.$$

This implies that a complex function $f(z)$ is analytic if and only if both real $Ref(z)$ and imaginary $Imf(z)$ are harmonic functions in the domain.

Example 1.3.4. Minimum surface equation.

In differential geometry, an interesting problem is to find a minimum surface in a closed rope. It is also the classical model for a soap bubble in mechanics.

Let $u(x)$ be the displacement of a surface at point $x = (x_1, x_2, \cdots, x_n) \in \Omega \subset R^n$. Then, $u(x)$ satisfies

$$\nabla(\frac{\nabla u}{\sqrt{1 + |\nabla u|^2}}) = 0, \qquad x \in \Omega. \tag{1.3.6}$$

Eq. (1.3.6) is nonlinear and the coefficient approaches 0 if $|\nabla u|$ approaches ∞. On the other hand, if $|\nabla u|$ is small, then Eq. (1.3.6) can be approximated by the Laplace equation (1.3.5).

In a more general setting, this type of PDE is called elliptic. We will give a precise classification of a PDE in two variables with constant coefficients in Section 1.4. In this book, we mainly study these three types of partial differential equations.

1.3.2 Some PDEs arising from various engineering fields

Example 1.3.5. The fundamental equation in quantum mechanics (Schrödinger's equation).

In quantum mechanics, the motion of a particle is described by a wave function $u(x, t)$ that satisfies the following fundamental equation:

$$ihu_t + \frac{h^2}{2m}\Delta u + v(x, t)u = 0,$$

where i represents the complex unit, h and m are the Planck constant and mass of the particles, respectively, while $v(x, t)$ is the potential function in a field.

It is very difficult to analyze the motion of a single particle. In reality, one often studies the motion for a large number of particles in a macroscopic way. This leads to a general transport equation that will be investigated in Chapter 9.

Example 1.3.6. The fundamental equations in electromagnetic fields (Maxwell's equations).

In electromagnetic fields, the interactions of electric and magnetic fields are governed by Maxwell's equations. Let $\mathbf{E}(x, t)$ and $\mathbf{H}(x, t)$ be the electric and magnetic fields at position $x \in \Omega \subset R^3$ and time t, where the bold letter represents a vector in R^3. Then, $\mathbf{E}(x,t)$ and $\mathbf{H}(x,t)$ satisfy the following system:

$$\varepsilon \mathbf{E}_t - \nabla \times \mathbf{H} = \mathbf{J}(x,t), \qquad x \in \Omega, t > 0, \tag{1.3.7}$$

$$\mu \mathbf{H}_t + \nabla \times \mathbf{E} = 0, \qquad x \in \Omega, t > 0, \tag{1.3.8}$$

where $\mathbf{J}(x,t)$ represents the known internal current, the physical parameters ε and μ are the electric permittivity and magnetic permeability.

Maxwell's equations play an essential role in electric engineering. Modern telecommunication equipment and many advanced device designs are based on the solution of Maxwell's equations.

Example 1.3.7. Fundamental equations in fluid mechanics (Euler's equations and Navier–Stokes equation).

In fluid mechanics, a challenging problem is to understand the motion of the fluid in a medium. Fluids are usually divided into compressible and incompressible. Let $\mathbf{v}(x,t)$ represent the velocity field for a fluid. Let ρ be the density of a fluid. The principle of mass conservation implies that

$$\rho_t + \nabla(\rho \mathbf{v}) = 0.$$

The principle of conservation of momentum yields

$$\rho[\mathbf{v}_t + (\mathbf{v} \cdot \nabla)\mathbf{v}] = -\nabla p + \mathbf{F}(x,t), \qquad x \in \Omega, t > 0,$$

where p represents the pressure and $\mathbf{F}$ is the external force, such as gravity.

For a compressible fluid such as natural gas, the pressure p is a function of ρ:

$$p = p(\rho).$$

The equations for $(\rho, \mathbf{v})$ are called Euler's equations.

For an incompressible fluid such as water, the density ρ is a constant. Moreover, there is a viscous force that can be expressed by $\Delta \mathbf{v}$. Then, $\mathbf{v}$ and $p(x,t)$ satisfy

$$\mathbf{v}_t - a\Delta \mathbf{v} + (\mathbf{v} \cdot \nabla)\mathbf{v} + \nabla p = \mathbf{F}(x,t), \qquad x \in \Omega, t > 0, \tag{1.3.9}$$

$$\nabla \cdot \mathbf{v} = 0, \qquad x \in \Omega, t > 0, \tag{1.3.10}$$

where $\mathbf{F}(x,t)$ represents the external force acting on the fluid. The constant $a > 0$ is the viscosity coefficient.

Both Euler's and Navier–Stokes equations are nonlinear. The global existence of a solution for the Navier–Stokes equation is still an open problem up to now. It is one of the millennium problems proposed by the Clay Mathematical Research Institute with a one million dollar prize if the problem is solved.

There are many other important equations in modern industries. An interesting example is the semiconductor equations that describe the ion transportation in electronic devices. We omit these equations here. The interested reader may find books dealing with these topics for device modeling and simulations.

1.3.3 Some model equations in biological and life sciences

In biological, ecological, health, and life sciences, there are many problems that can be modeled by partial differential equations. We give two examples here (also see [35] for an infectious disease model).

Example 1.3.8. Population growth model.

Suppose $u(x,t)$ is the population density. Due to the movement of the general population, a diffusion process occurs in which the flux is proportional to the gradient of the density (similar to Fourier's law). Then, the empirical data show that $u(x,t)$ satisfies the logarithmic growth model:

$$u_t - a^2 \Delta u = r(x,t)u(1 - \frac{u}{K}) - d(x,t)u, \tag{1.3.11}$$

where $r(x,t) > 0$ and $d(x,t)$ represent the combined growth and death rates, respectively. The constants a and K represent the diffusion coefficient and the maximum capacity of the population in the region.

Example 1.3.9. The predator–prey model.

In an ecological system, the balance between predator and prey is very important in order to avoid extinction of a species. Suppose $u(x,t)$ and $v(x,t)$ are the population densities of predator and prey, respectively. Then, u and v satisfy the modified population growth model, with an additional term due to an abundant prey population:

$$u_t - a\Delta u = r_1 u(1 - \frac{u}{K_1}) + \beta uv - d_1 u, \tag{1.3.12}$$

$$v_t - b\Delta v = r_2 v(1 - \frac{v}{K_2}) - \beta uv - d_2 v, \tag{1.3.13}$$

where $\beta > 0$ measures the additional growth due to the prey population.

There is an interesting dynamic when various parameters vary as time evolves.

1.3.4 Some PDEs arising in economics and finance

In modern economics, the PDE is used to price many types of financial assets, such as bonds and stock options. The model equation is typically given by a stochastic differential equation. By using the capital-asset pricing principle, it turns out that the equation is equivalent to a parabolic type of equation. Here we give two examples.

Example 1.3.10. Interest-rate model and bond-pricing model.

The interest rate is the most important factor in the world economy. There are many different mathematical models for an interest rate $r(t)$ in economics and finance. The uncertainty is that the interest rate is unknown in the future. Most interest-rate models are given by a stochastic form:

$$dr(t) = r_0 dt + \sigma dW(t), \tag{1.3.14}$$

where r_0 is a long-term interest rate given by the Federal Reserve Open Board Committee in the United States, $\sigma > 0$ is called the volatility and $W(t)$ is the standard Wiener process (Brownian motion).

Once $r(t)$ is determined, we can model a bond price. Let $B(s,t)$ represent the bond price for a Company's security s at time t. Under certain assumptions, $B(s,t)$ satisfies the following equation:

$$B_t + \frac{\sigma^2}{2} s^2 B_{ss} + rsB_s - rB = f(s,t), \tag{1.3.15}$$

where σ is the volatility of the bond price movement and $f(s,t)$ represents the cash flow produced by the company.

Throughout the history of the financial world, a financial crisis in a country is often induced by a sudden change of the interest rate $r(t)$. The reason is that a sudden change of $r(t)$ may drastically affect the price of a bond, which will affect the currency and security prices sharply, causing a financial crisis in a country.

Example 1.3.11. The fundamental equation of the option pricing model (Black–Scholes equation).

In finance engineering, a fundamental assumption is that a stock price follows the geometric Brownian motion. If $S(t)$ represents the stock price at time t, then $S(t)$ satisfies

$$\frac{dS(t)}{S(t)} = \mu dt + \sigma dW(t), \tag{1.3.16}$$

where μ represents the expected return rate and σ is the stock volatility.

Let $u(s,t)$ represent the option price for a stock s at time t. Then, $u(s,t)$ satisfies

$$u_t + \frac{1}{2}\sigma^2 s^2 u_{ss} + rsu_s - ru = 0, \qquad s \geq 0, t \geq 0, \tag{1.3.17}$$

where r represents the interest rate, σ is the volatility of the stock s. When σ changes dramatically, the option price will suffer a huge swing. The mathematical explanation is that a solution of Eq. (1.3.17) is very sensitive to the change of σ.

It can be shown by using suitable transforms that both models in Examples 1.3.10 and 1.3.11 are similar to the heat equation (1.3.1). However, unlike heat conduction, the models in economy and finance may not be valid in a certain time period when there is a disruptive event occurring (called the Black-Swan phenomenon). There are too many factors that can cause a drastic change that in turn the mathematical model for a bond or a stock price fails. However, the empirical financial data show that the model does hold over a long-term time period.

1.4 Classification of second-order partial differential equations

In this section we classify a general class of second-order partial differential equation with constant coefficients. These second-order PDEs can be broadly divided into three classes of canonical equations. There are some features and special mathematical tools for each type of equation. Moreover, each type of equation represents a large class of physical models. By introducing suitable new variables, we will see that second-order PDEs can be classified as one of the following canonical forms: the heat equation, wave equation or Laplace equation.

1.4.1 Classification of a second-order PDE with constant coefficients

To illustrate the basic idea, we begin with a second-order linear PDE with two variables.

Let $u(x, y)$ satisfy the following general second-order linear partial differential equation:

$$L[u] := au_{xx} + 2bu_{xy} + cu_{yy} + d_1u_x + d_2u_y + d_3u = f(x, y), \qquad (1.4.1)$$

where all coefficients in Eq. (1.4.1) are constants and $f(x, y)$ is a general function of (x, y).

Define a quadratic form

$$F(\xi, \eta) = a\xi^2 + 2b\eta\xi + c\eta^2.$$

The corresponding matrix, denoted by A, of the quadratic form is equal to

$$A = \begin{pmatrix} a & b \\ b & c \end{pmatrix}.$$

For the matrix A, the characteristic equation is equal to

$$p(\lambda) := |A - \lambda I| = \lambda^2 - (a + c)\lambda - (b^2 - ac) = 0.$$

Since

$$D := (a + c)^2 + 4(b^2 - ac) = (a - c)^2 + 4b^2 \geq 0,$$

there exist two real eigenvalues, λ_1 and λ_2. Moreover,

$$\lambda_1 + \lambda_2 = a + c, \qquad \lambda_1\lambda_2 = ac - b^2.$$

Case 1: $ac - b^2 > 0$.

Then, there exist two eigenvalues (may be repeated) with the same sign. We assume both eigenvalues are positive. In this case, for each of the real roots λ_1 and λ_2, we can find the corresponding eigenvectors, denoted by $\mathbf{v}_1$ and $\mathbf{v}_2$, respectively. Since A is symmetric, then the two eigenvectors $\mathbf{v}_1$ and $\mathbf{v}_2$ can be chosen to be orthonormal.

Let

$$\mathbf{v}_1 = (v_{11}, v_{12})^T, \mathbf{v}_2 = (v_{21}, v_{22})^T,$$

where the exponent T means the transpose of a vector.

Now, we introduce new variables

$$x = v_{11}\xi + v_{12}\eta,$$
$$y = v_{21}\xi + v_{22}\eta.$$

Define

$$v(\xi, \eta) := u(x, y) = u(v_{11}\xi + v_{12}\eta, v_{21}\xi + v_{22}\eta).$$

By the chain rule, we see $v(\xi, \eta)$ satisfies

$$\lambda_1 v_{\xi\xi} + \lambda_2 v_{\eta\eta} + \text{lower-order terms} = f.$$

We further introduce a new variable

$$\bar{\xi} = \frac{\xi}{\sqrt{\lambda_1}}, \bar{\eta} = \frac{\eta}{\sqrt{\lambda_2}}.$$

A simple calculation implies that $w(\bar{\xi}, \bar{\eta}) := v(\xi, \eta)$ satisfies the following equation:

$$w_{\bar{\xi}\bar{\xi}} + w_{\bar{\eta}\bar{\eta}} + \text{lower-order terms} = f(\bar{\xi}, \bar{\eta}). \tag{1.4.2}$$

Eq. (1.4.2) is an elliptic type of equation.

Case 2: $ac - b^2 < 0$.

In this case, the characteristic equation has two distinct eigenvalues with opposite sign, say, $\lambda_1 > 0$ and $\lambda_2 < 0$. By a similar approach, we can transform Eq. (1.4.1) into

$$w_{\bar{\xi}\bar{\xi}} - w_{\bar{\eta}\bar{\eta}} + \text{lower-order terms} = f. \tag{1.4.3}$$

Eq. (1.4.3) is the typical form of a wave equation if we use the familiar variables $t = \bar{\xi}$ and $x = \bar{\eta}$.

Case 3: $ac - b^2 = 0$.

In this case, at least one of the coefficients a and c is not equal to 0. We assume $a > 0$.

We can follow the same technique to transform Eq. (1.4.1) into the following equation:

$$w_{\xi\xi} + (d_1^* w_\xi + d_2^* w_\eta + d_3^* w) = f^*(\xi, \eta). \tag{1.4.4}$$

The coefficient d_2^* must be a nonzero constant. Otherwise, Eq. (1.4.4) contains only a partial derivative with respect to ξ, which is an ordinary differential equation.

Since $d_2^* \neq 0$, we can further introduce a new variable (say, $\bar{\eta} = -\frac{\eta}{d_2^*}$) to reduce the above equation to the following form:

$$w_{\bar{\eta}} - w_{\xi\xi} - (d_1^* w_\xi + d_3^* w) = -f(\xi, \bar{\eta})). \tag{1.4.5}$$

This is the classical heat equation if we consider η as the time variable t.

When the coefficients in Eq. (1.4.1) are continuous functions of x and y, we consider Eq. (1.4.1) in a neighborhood $B_r(P_0)$ of a point $P(x_0, y_0)$:

(a) if $a(x, y)c(x, y) - b^2(x, y) > 0$ in $B_r(P_0)$, Eq. (1.4.1) is called an elliptic equation;

(b) if $a(x, y)c(x, y) - b^2(x, y) < 0$ in $B_r(P_0)$, Eq. (1.4.1) is called a hyperbolic equation;

(c) If $a(x, y)c(x, y) - b^2(x, y) = 0$ in $B_r(P_0)$, Eq. (1.4.1) is called a parabolic equation.

If in a region $\Omega \in R^2$, Eq. (1.4.1) is elliptic in one subregion while hyperbolic or parabolic in another subregion, then we say Eq. (1.4.1) is of mixed type in Ω. An interesting example is the Tricomi equation:

$$u_{xx} + xu_{yy} = 0, \qquad (x, y) \in \Omega = \{(x, y) : -\infty < x < \infty, y > 0\},$$

which is elliptic if $x > 0$ and hyperbolic if $x < 0$.

1.4.2 Classification of general second-order linear PDEs

Now, we extend the idea in the above subsection to classify a general second-order linear PDE in n dimensions. For convenience, a point in R^n is denoted by $x = (x_1, \cdots, x_n)$.

$$L[u] := \sum_{i,j=1}^{n} a_{ij}(x)u_{x_i x_j} + \sum_{i=1}^{n} b_i(x)u_{x_i} + c(x)u = f(x),$$

where $a_{ij}(x)$, $b_i(x)$, and $c(x)$ are known coefficients.

We may assume

$$a_{ij}(x) = a_{ji}(x), \qquad i, j = 1, \cdots, n.$$

Otherwise, we may define

$$a_{ij}^* = \frac{a_{ij} + a_{ji}}{2},$$

then,

$$\sum_{i,j=1}^{n} a_{ij}^* u_{x_i x_j} = \sum_{i,j=1}^{n} a_{ij} u_{x_i x_j},$$

hence, the operator L is the same as before.

Introduce a quadratic form

$$Q(\xi) := \sum_{i,j=1}^{n} a_{ij}\xi_i\xi_j.$$

The corresponding matrix of the quadratic form is

$$A = (a_{ij})_{n\times n}.$$

Since A is symmetric, the corresponding characteristic equation

$$p(\lambda) := |A - \lambda I| = 0$$

has n real eigenvalues and n linearly independent orthonormal eigenvectors $\mathbf{v_1}, \cdots, \mathbf{v_n}$. Let V be the matrix with the ith-column vector $\mathbf{v_i}$, $i = 1, 2, \cdots, n$. From the theory of linear algebra, we can have the following cases:

Case 1: All eigenvalues are positive (or all negative).

From linear algebra, one can introduce new variables

$$\eta = V\xi,$$

to transform the quadratic form $Q(\xi)$ into the following canonical form:

$$Q(\xi) = \eta_1^2 + \eta_2^2 + \cdots + \eta_n^2.$$

Under the new variables, the corresponding PDE becomes

$$u_{\eta_1\eta_1} + \cdots + u_{\eta_n\eta_n} + \text{lower-order terms} = f,$$

which is a typical elliptic equation.

Case 2: There exists one eigenvalue that has the opposite sign to the rest of the eigenvalues.

We assume there exists one eigenvalue that has the opposite sign to the rest of the eigenvalues. For this case, we use the same technique to obtain the following canonical form

$$Q(\xi) = \eta_1^2 - (\eta_2^2 + \cdots + \eta_n^2).$$

Under the new variables with $t = \eta_1$, $y_k = \eta_k$, $k = 2, \cdots, n$, the corresponding PDE becomes

$$u_{tt} - (u_{y_2y_2} + \cdots + u_{y_ny_n}) + \text{lower order terms} = f,$$

which is a typical hyperbolic equation.

When there exists more than one eigenvalue with a different sign to the rest, we still call this type of equation a hyperbolic type.

Case 3: There exists at least one eigenvalue that is equal to 0.

In this case, just like the case with two variables we again can introduce new variables such that the PDE becomes

$$u_t - (u_{y_2 y_2} + \cdots + u_{y_n y_n}) + \text{lower order terms} = f,$$

which is a typical parabolic equation.

Case 4: There exist more eigenvalues that are not included in the previous cases.

In this case, the PDE problem becomes much more difficult to deal with the basic well-posedness questions. These are the research topics in differential operator theory (see [13]).

1.4.3 A general nonlinear second-order PDE

The classification for the second-order linear PDE can be easily extended to a general nonlinear second-order PDE in n-space dimension.

Let $u(x)$ be an unknown function in R^n. Introduce

$$D_i u = \frac{\partial u}{\partial x_i}, \qquad D_{ij} u = \frac{\partial^2 u}{\partial x_i x_j}, \qquad i, j = 1, \cdots, n.$$

Let $F(p, q, s, x)$ be a differentiable function for $(p, q, s, x) \in R^{n^2} \times R^n \times R^1 \times R^n$. The most general form of the second-order partial differential equation in R^n can be written as follows.

$$F(D_{ij}u, D_i u, u, x) = 0. \tag{1.4.6}$$

Eq. (1.4.6) is called a *fully nonlinear equation*. If $F(p, q, s, x)$ is linear with respect to p_{ij} for all $1 \le i, j \le n$, Eq. (1.4.6) is called *quasilinear*. If $F(p, q, s, x)$ is linear with respect to p_{ij}, q_i, then Eq. (1.4.6) is called *semilinear*.

Define an $n \times n$ matrix:

$$A = \left(\frac{\partial F(p, q, s, x)}{\partial p_{ij}} \right)_{n \times n}.$$

Eq. (1.4.6) is called the elliptic type if all eigenvalues of the matrix A are positive- (negative-) definite. If there exists one eigenvalue that is positive and the rest are negative, then Eq. (1.4.6) is called hyperbolic. If there exists one eigenvalue that is equal to 0 and the rest are all positive (negative), then Eq. (1.4.6) is called parabolic. All the rest of the cases are called mixed type or degenerate type.

There is no general method to deal with a fully nonlinear equation (1.4.6). However, there are many research papers in the literature dealing with each type of nonlinear PDEs. The study for this equation needs much deeper mathematical tools and techniques. The discussion about this equation is beyond the scope of this textbook.

1.5 Some elementary formulas and inequalities

In this section we give some identities and inequlities. Their proofs can be found in calculus and analysis.

1.5.1 Young's inequality

Let $a, b \geq 0$. Then, for any $\varepsilon > 0$

$$ab \leq \frac{\varepsilon a^p}{p} + \frac{\varepsilon^{-\frac{q}{p}} b^q}{q}, \tag{1.5.1}$$

where $p, q > 1$ and

$$\frac{1}{p} + \frac{1}{q} = 1.$$

A special case for $p = q = 2$ is the Cauchy–Schwarz inequality:

$$ab \leq \frac{\varepsilon a^2}{2} + \frac{b^2}{2\varepsilon}. \tag{1.5.2}$$

The proof can be found in a calculus book.

1.5.2 Gauss's divergence theorem and Green's identity

Gauss's divergence theorem. *Let Ω be a domain in R^n with C^1-boundary $\partial\Omega$ and $\mathbf{F}(x)$ be a vector field. Then,*

$$\int_\Omega \nabla \cdot \mathbf{F} dx = \int_{\partial\Omega} [\mathbf{F} \cdot \nu] ds, \tag{1.5.3}$$

where ν is the outward unit normal on $\partial\Omega$.

An immediate consequence is the following integral by parts in n dimensions: Let $u(x), v(x)$ be differentiable in Ω,

$$\int_\Omega u(x)v(x)_{x_i} dx = -\int_\Omega u(x)_{x_i} v(x) dx + \int_{\partial\Omega} [u(x)v(x)\nu_i] ds,$$

where ν_i is the ith component of the normal vector ν on $\partial\Omega$.

By using the above integration by parts, we have the following property.

Proposition 1.5.1. *(Green's identity)*

Let $u(x)$ and $v(x)$ be differentiable in Ω. Then,

$$\int_\Omega [u(\Delta v) - (\Delta u)v] dx = \int_{\partial\Omega} [u(\nabla_\nu v) - (\nabla_\nu u)v] ds. \tag{1.5.4}$$

The proof is left as an exercise.

1.5.3 Spherical coordinates in R^n and the singular integral

Let $x = (x_1, \cdots, x_n) \in R^n$. Introduce spherical coordinates $(\rho, \theta_1, \cdots, \theta_{n-1}, \psi)$ in R^n:

$$\begin{aligned} x_1 &= \rho \cos\theta_1 \cdots \cos\theta_{n-1} \sin\psi, \\ x_2 &= \rho \cos\theta_1 \cdots \sin\theta_{n-1} \sin\psi, \\ &\cdots, \\ x_n &= \rho \cos\psi, \end{aligned}$$

where $0 \le \rho < \infty, 0 \le \theta \le 2\pi, 0 \le \psi \le \pi$.

A sphere centered at the origin with radius r, denoted by $B_r(0) = \{x \in R^n : |x| < r\}$, can be expressed in spherical coordinates by

$$\rho = r, \qquad 0 \le \theta_i \le 2\pi, 0 \le \psi \le \pi, i = 1, 2, \cdots, n-1.$$

The volume of the sphere $B_r(0)$ is equal to

$$V = \omega(n) r^n,$$

where $\omega(n)$ represents the surface area of the unit ball in R^n.

Another interesting example is the following proposition about a singular integral.

Proposition 1.5.2. *Let $B_r(0)$ be a ball centered at the origin with radius $r > 0$. Then,*

$$\int_{B_r(0)} \frac{1}{|x|^p} dx \le C(n, p) r^{n-p},$$

where $C(n, p)$ depends only on p and n.

1.5.4 Gronwall's inequality

Proposition 1.5.3. *Let $\alpha(t)$ and $\beta(t)$ be integrable on $[0, \infty)$. Let $y(t) \ge 0$ satisfy*

$$y'(t) \le \alpha(t) y(t) + \beta(t), \qquad y(0) = y_0. \tag{1.5.5}$$

Then,

$$y(t) \le exp\{\int_0^t \alpha(\tau) d\tau\} \left[y_0 + \int_0^t exp\{-\int_0^\tau \alpha(\xi) d\xi\} \beta(\tau) d\tau \right], t \ge 0.$$

Proof. Introduce an integrating factor

$$exp\{-\int_0^t \alpha(\tau) d\tau\}.$$

We can rewrite the inequality (1.5.5) as follows:

$$\frac{d}{dt} \left[y(t) exp\{-\int_0^t \alpha(\tau) d\tau\} \right] \le \beta(t) exp\{-\int_0^t \alpha(\tau) d\tau\}.$$

The desired inequality follows immediately by integrating the above inequality over $[0, t]$. □

In applications, there is a different version of Gronwall's inequality.

Proposition 1.5.4. *Let $m_1(t)$ and $m_2(t)$ be nonnegative and integrable on $[0, \infty)$. Let $y(t) \geq 0$ satisfy*

$$y(t) \leq \int_0^t m_1(t-s)y(s)ds + m_2(t), \qquad t \geq 0. \tag{1.5.6}$$

Then, for any $T > 0$,

$$y(t) \leq C(T), \qquad t \in (0, T],$$

where C depends on the $L^1(0, T)$-norm of $m_1(t)$ and $m_2(t)$.

Proof. Instead of dealing with the inequality (1.5.6) we first consider the following integral equality:

$$y(t) = \int_0^t m_1(t-s)y(s)ds + m_2(t), \qquad t \geq 0. \tag{1.5.7}$$

Set $y_0(t) = 0$. Define a successive sequence $\{y_n(t)\}$ as follows:

$$y_{n+1}(t) = \int_0^t m_1(t-s)y_n(s)ds + m_2(t), \qquad t \geq 0, n \geq 0.$$

The Volterra integral equation (1.5.7) has a unique solution $y(t)$ in a small interval $[0, T_0]$:

$$y(t) = \lim_{n\to\infty} y_n(t), \qquad 0 \leq t \leq T_0.$$

Since $m_1(s, t) \geq 0$, $m_2(s) \geq 0$, it follows that

$$y(t) \leq C(T_0).$$

We can start with $t = T_0$ as an initial value and continue to obtain the estimate on $[T_0, 2T_0]$. This step can done due to the fact that there exists a priori bound for $y(t)$ in $[0, T]$ (see Chapter 2, Example 2.5.2). For any $T > 0$, by repeating this process a finite number of times, we obtain the desired estimate

$$0 \leq y(t) \leq C(T), \qquad t \in (0, T],$$

where $C(T)$ depends on known data and an upper bound of T.

Since $m_1, m_2 \geq 0$, we see the desired estimate holds if the equality (1.5.7) is replaced by inequality (1.5.6). □

An interesting example frequently used in applications is that

$$m_1(t-s) = \frac{1}{(t-s)^\alpha}, \qquad \alpha \in (0, 1).$$

1.5.5 Poincare's inequality

In this book we do not deal with weak derivatives and the theory of Sobolev spaces (see [1,23]). However, we may need some inequalities in the proofs of some results. The reader may consider a classical C^1-function for these inequalities as a first step. For the readers' convenience, a brief introduction about the theory of Sobolev spaces is given in Appendix B.

Poincare's inequality I. Let Ω be a bounded domain in R^n with Lipschitz boundary $\partial\Omega$. Suppose $u(x) \in H_0^1(\Omega) = W_0^{1,2}(\Omega)$. Then, there exists a constant c_0 such that

$$\int_\Omega u^2 dx \le c_0 \int_\Omega |\nabla u|^2 dx, \qquad \forall u \in H_0^1(\Omega), \tag{1.5.8}$$

where c_0 depends only on Ω and n.

A different version of Poincare's inequality is the following one.

Poincare's inequality II. Let Ω be a bounded domain in R^n with Lipschitz boundary $\partial\Omega$. Suppose $u(x) \in H^1(\Omega)$. Then, there exists a constant c_1 such that

$$\int_\Omega (u - u_0)^2 dx \le c_1 \int_\Omega |\nabla u|^2 dx, \qquad \forall u \in H^1(\Omega), \tag{1.5.9}$$

where c_1 depends only on Ω, n and

$$u_0 := \frac{1}{|\Omega|} \int_\Omega u(x) dx.$$

In dealing with some boundary conditions in the investigation of a PDE problem, we may need the following trace inequality.

Trace inequality. Let Ω be a bounded domain in R^n with Lipschitz boundary $\partial\Omega$. Suppose $u(x) \in H^1(\Omega)$. Then, there exists a constant C such that

$$\int_{\partial\Omega} u^2 ds \le \varepsilon \int_\Omega |\nabla u|^2 dx + C(\varepsilon) \int_\Omega u^2 dx, \qquad \forall u \in H^1(\Omega), \tag{1.5.10}$$

where $\varepsilon > 0$ is arbitrary and $C(\varepsilon)$ depends only on ε, Ω and n.

Poincare's inequalities and trace inequality can be extended to functions in Sobolev space $W_0^{1,p}(\Omega)$ with $p > 1$ (see Appendix B). There is a rich theory about Sobolev spaces. The interested reader may consult with the monographs on Sobolev spaces such as [1,23].

1.5.6 Some notations

$x = (x_1, \cdots, x_n) \in R^n.$

$\Omega =$ a domain in R^n .

$$
\begin{aligned}
\partial\Omega &= \text{the boundary of } \Omega \\
\nu(x) &=< \nu_1(x), \cdots, \nu_n(x) >= \text{ the outward unit normal at } x \in \partial\Omega. \\
B_a(x_0) &= \{x \in R^n : |x - x_0| < a\}; \\
Q_T &= \Omega \times (0, T], \text{ where } T > 0. \\
Q &= \Omega \times (0, \infty). \\
\partial_p Q_T &= \partial\Omega \times (0, T] \bigcup \Omega \times \{t = 0\} \\
&= \text{parabolic boundary for the domain } Q_T. \\
R^n_+ &= \{(x = (x_1, \cdots, x_n) \in R^n : x_n > 0\}.
\end{aligned}
$$

For a multiindex $\alpha = (\alpha_1, \cdots, \alpha_n)$, where α_i is nonnegative integer,

$$
\begin{aligned}
|\alpha| &= \sum_{i=1}^{n} \alpha_i, \\
D^\alpha &= \frac{\partial^{\alpha_1} \cdots \partial^{\alpha_n}}{\partial x_1^{\alpha_1} \cdots \partial x_n^{\alpha_n}}, \\
\nabla &= (\frac{\partial}{\partial x_1}, \cdots, \frac{\partial}{\partial x_n}).
\end{aligned}
$$

In many estimates, we use $C, C_1, C_2, \cdots$, to be generic constants. They may be different from one line to the next. Their precise dependency will be specified at the final step.

1.6 Notes and remarks

In this chapter we introduced some basic concepts in partial differential equations. Some important examples arising from the applied fields are presented in this chapter as motivations for students with different background and research interests. They include the model equation in population dynamics, the fundamental equations in fluid mechanics, electromagnetic fields, and in economics and finance (the Black–Scholes equation). Some of those equations will be derived and investigated in detail in later chapters. We also present some very different phenomena for linear and nonlinear PDE problems. These examples illustrate why the analysis for PDEs plays an important role in applied sciences. Some nonlinear problems such as the Yamabe problem needs advanced mathematical theory and delicate analysis.

We emphasize that most PDE problems studied in this book are derived from concrete physical models. It is a challenging step to establish a suitable mathematical model for a practical problem by using PDEs. This step shows the importance that we need to study the well-posedness for a PDE problem. The theoretical study for a PDE problem is also an interesting art of sciences, which is a large branch of the research field in mathematical sciences. The classification of second-order PDEs gives students the reason why we mainly focus on three types of basic equations in this book.

1.7 Exercises

1. Verify the following properties: Let L be a linear differential operator.
(a) **Superposition principle**: Let $u_1, u_2, \cdots u_m$ be solutions of a homogeneous linear equation $L[u]=0$, then the linear combination $k_1u_1+k_2u_2+\cdots+k_mu_m$ is also a solution of the equation $L[u]=0$. Can you find a counterexample in which the superposition principle fails if $m=\infty$?
(b) **Addition principle**: Let $u=\sum_{k=1}^{m} u_k$ be a general solution of a homogeneous linear equation $L[u]=0$ and u_c be a particular solution of a nonhomogeneous equation $L[u]=f$. Then, $u+u_c$ is a general solution of $L[u]=f$.
2. Let c be a constant. Express the equation

$$u_{tt}-c^2u_{xx}+u^3-t^2e^x=0, x\in R^1, t\geq 0.$$

as an operator equation form $L[u]=f$. Is it homogeneous or nonhomogeneous? Is the operator L linear or nonlinear?
3. For Schrödinger's equation in Example 1.3.5, if $u(x,t)=Re(u(x,t))+iIm(u(x,t)):=v(x,t)+iw(x,t)$, find the system of equations for real functions $v(x,t)$ and $w(x,t)$.
4. (a) Verify $u(x,t)=\frac{1}{\sqrt{2\pi t}}e^{-\frac{x^2}{4t}}$, $x\in R^1$, $t>0$ satisfies the following partial differential equation:

$$u_t-u_{xx}=0, \qquad x\in R^1, t>0.$$

(b) Find the following integral

$$\int_0^{\infty} u(x,t)dx.$$

5. Let $u(x,y)$ satisfy the following equation

$$u_{xx}+(x^2+y^2-1)u_{yy}=0, \qquad (x,y)\in R^2.$$

Find the region in R^2 in which the equation is elliptic and the region in which the equation is hyperbolic.
6. Suppose $u(x,y)$ satisfies the equation

$$u_{xx}-u_{yy}=0, \qquad (x,y)\in R^2.$$

Introduce new variables $\eta=x+y$ and $\xi=x-y$. Let $w(\eta,\xi):=u(x,y)$. Find the equation for $w(\eta,\xi)$.

7. Let $f(x,s)$ be a differentiable function in $R^n \times R^1$. Consider the following equation

$$\Delta u + f(x,u) = 0.$$

Find conditions for f such that the equation becomes linear and homogeneous.

8. Consider the following boundary value problem (BVP) for an ODE:

$$y''(x) = a, \qquad 0 < x < 1,$$

where a is a constant.
(a) Find the solution with boundary conditions $y(0) = b_1$, $y(1) = b_2$, where b_1 and b_2 are arbitrary constants.
(b) Is there any solution with boundary condition $y'(0) = b_1$, $y'(1) = b_2$? What is the necessary condition for the problem to have a solution?

9. Let $p > 0$. Find the solution for the nonlinear ODE:

$$y' + ay = y^p, \qquad y(0) = y_0,$$

where a is a constant. (Hint: Introduce $u = y^{1-p}$ when $p \neq 1$.)

10. Consider the heat equation in one-space dimension:

$$u_t - u_{xx} = 0, \qquad x \in R^1, t > 0.$$

Let $u(x,t) := v(\xi) = v(x-t)$. What is the equation for $v(\xi)$? Find the general solution for $v(\xi)$.

11. Let Ω be a bounded domain with a smooth boundary $\partial\Omega$. Prove Green's identity: for any $u(x)$ and $v(x)$ in $C^1(\bar{\Omega})$, then

$$\int_\Omega [u(\Delta v) - (\Delta u)v]\,dx = \int_{\partial\Omega} [u(\nabla_\nu v) - (\nabla_\nu u)v]\,ds.$$

12. Let $y(t) \geq 0$ satisfy

$$y(t) = \int_0^t \frac{y(\tau)}{\sqrt{t-\tau}}\,d\tau + 1, \qquad t \geq 0.$$

Use the iteration method to find $y(t)$.

13. Let $0 < \alpha < 1$ and $f(t) \in L^1(0,\infty)$. Prove the following integral equation has a solution in $[0,\infty)$:

$$y(t) = \int_0^t \frac{y(\tau)}{(t-\tau)^\alpha}\,d\tau + f(t), \qquad t \geq 0.$$

14. Let $\alpha > 0$ be a constant and $y(t)$ satisfy

$$y'(t) + \alpha y(t) = g(t), \; y(0) = y_0$$

where $g(t)$ is continuous on $[0, \infty)$ and

$$\lim_{t\to\infty} g(t) = 0.$$

Prove

$$\lim_{t\to\infty} y(t) = 0.$$

15. Let $u(x, y)$ satisfy the Laplace equation

$$\Delta u = 0.$$

Derive the equation for $v(r, \theta) := u(x, y)$ with polar coordinates $x = r\cos\theta$, $y = r\sin\theta$.

CHAPTER

2 Function spaces and the Fredholm Alternative

2.1 Banach spaces and Hilbert spaces

This section serves as a brief introduction as well as a review for basic spaces in functional analysis.

2.1.1 Normed vector spaces and Banach spaces

A linear space V is called a normed space if there is a mapping from V to R^1, denoted by $||v||$ for $v \in V$, such that the following axioms hold:

$$
\begin{aligned}
&(a) \quad ||v|| \geq 0. \text{ The equality holds if and only if } v = 0.\\
&(b) \quad ||cv|| = |c|||v||, \text{ for any } c \in R^1.\\
&(c) \quad ||v_1 + v_2|| \leq ||v_1|| + ||v_2|| \text{ for any } v_1, v_2 \in V.
\end{aligned}
$$

For a normed space V, we can define the convergence for a sequence v_n in V. We say a sequence v_n in V is convergent to $v_0 \in V$ if

$$||v_n - v_0|| \to 0, \qquad \text{as } n \to \infty.$$

All common limit properties from calculus hold for sequences in a normed space.

Similar to calculus, we say a sequence v_n in a normed space V is a Cauchy sequence if for any $\varepsilon > 0$, there exists an integer $N > 0$ such that

$$||v_m - v_n|| < \varepsilon,$$

whenever $m, n > N$.

Definition 2.1.1. (a) A normed space V is called a Banach space if every Cauchy sequence in V converges to an element in V.

(b) A Banach space is called separable if there exists a sequence $\{v_n\}_{n=1}^{\infty} \subset V$ such that the set of all limit points of the set $\{v_n, n = 1, 2, \cdots.\}$ is equal to V. Namely, $\{v_n, n = 1, 2, \cdots\}$ is a dense subset of V.

Example 2.1.1. In R^n, we define a norm for every $x = (x_1, x_2, \cdots, x_n)$ by

$$||x||_2 = \sqrt{x_1^2 + \cdots + x_n^2}.$$

Partial Differential Equations and Applications. https://doi.org/10.1016/B978-0-44-318705-6.00008-2

One can easily verify that $||x||_2$ is a norm. Moreover, it is a separable Banach space since each real number can be obtained as a limit of a sequence of rational numbers.

Example 2.1.2. Let $n \geq 2$,

$$M_n = \{\text{all } n \times n \text{ square matrices with entries in } R^1\}.$$

For any $A = (a_{ij})_{n \times n} \in M_n$, define

$$||A|| = \max_{1 \leq i, j \leq n} |a_{ij}|.$$

It can be shown that M_n is a Banach space and is separable.

Example 2.1.3. Let

$$l^\infty = \{\text{all sequences } l = (a_1, a_2, \cdots, a_n, \cdots) \text{ with } a_i \in [0, 1]\}.$$

For any sequence $l \in l^\infty$, define

$$||l|| = \max_{1 \leq i < \infty} |a_i|.$$

It can be shown that l^∞ is a Banach space and is not separable (see Exercise 2.19).

2.1.2 Hilbert spaces

A special class of Banach spaces V is called Hilbert spaces if there exists an inner production defined in V.

Definition 2.1.2. Let V be a vector space. Suppose there is a bilinear mapping from $V \times V$ to R^1, denoted by $< \cdot, \cdot >$ that satisfies the following axioms:

$$(a) \quad < v, v > \geq 0. \text{ The equality holds if and only if } v = 0.$$

$$(b) \quad < c_1 u_1 + c_2 u_2, v > = c_1 < u_1, v > + c_1 < u_2, v >,$$

$$\text{for any } c_1, c_2 \in R^1,\ u_1, u_2, v \in V.$$

$$(c) \quad < u, v > = < v, u > .$$

Then, V is called an inner product space.

We can define a norm for V from an inner product in V by

$$||v|| = \sqrt{< v, v >}, \qquad v \in V.$$

One can verify from Cauchy–Schwarz's inequality (see below) along with the linearity property that the above-defined mapping from V to R^1 is indeed a norm, which is called an induced norm. A completed inner product space is called a Hilbert space.

Example 2.1.4. In R^n, for any $x = (x_1, \cdots x_n)$ and $y = (y_1, \cdots, y_n)$, define

$$< x, y >:= \sum_{k=1}^{n} x_k y_k.$$

One can verify that the above-defined mapping is an inner product. We leave this as an exercise (Exercise 2.4). Hence, R^n is a Hilbert space.

Cauchy–Schwarz's inequality. *Let V be an inner product space. Then,*

$$| < u, v > | \leq ||u||||v||, \qquad \forall u, v \in V. \tag{2.1.1}$$

Proof. Let $u, v \in V$. Define a function

$$f(t) = < u + tv, u + tv >, t \in R^1.$$

The linear property yields that

$$f(t) = t^2||v||^2 + 2t < u, v > + ||u||^2.$$

Since $f(t) \geq 0$ for all $t \in R^1$, it follows that

$$| < u, v > | \leq ||u|| \cdot ||v||.$$

Moreover, the equality holds if and only if one vector is a multiple of another vector. □

There are many interesting results about Banach spaces and Hilbert spaces. They are included in the basic materials of *Functional Analysis* (see [6]). For this book, we are interested in function spaces.

2.2 Function spaces

In this section we introduce some function spaces that will be used in this book.

2.2.1 Continuous and Hölder continuous spaces: $C(\bar{\Omega})$ and $C^\alpha(\bar{\Omega})$

Let Ω be a domain in R^n with Lipschitz boundary. Let

$$C(\bar{\Omega}) = \{\text{all continuous functions defined in } \bar{\Omega}\}.$$

The norm for $C(\bar{\Omega})$ for a bounded domain Ω is defined by

$$||f||_0 = \max_{x \in \bar{\Omega}} |f(x)|.$$

Let α be any fixed number in (0, 1],

$$C^{\alpha}(\bar{\Omega}) = \{f(x) \in C(\bar{\Omega}) : [f]_{\alpha} < \infty\},$$

where

$$[f]_{\alpha} = \sup_{x,y \in \bar{\Omega}, x \neq y} \frac{|f(x) - f(y)|}{|x - y|^{\alpha}}.$$

The norm for $C^{\alpha}(\bar{\Omega})$ is defined by

$$||f||_{C^{\alpha}(\bar{\Omega})} = ||f||_0 + [f]_{\alpha}.$$

A function in $C^{\alpha}(\bar{\Omega})$ is called Hölder continuous with exponent α. When $\alpha = 1$, it is called *Lipschitz* continuous.

2.2.2 The function spaces $C^{k,\alpha}(\bar{\Omega})$, $C^{\infty}(\bar{\Omega})$, and $C_0^{\infty}(\bar{\Omega})$

Let $k \geq 1$ be an integer. Define

$$C^k(\bar{\Omega}) = \{\text{all functions } f \text{ with } D^k f \in C(\bar{\Omega})\},$$
$$C^{k,\alpha}(\bar{\Omega}) = \{f(x) \in C^k(\bar{\Omega}) : D^k f \in C^{\alpha}(\bar{\Omega})\},$$

where D^k represents all possible kth-order partial derivatives of f. The norm of $C^{k,\alpha}(\bar{\Omega})$ is defined by

$$||f||_{C^{k,\alpha}(\bar{\Omega})} = \sum_{|m| \leq k} ||D^m f||_0 + \sum_{|m|=k} [|D^m f|]_{\alpha},$$

where m is a multiindex $m = (m_1, \cdots, m_n)$, $|m| = \sum_{i=1}^{n} m_i$, m_i is a nonnegative integer, and

$$D^m = \frac{\partial^m}{\partial x_1^{m_1} \cdots \partial x_n^{m_n}}.$$

When $k = \infty$, it is denoted by $C^{\infty}(\bar{\Omega})$. $C_0^{\infty}(\Omega)$ is a subspace of $C^{\infty}(\Omega)$ where every function has a compact support. Namely, the set

$$supp(f) = \{x \in \Omega : f(x) \in C^{\infty}(\Omega), f(x) \neq 0\}$$

is bounded in R^n and $dist\{supp(f), \partial\Omega\} > 0$.

When Ω is bounded, it can be shown in functional analysis that $C(\bar{\Omega})$, $C^{\alpha}(\bar{\Omega})$, $C^{k,\alpha}(\bar{\Omega})$ are Banach spaces. Moreover, the dimension for each of these spaces is infinite. The reader can find these results in functional analysis such as in [5].

2.2.3 L^p-space

A more interesting function space is called an L^p-space. A rigorous definition for this space requires the Lebesgue measure theory and Lebesgue integrals. However, we can think of these functions as approximations of continuous functions under an L^p-norm.

Let $p \geq 1$ and Ω be a domain in R^n.

Definition 2.2.1. A function $f(x)$ defined on Ω is said to be p-integrable on Ω if

$$\int_\Omega |f(x)|^p dx < \infty.$$

The set of all p-integrable functions is denoted by $L^p(\Omega)$. The norm is defined by

$$||f||_p = \left(\int_\Omega |f|^p dx\right)^{\frac{1}{p}}.$$

One can show from the functional analysis that $L^p(\Omega)$ is a Banach space under the above norm.

When $p = \infty$, the space is denoted by $L^\infty(\Omega)$ equipped with the norm

$$||f||_\infty = ess. \sup_{x \in \Omega} |f(x)|.$$

When Ω is unbounded, we use $L^p_{loc}(\Omega)$ as a space in which every function is p-integrable over any bounded subset of Ω. One can show from the functional analysis that $L^p_{loc}(\Omega)$ is a Banach space under the above norm (see [5]).

In the above definition, the integration is defined as a Lebesgue integral. If one does not have the knowledge of the measure theory, one may use the Riemann integral instead. It is clear that all piecewise-bounded continuous functions are in $L^p(\Omega)$. Note that every function in $L^p(\Omega)$ represents a class of functions as long as their norms are the same. In the measure theory, we say two functions are equal except for a set of points where the Lebesgue measure of the set is equal to 0.

Unless indicated in the context, in this book a function sequence $f_n(x)$ in a Banach space that is convergent to $f(x)$ means

$$||f_n - f|| \to 0, \qquad \text{as } n \to \infty.$$

For example, in $C(\bar{\Omega})$, the convergence of a function sequence means a uniform convergence on $\bar{\Omega}$.

There are some important inequalities that are used frequently in the analysis of PDEs.

Proposition 2.2.1. *(Hölder's inequality and Minkowski's inequality)*

(a) Let $p, q \geq 1$ and $f \in L^p(\Omega)$ and $g \in L^q(\Omega)$. Then,

$$||fg||_1 \leq ||f||_p ||g||_q,$$

where

$$\frac{1}{p}+\frac{1}{q}=1.$$

Moreover, the equality holds if and only if one function is a multiple of another function.

(b) Let $p_1, \cdots, p_k \geq 1$ *and*

$$\frac{1}{p_1}+\cdots+\frac{1}{p_k}=1.$$

Then, for $f_i \in L^{p_i}(\Omega)$, $i=1,\cdots,k$,

$$||f_1\cdots f_k||_1 \leq ||f_1||_{p_1}\cdots||f_k||_{p_k}.$$

(c) Let $p \geq 1$. *For any* $f(x), g(x) \in L^p(\Omega)$,

$$||f+g||_p \leq ||f||_p + ||g||_p.$$

The proof is left as an exercise.

When $p=2$, we can define an inner product. Let $f(x), g(x) \in L^2(\Omega)$, define

$$< f, g >:= \int_\Omega f(x)g(x)dx.$$

One can verify that the above inner product satisfies the axioms in Definition 2.1.1. Hence, $L^2(\Omega)$ is a Hilbert space. The proof can be found in a standard textbook of functional analysis such as [6].

Definition 2.2.2. (a) Two functions $f(x), g(x) \in L^2(\Omega)$ are said to be orthogonal, if

$$< f, g >= \int_\Omega f(x)g(x)dx = 0.$$

(b) A subset M of $L^2(\Omega)$ is said to be mutually orthogonal if

$$< f, g >= \int_\Omega f(x)g(x)dx = 0$$

for any distinct functions $f(x), g(x) \in M \subset L^2(\Omega)$.

(c) A function sequence $f_1(x), \cdots, f_n(x), \cdots$, in $L^2(\Omega)$ is said to be orthonormal if the set

$$M = \{f_1(x), f_2(x), \cdots f_n(x), \cdots\}$$

is mutually orthogonal in $L^2(\Omega)$ and $||f_n|| = 1, n = 1, 2, \cdots$.

Example 2.2.1. Prove each of the following sets is mutually orthogonal.

$$M_1 = \{1, \cos(\frac{n\pi x}{a}), \sin(\frac{n\pi x}{a}), n = 1, 2, \cdots .\} \text{ are orthogonal in } L^2(-a, a).$$

$$M_2 = \{\sin(\frac{n\pi x}{a}), n = 1, 2, \cdots\} \text{are orthogonal in } L^2(0, a).$$

$$M_3 = \{\cos(\frac{n\pi x}{a}), \ n = 0, 1, 2, \cdots\} \text{ are orthogonal in } L^2(0, a).$$

Proof. We recall the following trigonometric identities:

$$\sin\alpha \sin\beta = \frac{1}{2}[\cos(\alpha-\beta) - \cos(\alpha+\beta)];$$

$$\sin\alpha \cos\beta = \frac{1}{2}[\sin(\alpha-\beta) + \sin(\alpha+\beta)];$$

$$\cos\alpha \cos\beta = \frac{1}{2}[\cos(\alpha-\beta) + \cos(\alpha+\beta)].$$

By using these identities one can easily verify the desired results. □

Remark 2.2.1. It will be seen in Section 2.4 that M_1 forms an orthogonal basis for the Hilbert space $L^2(-a, a)$, while M_2 or M_3 forms an orthogonal basis for $L^2(0, a)$.

In some applications, one may need to extend the inner product space $L^2(\Omega)$ with a weight function. Let $w(x) \geq 0$ be positive except for a small set E with zero-Lebesgue measure. For any $f, g \in L^2(\Omega)$, define the inner product

$$< f, g >_w = \int_\Omega f(x)g(x)w(x)dx.$$

One can verify that $< \cdot, \cdot >_w$ is indeed an inner product (see Exercise 2.5). This space is denoted by $L^2(\Omega, w)$.

The next example comes from special function theory. These special functions are used in Chapter 3.

Example 2.2.2. Define a Bessel function sequence $J_m(x)$ over $[0, \infty)$ as the following:

$$J_m(x) = \sum_{n=0}^{\infty} \frac{(-1)^n}{n!\Gamma(1+m+n)} \left(\frac{x}{2}\right)^{2n+m},$$

where the Γ-function is defined by

$$\Gamma(m) = \int_0^\infty x^m e^{-x} dx.$$

When $m \geq 0$, the series is convergent on $[0, \infty)$. When $m < 0$, then the series is convergent on $(0, \infty)$.

The Bessel function is derived from the series solution for the ordinary differential equation (see reference [16] for example):

$$x^2y'' + xy' + (x^2 - m^2)y = 0.$$

We will see in Chapter 5 that the result of Example 2.2.2 is used to find a solution of the wave equation in a disk.

The Bessel functions $\{J_m(x), m = 0, 1, 2, \cdots\}$ are mutually orthogonal on $[0, a]$ with a weight function $w(x) = x$. A direct proof is quite complicated. It will be seen that these special function sequences are eigenfunctions of certain eigenvalue problems and the orthogonality for eigenfunctions is just one of the properties (see the Sturm–Liouville theorem in Chapter 3).

2.2.4 Approximation for functions

As indicated in the previous subsection a function in $L^1(\Omega)$ may be approximated by a sequence of smooth functions. Define

$$\eta(x) = \begin{cases} k exp\{-\frac{1}{1-|x|^2}\}, & \text{if } |x| < 1, \\ 0, & \text{if } |x| \geq 1, \end{cases}$$

where the constant k is chosen such that

$$\int_{R^n} \eta(x)dx = 1.$$

The function $\eta(x)$ is called a mollifier. Let $\varepsilon > 0$. Define

$$\eta_\varepsilon(x) = \varepsilon^{-n}\eta(\frac{x}{\varepsilon}).$$

Then,

$$\int_{R^n} \eta_\varepsilon(x)dx = 1, \qquad \eta_\varepsilon(x) = 0, \qquad \text{if} \qquad |x| \geq \varepsilon.$$

For any function $f(x) \in L^1(R^n)$, we define the convolution

$$(\eta_\varepsilon * f)(x) = \int_{R^n} \eta_\varepsilon(x - y)f(y)dy.$$

Clearly, $(\eta_\varepsilon * f)(x)$ has a compact support if $f(x)$ has a compact support in R^n.

Theorem 2.2.1. *Let $f(x) \in L^1_{loc}(R^n)$. Then,*

$$\begin{array}{lll} (i) & (\eta_\varepsilon * f)(x) \in C^\infty(R^n); & \\ (ii) & \eta_\varepsilon * f(x) \in L^p(\Omega) & \text{if} \quad f(x) \in L^p(\Omega), \\ & \|\eta_\varepsilon * f(x) - f\|_p \to 0 & \text{as } \varepsilon \to 0; \\ (iii) & \eta_\varepsilon * f(x) \to f(x) & \text{uniformly in } K \text{ for any compact } K \subset \Omega, \\ & \text{if } f(x) \text{ is continuous in } \Omega. & \end{array}$$

Proof. We give a proof for (iii). The rest of the proofs are left as exercises. Let $K \subset \Omega$ be compact. Then,

$$|(\eta_\varepsilon * f)(x) - f(x)| = \left| \int_{R^n} (\eta_\varepsilon(x-y)[f(y) - f(x)]dy \right|$$
$$\leq \sup_{|x-y|<\varepsilon} |f(x) - f(y)| \to 0$$

uniformly as $\varepsilon \to 0$. □

2.2.5 Sobolev spaces $W^{k,p}(\Omega)$, $H^k(\Omega)$, and $L^q((0,T);X)$

Sobolev space plays an essential role in the study of advanced partial differential equations. There are several books about the theory of this space (for examples, see monographs [1,23]). However, in this book we use it occasionally in dealing with the existence theory for a PDE problem. The reader may find more materials in Appendix B for the basic concept. The theory of Sobolev space is very interesting. This part may be skipped for beginners.

Let $k \geq 1$ be an integer and $p \geq 1$. Let Ω be a bounded domain with a smooth boundary. Define

$$W^{k,p}(\Omega) := \{f(x) \in L^p(\Omega) : ||f||_{W^{k,p}(\Omega)} < \infty\},$$

where

$$||f||_{W^{k,p}(\Omega)} := \sum_{|m|\leq k} ||D^m f||_{L^p(\Omega)}.$$

$W^{k,p}(\Omega)$ can be considered as the closure of $C^k(\bar{\Omega})$ under the $W^{k,m}$-norm.

$$W_0^{k,p}(\Omega) = \text{closure of } C_0^k(\Omega) \text{ under the } W^{k,p}\text{-norm}.$$

When $p = 2$, then $W^{k,2}(\Omega)$ is a Hilbert space with the inner product

$$< f, g >:= \sum_{|m|\leq k} < D^m f, D^m g > .$$

In this case, we use

$$H^k(\Omega) := W^{k,2}(\Omega), \qquad H_0^k(\Omega) := W_0^{k,2}(\Omega).$$

In dealing with time-dependent evolution equations, one needs to define a function from $[0, T]$ to a Banach space X. Here is an example. Let X be a Banach space and $p \geq 1$.

$$L^p((0,T); X) := \{\text{all functions from } [0,T] \text{ to } X \text{ with } ||f||_{L^p((0,T);X)} < \infty\},$$

where

$$||f||_{L^p((0,T);X)} := \{\int_0^T ||f||_X^p dt\}^{\frac{1}{p}}.$$

When $X = L^p(\Omega)$, we simply use

$$L^p(Q_T) := L^p((0,T); L^p(\Omega)),$$

where $Q_T = \Omega \times (0, T]$.

The following compactness result is often used in the book.

Theorem 2.2.2. *(Aubin–Lions lemma) Let X_0, X, and X_1 be three Banach spaces with $X_0 \subset X \subset X_1$. Suppose that X_0 is compactly embedded in X and that X is continuously embedded in X_1. For $1 \leq p, q \leq \infty$. Let*

$$W = \{u \in L^p([0,T]; X_0) \mid \dot{u} \in L^q([0,T]; X_1)\}.$$

(i) If $p < \infty$, then the embedding of W into $L^p([0,T]; X)$ is compact.
(ii) If $p = \infty$ and $q > 1$ then the embedding of W into $C([0,T]; X))$ is compact.

The reader may find its proof in [28].

2.2.6 Linear independence and the Gram–Schmidt process

We extend the concept of linear independence in linear algebra into infinite-dimensional function space.

Definition 2.2.3. We say a set of functions

$$M = \{f_k(x), x \in \Omega : k = 1, 2, \cdots\}$$

is linearly independent on Ω, if for any $m \geq 1$ and

$$k_1 f_1(x) + k_2 f_2(x) + \cdots + k_m f_m(x) = 0, \qquad x \in \Omega,$$

implies $k_1 = k_2 = \cdots = k_m = 0$.

Example 2.2.3. The function set $M = \{1, x, x^2, \cdots, x^m, \cdots\}$ is linearly independent at any interval (a, b).

Proof. Suppose there exist constants $k_1, \cdots, k_m$ such that

$$k_1 + k_2 x + \cdots + k_m x^m = 0, \qquad x \in (a, b).$$

From the fundamental theorem of algebra we know that an mth-order polynomial has m roots. However, we have an infinite number of $x \in (a, b)$ that satisfy the equation. This is a contradiction unless all coefficients are equal to 0. It follows that the functions in M are linearly independent.

This example shows that the dimension of a function space such as $C^\infty(\bar{\Omega})$ is infinite. Similar to the property in linear algebra, one can show that an orthogonal set in a Hilbert space must be linearly independent.

Proposition 2.2.2. *Let $M\backslash\{0\}$ be a set of functions that are mutually orthogonal in a Hilbert space H. Then, any sequence in M is linearly independent.*

Proof. Let $K = \{\phi_k(x), x \in \Omega\}_{k=1}^m$ be any mutually orthogonal subset of M. Suppose

$$k_1\phi_1(x) + \cdots + k_m\phi_m(x) = 0, \qquad x \in \Omega.$$

If we take an inner product to the equation with ϕ_i for each $i = 1, 2, \cdots, m$, we obtain

$$k_i < \phi_i, \phi_i >= 0.$$

It follows that $k_i = 0, i = 1, 2, \cdots, m$, which implies that K is linearly independent. Consequently, M is linearly independent. □

Example 2.2.4. Any subsequence in M_1, M_2 or M_3 from Example 2.2.1 is linearly independent.

From linear algebra, we can construct a set of orthonormal vectors from a set of linearly independent vectors. This process is called the Gram–Schmidt process. For a Hilbert space, we can use the same process to construct an orthonormal set from a linearly independent set of functions.

Gram–Schmidt process. Let $M = \{\phi_1(x), \cdots, \phi_m(x), \cdots .\}$ be linearly independent in a Hilbert space H.

Step 1: Let $\psi_1(x) = \phi_1(x)$.

Step 2: Set $\psi_2(x) = \phi_2(x) - c_1\psi_1(x)$, where c_1 is chosen such that $< \psi_2, \psi_1 >= 0$.
It follows that

$$c_1 = \frac{< \psi_1, \phi_2 >}{< \psi_1, \psi_1 >}.$$

Step 3: Set $\psi_3 = \phi_3 - c_1\psi_1(x) - c_2\psi_2(x)$, where c_1 and c_2 are chosen such that $< \psi_1, \psi_3 >=< \psi_2, \psi_3 >= 0$.

We can easily solve for c_1 and c_2:

$$c_1 = \frac{< \psi_1, \phi_3 >}{< \psi_1, \psi_1 >}; \ c_2 = \frac{< \psi_2, \phi_3 >}{< \psi_2, \psi_2 >}.$$

Step 4: Continuing this process, we obtain a new set of mutually orthogonal sets

$$M^* = \{\psi_1(x), \cdots, \psi_m(x), \cdots\},$$

where

$$\psi_k(x) = \phi_k(x) - \sum_{i=1}^{k-1} c_i\phi_i(x), \qquad c_i = \frac{< \psi_i, \phi_k >}{< \psi_i, \psi_i >}, k = 1, 2, \cdots .$$

Step 5: Dividing the norm of each function in M^* yields an orthonormal set.

Example 2.2.5. Use the Gram–Schmidt process to construct an orthonormal function sequence from the polynomial set $M := \{1, x, x^2, x^3, \cdots .\}$ in $[0, 1]$.

Solution. Let $f_i(x) = x^{i-1}$ for all $i \geq 1$.

Step 1: We choose

$$g_1(x) = f_1(x) = 1, \qquad x \in [0, 1].$$

Step 2: We set

$$g_2(x) = f_2(x) - c_1 g_1,$$

where c_1 is chosen such that $< g_1, g_2 >= 0$.

That is,

$$\int_0^1 g_1(x) g_2(x) dx = 0.$$

It follows that

$$\int_0^1 [x - c_1] dx = 0.$$

Hence, $c_1 = \frac{1}{2}$ and $g_2(x) = x - \frac{1}{2}$.

Step 3: We set

$$g_3(x) = f_3(x) - c_1 g_1(x) - c_2 g_2(x),$$

where c_1, c_2 are chosen such that

$$< g_1, g_3 >= 0, < g_2, g_3 >= 0.$$

An elementary calculation gives

$$c_1 = \frac{1}{3}, c_2 = 1$$

and

$$g_3(x) = x^2 - (x - \frac{1}{2}) - \frac{1}{3} = x^2 - x + \frac{1}{6}.$$

We can continue this process to obtain an orthogonal set of functions $\{g_m(x)\}$. An orthonormal sequence can be obtained immediately by dividing the norm of each $g_k(x)$.

2.3 Completeness and series representation

In linear algebra, an n-dimensional vector can be expressed as a linear combination of any n basis vectors. In this section we extend this concept into infinite-dimensional function spaces. A Banach space may not have accountable basis in general. However, a separable Banach space does have an accountable basis.

2.3.1 Bessel's inequality and Parseval's equality

A Banach space is said to be separable if there exists an accountable set $E \subset B$ such that the closure $\bar{E} = B$. Most function spaces in applications such as $L^p(\Omega)$ are separable. However, l^∞ is not separable.

Definition 2.3.1. We say $M := \{v_1, \cdots, v_m, \cdots\} \subset V$ is a basis (also called complete) of a Banach space V if M is linearly independent and every $u \in V$ can be approximated by a sequence u_n in subspace $V_n = span\{v_1, v_2, \cdots, v_n\}$:

$$\lim_{n\to\infty} ||u - u_n|| = 0,$$

where

$$u_n := \sum_{k=1}^{n} c_k v_k, \qquad c_k \in R^1, n = 1, 2, \cdots.$$

If $M \subset V$ is complete, then any $u \in V$ can be expressed as a series:

$$u = \sum_{k=1}^{\infty} c_k v_k, \qquad c_k \in R^1.$$

For a Hilbert space H, if $M \subset H$ is an orthonormal basis of H, then

$$c_k =< u, v_k >, \qquad k = 1, 2, \cdots.$$

Let Ω be a bounded domain in R^n.

Let $M = \{\phi_1(x), \cdots, \phi_n(x), \cdots\}$ be an orthonormal sequence in $L^2(\Omega)$. A natural question to ask is whether every function in $L^2(\Omega)$ can be expressed as a linear combination of $\{\phi_n(x), n = 1, 2, \cdots\}$. A different way is to ask whether or not M forms a basis for the space $L^2(\Omega)$.

To answer the question, we consider the following function with parameters $d_1, d_2, \cdots, d_m$:

$$D(d_1, d_2, \cdots, d_m) := \int_\Omega [f(x) - \sum_{i=1}^{m} d_i \phi_i(x)]^2 dx,$$

where $d_i \in R^1, i = 1, 2, \cdots, m$.

Using the orthogonality, we see that

$$D(d_1, d_2, \cdots, d_m) = \int_\Omega f(x)^2 dx - 2\sum_{i=1}^{m} d_i \int_\Omega f(x)\phi_i(x) dx + \sum_{i=1}^{m} d_i^2.$$

It follows that

$$D(d_1, d_2, \cdots, d_m) = \int_\Omega f(x)^2 dx + \sum_{i=1}^{m} [(d_i - c_i)^2 - c_i^2],$$

where

$$c_k = \int_\Omega f(x)\phi_k(x)dx, \qquad k = 1, 2, \cdots, m.$$

Consequently, the function $D(d_1, \cdots, d_m)$ attains its minimum if and only if

$$d_k = c_k, \qquad k = 1, \cdots, m.$$

Proposition 2.3.1. *Let $M = \{\phi_k\}_{k=1}^{\infty}$ be an orthonormal sequence in $L^2(\Omega)$. Then, for any $f(x) \in L^2(\Omega)$, the following inequality holds:*

$$\sum_{i=1}^{\infty} c_i^2 \le \int_\Omega f(x)^2 dx, \qquad \textit{(Bessel's inequality)} \tag{2.3.1}$$

where

$$c_k = \int_\Omega f(x)\phi_k(x)dx, \qquad k = 1, 2, \cdots.$$

Proof. Since $D(d_1, d_2, \cdots, m)$ is nonnegative, it follows that

$$\sum_{i=1}^{m} c_i^2 \le \int_\Omega f(x)^2 dx.$$

Since the right-hand side of the above inequality is independent of m, let $m \to \infty$, we see that

$$\sum_{i=1}^{\infty} c_i^2 \le \int_\Omega f(x)^2 dx. \qquad \square$$

Theorem 2.3.1. *Let $\{\phi_k(x)\}_{k=1}^{\infty}$ be a function sequence that is orthonormal in $L^2(\Omega)$. Then, the function sequence $\{\phi_k(x)\}_{k=1}^{\infty}$ forms a basis of $L^2(\Omega)$ (complete) if and only if for any function $f(x)$ in $L^2(\Omega)$,*

$$\int_\Omega f(x)^2 dx = \sum_{k=1}^{\infty} c_k^2, \qquad \textit{(Parseval's equality)}, \tag{2.3.2}$$

where

$$c_k = \int_\Omega f(x)\phi_k(x)dx, \qquad k = 1, 2, \cdots.$$

Proof. Let $f(x) \in L^2(\Omega)$ and $f_n(x) = \sum_{k=1}^{n} c_k\phi_k(x)$ with c_k defined as in the theorem. Then,

$$\int_\Omega (f(x) - f_n(x))^2 dx = \int_\Omega f(x)^2 dx - \sum_{k=1}^{n} c_k^2.$$

It follows that $f_n(x)$ converges to $f(x)$ in the sense of an $L^2(\Omega)$-norm if and only if

$$\int_\Omega f(x)^2 dx = \sum_{k=1}^{\infty} c_k^2. \qquad \square$$

Proposition 2.3.2. *There exists an orthonormal basis for $L^2(\Omega)$.*

The proof of the proposition is due to the fact that $L^2(\Omega)$ is separable and every function in $L^2(\Omega)$ can be approximated by a polynomial in L^2-sense (see [5]).

2.3.2 Series representation of functions

The motivation for a function representation comes from a superposition principle. To find a solution of a PDE problem, we first seek a series function just like a series solution in terms of power series for ODEs. For a partial differential equation it turns out that a power series is not a suitable series solution due to some boundary conditions. A new series in terms of trigonometric functions or eigenfunctions is used instead of a power-series solution.

It is often difficult to prove that a set of orthogonal functions forms a basis for $L^2(\Omega)$. However, we will prove a general theorem that implies all linearly independent eigenfunctions for a self-adjoint elliptic operator form a basis for $L^2(\Omega)$. This result is very useful in the study of partial differential equations.

Suppose

$$M = \{\phi_k(x),\ x \in \Omega\}_{k=1}^{\infty}$$

forms an orthonormal basis in $L^2(\Omega)$. Then, for any function $f(x) \in L^2(\Omega)$, we can express $f(x)$ in terms of the basis functions in M. Let

$$f(x) = \sum_{n=1}^{\infty} c_n \phi_n(x), \qquad x \in \Omega.$$

If we take an inner product by $\phi_k(x)$ in the representation, we see that

$$c_k = \int_\Omega f(x)\phi_k(x)dx, \qquad k = 1, 2, \cdots.$$

We would like to emphasize that the above equality holds in L^2-sense.

Remark 2.3.1. In the 1990s, the wavelet theory played a special role in modern telecommunication (digital revolution). The idea behind this revolution is due to the introduction of a special class of function basis (called wavelets). This class of basis functions can be used to approximate high-frequency signals much more accurately, which in turn makes television pictures and other graphics much clearer than analog signals (Fourier modes).

2.4 Fourier series

In this section we focus on a special class of functions that forms a basis of the function space $L^2(\Omega)$. The set of basis functions consists of simple trigonometric functions. For this special class of basis, the representation of a function is called a *Fourier series*.

2.4.1 Fourier series

We begin with a simple case where the dimension is equal to one. For simplicity, we choose $a = \pi$ in the function sets M_1, M_2, and M_3 in Example 2.2.1. Let

$$M_1 = \{1, \cos(nx), \sin(nx) : n = 1, 2, \cdots\}.$$

We already know from the previous section that M_1 is mutually orthogonal in $(-\pi, \pi)$. We will see in Chapter 3 that M_1 forms a basis for $L^2(-\pi, \pi)$. It follows that any function in $L^2(-\pi, \pi)$ can be expressed by the trigonometric functions in M_1.

Let

$$f(x) = a_0 + \sum_{n=1}^{\infty} [a_n \cos(nx) + b_n \sin(nx)]. \tag{2.4.1}$$

From the previous theorem, we know the coefficients are

$$\begin{aligned} a_0 &= \frac{1}{2\pi} \int_{-\pi}^{\pi} f(x) dx, \\ a_n &= \frac{1}{\pi} \int_{-\pi}^{\pi} f(x) \cos(nx) dx, \\ b_n &= \frac{1}{\pi} \int_{-\pi}^{\pi} f(x) \sin(nx) dx, \qquad n = 1, 2, \cdots. \end{aligned}$$

The coefficients $\{a_n, b_n\}$ are called the *Fourier coefficients* of function $f(x)$. Moreover, from Parseval's identity, we have the following property: for any $f(x) \in L^2(-\pi, \pi)$,

$$\lim_{n\to\infty} \int_{-\pi}^{\pi} f(x) \sin(nx) dx = \lim_{n\to\infty} \int_{-\pi}^{\pi} f(x) \cos(nx) dx = 0.$$

Let

$$f_n(x) = a_0 + \sum_{k=1}^{n} [a_k \cos(kx) + b_k \sin(kx)].$$

Theorem 2.4.1. *Every function $f(x) \in L^2(-\pi, \pi)$ can be expressed as a Fourier series (2.4.1). Moreover, the Fourier series $f_n(x)$ converges to $f(x)$ in L^2-sense.*

Proof. From the definition of the Fourier coefficients, we see that

$$\begin{aligned} f_n(x) &= \frac{1}{2\pi}\int_{-\pi}^{\pi} f(t)dt + \frac{1}{\pi}\sum_{m=1}^{n}\int_{-\pi}^{\pi} f(t)[\cos mt \cos mx + \sin mt \sin mx]dt \\ &= \frac{1}{2\pi}\int_{-\pi}^{\pi} f(t)dt + \frac{1}{\pi}\sum_{m=1}^{n}\int_{-\pi}^{\pi} f(t)\cos m(t-x)dt \\ &= \frac{1}{\pi}\int_{-\pi}^{\pi} f(t)[\frac{1}{2} + \sum_{m=1}^{n}\cos m(t-x)]dt. \end{aligned}$$

Note that

$$\frac{1}{2} + \cos\alpha + \cdots + \cos(n\alpha) = \frac{\sin(n+\frac{1}{2}\alpha)}{2\sin(\frac{1}{2}\alpha)}. \tag{2.4.2}$$

It follows that

$$f_n(t) - \frac{1}{\pi}\int_{-\pi}^{\pi} f(t)D_n(x,t)dt,$$

where

$$D_n(x,t) = \frac{\sin((n+\frac{1}{2})(t-x))}{2\sin(\frac{t-x}{2})}.$$

From (2.4.2), we see that

$$\int_{-\pi}^{\pi} D_n(x,t)dt = 1.$$

It follows by Bessel's inequality that

$$\int_{-\pi}^{\pi} |f(x) - f_n(x)|^2 dx \le [\int_{-\pi}^{\pi} |f(x) - f(t)||D_n(x,t)|dt]^2 \to 0, \text{ as } n \to \infty. \qquad \square$$

A central question in Fourier analysis is whether or not the Fourier series converges to the original function in a stronger norm than the L^2-norm. Let

$$f_n(x) = a_0 + \sum_{k=1}^{n}[a_k\cos(kx) + b_k\sin(kx)].$$

Theorem 2.4.2. *Let $f(x) \in L^2(-\pi,\pi)$.*

(a) If $f(x)$ has a jump discontinuity at a point $c \in (-\pi,\pi)$, then $f_n(c)$ converges to

$$\frac{f(c-) + f(c+)}{2}.$$

(b) If $f(x) \in C[-\pi,\pi]$, then $f_n(x)$ converges $f(x)$ uniformly on $[-\pi,\pi]$.

Proof. We prove (b) and leave (a) as an exercise. For $x \in [-\pi, \pi]$, we note for all $x \in [-\pi, \pi]$,

$$\begin{aligned} |f(x) - f_n(x)| &= |\int_{-\pi}^{\pi} D_n(x,t)[f(x) - f_n(t)]dx| \\ &\to 0, \qquad n \to \infty. \end{aligned}$$

□

2.4.2 Fourier sine series and Fourier cosine series

In some applications we may need some special Fourier series.

Let $f(x)$ be defined on $[0, \pi]$. If we take an even extension for $f(x)$ to $(-\pi, \pi)$ by

$$\hat{f}(x) = f(-x), \qquad x \in (-\pi, 0).$$

Then, $\hat{f}(x)$ is defined on $(-\pi, \pi)$. We can express $\hat{f}(x)$ as a Fourier series:

$$\hat{f}(x) = a_0 + \sum_{n=1}^{\infty}[a_n \cos(nx) + b_n \sin(nx)].$$

Note that $\hat{f}(x)$ is an even function on $(-\pi, \pi)$, we see that

$$b_n = 0, \qquad n = 1, 2, \cdots.$$

It follows that

$$f(x) = a_0 + \sum_{n=1}^{\infty} a_n \cos(nx), \qquad x \in (0, \pi),$$

where

$$a_0 = \frac{1}{\pi}\int_0^{\pi} f(x)dx, a_n = \frac{2}{\pi}\int_0^{\pi} f(x)\cos(nx)dx, \qquad n = 1, \cdots.$$

The above representation $f(x)$ is called a *Fourier cosine series.*

Similarly, we take an odd extension of $f(x)$ to $(-\pi, 0)$ by

$$\hat{f}(x) = -f(-x), \qquad x \in (-\pi, 0).$$

Then,

$$a_n = 0, \qquad n = 0, 1, 2, \cdots.$$

It follows that

$$f(x) = \sum_{n=1}^{\infty} b_n \sin(nx), \qquad x \in [0, \pi],$$

where

$$b_n = \frac{2}{\pi}\int_0^{\pi} f(x)\sin(nx)dx, \qquad n = 1, \cdots.$$

This representation of $f(x)$ is called a *Fourier sine series*. One can easily extend the Fourier series into any interval $(-L, L)$ by using a new variable

$$y = \frac{L}{\pi}x.$$

Example 2.4.1. Let $f(x) = x + x^2,\ x \in [-1, 1]$.
(a) Find the Fourier series of $f(x)$ in $(-1, 1)$.
(b) Find the Fourier cosine series of $f(x)$ in $[0, 1]$.
(c) Find the Fourier sine series of $f(x)$ in $[0, 1]$.

Solution. (a) By the definition, we find

$$a_0 = \frac{1}{3}, a_n = \frac{4(-1)^n}{(n\pi)^2}, b_n = \frac{2(-1)^{n+1}}{n\pi},$$

$$x + x^2 = \frac{1}{3} + \sum_{n=1}^{\infty}\left[\frac{4(-1)^n}{(n\pi)^2}\cos(n\pi x) + \frac{2(-1)^{n+1}}{n\pi}\sin(n\pi x)\right].$$

Similarly, for (b) and (c) we find

$$x + x^2 = \frac{5}{6} + \sum_{n=1}^{\infty}\left[\frac{2[3(-1)^n - 1]}{(n\pi)^2}\cos(n\pi x)\right];$$

$$x + x^2 = \sum_{n=1}^{\infty}\left[\frac{(-1)^{n+1}4}{n\pi} + \frac{4[(-1)^n - 1]}{(n\pi)^3}\right]\sin(n\pi x).$$

2.4.3 Fourier series in higher dimension

The Fourier series can be extended to higher dimension. We use two-dimensional sine series as an example.

Theorem 2.4.3. *Let $D = [0, L] \times [0, H] \in R^2$ be a rectangle. Then, the function set*

$$M_4 = \{\sin(\frac{n\pi x}{L})\sin(\frac{m\pi y}{H}), n, m = 1, 2, \cdots\}$$

forms an orthogonal basis for $L^2(D)$.

Proof. We know that $\{\sin(\frac{n\pi x}{L}), n = 1, 2, \cdots\}$ forms a basis for $L^2(0, L)$. It follows that for any $y \in [0, H]$, $f(x, y)$ can be expressed as a Fourier sine series:

$$f(x, y) = \sum_{n=1}^{\infty} a_n(y)\sin(\frac{n\pi x}{L}), \qquad x \in [0, L],$$

where

$$a_n(y) = \frac{2}{L}\int_0^L f(x, y)\sin(\frac{n\pi x}{L})dx, \qquad n = 1, 2, \cdots.$$

For every n, $a_n(y)$ can be expressed as a Fourier sine series on $[0, H]$:

$$a_n(y) = \sum_{m=1}^{\infty} A_{nm}\sin(\frac{n\pi y}{H}), \qquad m = 1, 2, \cdots,$$

where

$$A_{nm} = \frac{2}{H}\int_0^H a_n(y)\sin(\frac{n\pi y}{H})dy, \; m = 1, 2, \cdots.$$

It follows that

$$f(x, y) = \sum_{n=1}^{\infty}\sum_{m=1}^{\infty} A_{nm}\sin(\frac{n\pi x}{L})\sin(\frac{n\pi y}{H}),$$

where

$$A_{nm} = \frac{4}{HL}\int_0^H\int_0^L f(x, y)\sin(\frac{n\pi x}{L})\sin(\frac{m\pi y}{H})dxdy, \qquad \forall n, m \geq 1.$$

Define

$$f_{kl}(x, y) := \sum_{n=1}^{k}\sum_{m=1}^{l} A_{nm}\sin(\frac{n\pi x}{L})\sin(\frac{m\pi y}{H}), \qquad (x, y) \in D.$$

We can follow the same idea as in the one-dimensional case to obtain

$$||f_{kl}(x, y) - f(x, y)||_{L^2(D)} \to 0, \; k, l \to \infty. \qquad \square$$

By using the same idea, we can easily construct a Fourier basis for a rectangular box $E = [0, L] \times [0, H] \times [0, K] \subset R^3$. However, when E is a sphere, the construction of a Fourier basis is much more complicated (see Chapter 3).

2.5 The contraction mapping principle and applications

In the study of partial differential equations, the solvability for a PDE problem is one of the central questions for the justification of a physical model. The contraction mapping principle is a powerful tool used in research. The method can be used to deal with some nonlinear problems.

2.5.1 The contraction mapping principle

A mapping M from a Banach space V into V is called a contraction mapping if there exists a constant $\theta \in (0, 1)$ such that

$$||M[u] - M[v]|| \leq \theta ||u - v||, \qquad \forall u, v \in V.$$

Clearly, a contraction mapping must be continuous.

Theorem 2.5.1. *There exists a unique fixed point for every contraction mapping in a Banach space.*

Proof. Let $u_0 \in V$ be any fixed point in V. We define a sequence $\{u_n\}$ by

$$u_{n+1} = M[u_n], \qquad n = 0, 1, 2, \cdots.$$

We claim $\{u_n\}$ is a Cauchy sequence. Indeed, from the definition for any $n \geq 1$ we see that

$$||u_{n+1} - u_n|| = ||M[u_n] - M[u_{n-1}]|| \leq \theta ||u_n - u_{n-1}|| \leq \cdots \leq \theta^n ||u_1 - u_0||.$$

Next, for any $n, m \geq 1$ with $n \geq m$, we use the triangle inequality to obtain

$$\begin{aligned} ||u_n - u_m|| &\leq \sum_{j=m+1}^{n} ||u_j - u_{j-1}|| \\ &\leq \sum_{j=m+1}^{n} \theta^{j-1} ||u_1 - u_0|| \\ &\leq \frac{||u_1 - u_0|| \theta^m}{1 - \theta} \to 0, \qquad \text{as } m \to \infty. \end{aligned}$$

Since V is a Banach space, we see $\{u_n\}$ must converge to an element, denoted by $u \in V$. Since the mapping M is continuous, it follows that

$$u = \lim_{n \to \infty} u_{n+1} = \lim_{n \to \infty} M[u_n] = M[u],$$

i.e., u is a fixed point of the mapping M. Uniqueness is obvious. □

The method used in the proof of Theorem 2.5.1 is called the successive iteration method. One can use this method to find a numerical approximation for a nonlinear PDE problem. The method also gives a way to estimate the error between the approximate solution u_n and the actual solution u:

$$||u - u_n|| \leq \frac{\theta^n}{1 - \theta} ||u_1 - u_0||.$$

This is particularly useful when dealing with a nonlinear problem.

2.5.2 Some applications

To illustrate the application of the contraction mapping principle, we give two examples here. The first one deals with the solvability of a linear integral equation. The other is concerned with a system of nonlinear ODEs.

Example 2.5.1. Solvability for a linear integral equation.

Let $f(x) \in L^2(\Omega)$. Consider an integral equation: Find $u(x) \in L^2(\Omega)$ such that

$$u(x) = f(x) + \lambda \int_\Omega K(x, y)u(y)dy, \tag{2.5.1}$$

where $K(x, y)$ is given (called a kernel) in $L^2(\Omega \times \Omega)$ and λ is a constant.

We choose $V = L^2(\Omega)$. For any $u \in L^2(\Omega)$, define

$$M[u] := f(x) + \lambda \int_\Omega K(x, y)u(y)dy.$$

Then, M is a contraction mapping if

$$\theta := \lambda \int_\Omega \int_\Omega K(x, y)^2 dxdy < 1.$$

It follows that the integral equation (2.5.1) has a unique solution $u(x) \in L^2(\Omega)$ if $\theta < 1$. Moreover, one can construct a successive sequence $u_n(x)$ as follows: Choose $u_0(x) = f(x)$ and

$$u_{n+1}(x) = M[u_n] = f(x) + \int_\Omega K(x, y)u_n(y)dy, \qquad n = 0, 1, \cdots.$$

We know that $u_n(x)$ must converge to a function $u(x) \in L^2(\Omega)$ that is the unique solution of the integral equation (2.5.1). Moreover, one can obtain the error estimate $||u_n - u||_2$ as long as θ can be estimated.

Example 2.5.2. Solvability for a nonlinear system of ODEs.

Let $f(t, u)$ be defined on $R^1_+ \times R^1$ be continuous with respect to t and Lipschitz continuous with respect to u:

$$|f(t, u) - f(t, v)| \leq M_0|u - v|, \qquad \forall u, v \in R^1, t \in R^1_+,$$

where M_0 is a constant.

Consider the following initial-value problem:

$$y'(t) = f(t, y(t)), \qquad t > 0, \tag{2.5.2}$$

$$y(0) = y_0. \tag{2.5.3}$$

It is clear that the solution $y(t)$ satisfies the following integral equation:

$$y(t) = y_0 + \int_0^t f(\tau, y(\tau))d\tau. \tag{2.5.4}$$

We choose

$$V = L^{\infty}(0, K),$$

where K will be determined later.

Define a mapping M from V into V as follows:

$$M[y(t)] = y_0 + \int_0^t f(\tau, y(\tau))d\tau.$$

We choose T_0 such that $\theta := M_0 T_0 < 1$. Then, for all $u(t), v(t) \in V$,

$$||M[u] - M[v]|| \le M_0 T_0 ||u(t) - v(t)||.$$

Now, we choose

$$K = |y_0| + \frac{1}{1-\theta}||y_1 - y_0||.$$

Then, $M[y_n] \in V$ for all $n \geq 1$. The contraction mapping principle implies that the nonlinear problem (2.5.2)–(2.5.3) has a unique solution $y(t)$ on $[0, T_0]$. Since $y(t) \in V$ and $f(t, u)$ are continuous, from the integral equation of $y(t)$ we see that $y(t)$ must be continuous on $[0, T_0]$. If we go back to the integral equation (2.5.4) for $y(t)$, we see that $y(t)$ must be differentiable on $[0, T_0]$. Hence, $y(t)$ satisfies the differential equation (2.5.2). For $T = \frac{1}{2M_0}$,

$$|y|_{L^{\infty}(0,T)} \le |y_0| + T M_0 |y|_{L^{\infty}(0,T)} + F_0 T,$$

where $F_0 = \max_{t \in [0,T]} |f(t, 0)|$.

It follows that

$$|y|_{L^{\infty}(0,T)} \le 2(|y_0| + \frac{F_0}{M_0}).$$

We can start with $t = T_0$ as an initial time and solve the ODE (2.5.2)–(2.5.3) to obtain a solution $[0, 2T_0]$. By using the Gronwall's inequality, we see that $y(t)$ is bounded in any bounded interval $[0, T]$ for $T > 0$. By repeating this process, we can obtain a global solution $y(t) \in C^1[0, \infty)$.

Define a successive sequence $y_n(t)$ as follows:

$$y_{n+1}(t) = y_0 + \int_0^t f(\tau, y_n(\tau))d\tau, \qquad n = 0, 1, \cdots.$$

One can obtain an approximate solution $y_n(t)$ and an error estimate as long as θ can be estimated. It is easy to see that the method can be extended to deal with a system of nonlinear ODEs (see Chapter 9).

2.6 The continuity method

In this section we introduce the method of continuity as another application of the contraction mapping principle. The basic idea of the method is that one can answer the solvability question for a linear PDE problem if the problem is connected continuously to the other problem in which the solvability of the problem is known, as long as some *a priori* estimates can be proved for the PDE problem. The generalization of this method is Leray–Schauder's fixed-point theorem, which is a powerful tool used in the study of solvability for nonlinear PDE problems.

Let $(V_1 || \cdot ||_1)$ and $(V_2, || \cdot ||_2)$ be Banach spaces. A mapping M from V_1 to V_2 is said to be linear if

$$M[k_1u_1 + k_2u_2] = k_1M[u_1] + k_2M[u_2], \qquad \forall u_1, u_2 \in V_1, k_1, k_2 \in R^1.$$

M is said to be bounded if there exists a constant K such that

$$||M[u]||_2 \leq K||u||_1, \qquad \forall u \in V_1.$$

Similar to the definition in linear algebra, M is said to be one-to-one if $M[u_1] = M[u_2]$ implies $u_1 = u_2$. M is said to be onto if for every $v \in V_2$ there exists an element $u \in V_1$ such that $M[u] = v$. If M is one-to-one and onto (bijection), then the inverse mapping of M exists. All of the bounded linear mapping forms a normed space with a norm defined by

$$||M|| = \sup_{u \in V_1, u \neq 0} \frac{||M[u]||_2}{||u||_1}.$$

One can easily verify that a bounded linear mapping must be continuous.

Let L_0 and L_1 be bounded linear operators from V_1 to V_2. Define a linear operator L_t as follows:

$$L_t = (1-t)L_0 + tL_1, \qquad 0 \leq t \leq 1.$$

It is clear that L_t is a bounded linear operator from V_1 to V_2 for all $t \in [0, 1]$. Suppose we know that for every $f \in V_2$, a linear equation

$$L_0[u] = f$$

has a solution $u \in V_1$, what conditions are needed such that

$$L_1[u] = f$$

also has a solution? The following theorem answers this question.

Theorem 2.6.1. *Let the linear operator L_t be defined as above. Suppose that there exists a constant C_0 such that*

$$||u||_1 \leq C_0||L_t[u]||_2, \qquad \forall t \in [0, 1]. \tag{2.6.1}$$

Then, the operator L_1 maps V_1 onto V_2 if and only if L_0 maps V_1 onto V_2.

Proof. First, the estimate (2.6.1) implies that the operator L_t is one-to-one for all $t \in [0, 1]$. It follows that the inverse operator L_t^{-1} exists and is also a bounded linear one. Suppose that L_s is onto for some $s \in [0, 1]$. Then, for any $v \in V_2$, the equation

$$L_t[u] = v$$

is equivalent to

$$L_s[u] = v + (L_s - L_t)[u] = v + (t - s)L_0[u] - (t - s)L_1[u].$$

Since L_s^{-1} exists, we see that

$$u = L_s^{-1}[v] + (t - s)L_s^{-1}(L_0 - L_1)[u].$$

Define a mapping $M : V_2 \to V_1$ as follows:

$$M[u] := L_s^{-1}[v] + (t - s)L_s^{-1}(L_0 - L_1)[u].$$

Then,

$$||M[u_1] - M[u_2]||_2 \leq |t - s|C_0||u_1 - u_2||_1,$$

where

$$C_0 = ||L_s^{-1}(L_0 - L_1)|| < \infty.$$

Now, if t is sufficiently close to s such that

$$\theta := |t - s|C_0 < 1,$$

then the mapping M is contractive from V_2 into V_1, which implies that M has a fixed point. Consequently, L_t is invertible as long as $|t - s|C_0 < 1$. Therefore for $t \in [0, \frac{1}{\theta}]$, L_t is an onto mapping if L_0 is an onto mapping. After a finite number of steps, we see that L_1 is an onto mapping. □

To illustrate the application of the continuity method, we consider two Poisson equations:

$$L_0[u] := -\Delta u = f(x), \qquad x \in \Omega, \tag{2.6.2}$$

$$u(x) = 0, \qquad x \in \partial\Omega, \tag{2.6.3}$$

and

$$L_1[u] := -\sum_{i,j=1}^{n} a_{ij}(x)u_{x_i x_j} + \sum_{i=1}^{n} b_i(x)u_{x_i} + c(x)u = f(x), \qquad x \in \Omega, \tag{2.6.4}$$

$$u(x) = 0, \qquad x \in \partial\Omega, \tag{2.6.5}$$

where the coefficients of Eq. (2.6.4) are assumed to be in $C^\alpha(\bar{\Omega})$.

Choose $V_1 = C^{2+\alpha}(\bar{\Omega})$ and $V_2 = C^\alpha(\bar{\Omega})$ for some $\alpha \in (0, 1)$. It is clear that L_0 and L_1 are linear and bounded operators in V_1. Define

$$L_t = (1-t)L_0 + tL_1, \qquad 0 \le t \le 1.$$

Suppose we know that for every $f(x) \in V_2$, there exists a unique solution $u \in V_1$ for the problem (2.6.2)–(2.6.3). This implies that the operator L_0 from V_1 into V_2 is an onto mapping.

For all $t \in [0, 1]$, if we have the following estimate

$$||u||_{V_1} \le C||L_t[u]||_{V_2}, \tag{2.6.6}$$

where C is a constant that depends only on known data.

Then, the continuity method implies that the problem (2.6.4)–(2.6.5) has a unique solution $u(x) \in V_1$ for every $f(x) \in V_2$.

The crucial step in this example is to derive the estimate (2.6.6), which is called an *a priori* estimate. It is often difficult to derive such an *a priori* estimate. Many of these estimates need very delicate analysis. The interested reader may find such an estimate in Schauder's theory (see the advanced PDE book [12]).

2.7 The Fredholm Alternative and applications

In this section we state an important theorem related to a compact operator. Some examples are given to illustrate the application of the theorem.

2.7.1 The Fredholm Alternative

First, we need to define a compact operator.

Definition 2.7.1. Let X and Y be Banach spaces. An operator

$$K : X \to Y$$

is said to be compact if for every bounded sequence $\{u_n\} \subset X$, the set $\{Ku_n\}$ is precompact in Y, i.e., the sequence $\{Ku_n\}$ has a convergent subsequence in Y.

For a Hilbert space H, we can define an adjoint operator K^* by

$$< Ku, v >=< u, K^*v >, \qquad u, v, \in H.$$

K is called self-adjoint if $K = K^*$.

An interesting property is that K^* is compact if and only if K is compact.

Proposition 2.7.1. *Let $K : H \to H$ be a linear operator. Then, the operator K is compact if and only if the adjoint operator K^* is compact.*

The proof is directly from the definition. We leave it as an exercise.

Definition 2.7.2. (a) For an operator $K : H \to H$, if there exists a number λ and a nonzero vector $u \in H$ such that

$$Ku = \lambda u, \tag{2.7.1}$$

then we say λ is an eigenvalue of K and u is the corresponding eigenvector. Define

$$\sigma(K) = \{\text{all eigenvalues of the operator } K\}.$$
$$N_\lambda(K) = span\{\text{all eigenvectors corresponding to an eigenvalue } \lambda\}.$$

The set $\sigma(K)$ is called the spectrum of the operator K. $N_\lambda(K)$ is called the eigenspace corresponding to the eigenvalue λ. Clearly, $N_\lambda(K)$ is a subspace of H.

Let $I : H \to H$ be an identity mapping. Eq. (2.7.1) can be rewritten as

$$(K - \lambda I)u = 0. \tag{2.7.2}$$

Define the null space for the operator $\lambda I - K$:

$$Ker(\lambda I - K) = \{u \in H : (\lambda I - K)u = 0.\} = N_\lambda(K).$$

The following theorem is a generalization from linear algebra.

Theorem 2.7.1. *Let H be a Hilbert space and $K : H \to H$ be a bounded linear operator and compact. Then, for any $\lambda \in R^1$, either $(\lambda I - K)$ is invertible and for every $f \in H$ the equation*

$$(\lambda I - K)u = f$$

has a unique solution $u = (\lambda I - K)^{-1} f$ or $\lambda \in \sigma(K)$ and the equation

$$(\lambda I - K)u = f$$

has a solution if and only if $f \in N_\lambda(K)^\perp$. Moreover, when $\lambda \in \sigma(K)$, the following statements hold:

(a) *$Ker(I - \lambda K)$ is finite dimensional;*
(b) *$R(\lambda I - K)$ is closed;*
(c) *$R(\lambda I - K) = N(\lambda I - K)^\perp$;*
(d) *$Ker(\lambda I - K^*) = 0$ if and only if $R(\lambda I - K) = H$;*
(e) *$dimN(\lambda I - K) = dimN(\lambda I - K^*)$.*

The proof can be found in Functional Analysis, such as in [6].
From Theorem 2.7.1, we immediately have the following consequence.

Corollary 2.7.1. *(a) If K is compact, then $\sigma(K)$ is a compact subset of R^1 and the only possible limit point in $\sigma(K)$ is 0. Moreover, $\sigma(K) \subset [-||K||, ||K||]$.*
(b) $dimN_\lambda(K)$ is finite.

For any number $\mu \in R\backslash\sigma(K)$, the operator $(K - \mu I)$ is invertible in H. It follows that for every $f \in H$, there exists a unique $u \in H$ such that

$$u = (K - \mu I)^{-1} f.$$

One can see the compactness of the operator K plays an essential role in solving an equation in an infinite-dimensional space. The existence for an ODE or PDE problem relies on the compactness property of the corresponding inverse operator.

2.7.2 Applications

In this subsection we give two examples to demonstrate how the Fredholm Alternative can be applied to solve linear equations.

Example 2.7.1. Let $f(x) \in L^2(\Omega)$. Consider an integral equation: Find $u(x) \in L^2(\Omega)$ such that

$$u(x) = f(x) + \lambda \int_\Omega K(x, y)u(y)dy, \tag{2.7.3}$$

where $K(x, y) = K(y, x)$ are given in $L^2(\Omega \times \Omega)$ and λ is a constant.

It is known from Section 2.5 that the integral equation (2.7.3) has a unique solution $u(x)$ if $\lambda K_0 < 1$, where

$$K_0 = \int_\Omega \int_\Omega K(x, y)^2 dxdy.$$

When this condition is violated, does the integral equation have a solution for any $f(x) \in L^2(\Omega)$?

To answer the question, we define an operator M_0 as follows:

$$M_0 : u(x) \in L^2(\Omega) \to v(x) = M_0[u] = \int_\Omega K(x, y)u(y)dy.$$

It is easy to verify that M_0 is self-adjoint since $K(x, y) = K(y, x)$.

Suppose the operator M_0 is compact, then by the Fredholm Alternative, the integral equation (2.7.3) has a solution $u(x)$ for any $f(x) \in L^2(\Omega)$ if λ is not an eigenvalue of M_0. On the other hand, if λ is an eigenvalue of K, then the integral equation (2.7.3) has a solution for every $f(x) \in N_\lambda^\perp$, where

$$N_\lambda := \{u(x) \in L^2(\Omega) : M_0[u] = \lambda u\}.$$

This gives a completed answer for the solvability of the integral equation (2.7.3) as long as K is compact.

There are several different sets of conditions that can ensure the compactness of the operator M_0. For example, when $K(x, y) = K(y, x)$ is continuous in $\bar{\Omega} \times \bar{\Omega}$, then the operator M_0 is compact by using Arzela–Ascoli's theorem (see [5]).

Example 2.7.2. Consider the Laplace equation:

$$L_0[u] := -\Delta u = \lambda u + f(x), \qquad x \in \Omega, \tag{2.7.4}$$

$$u(x) = 0, \qquad x \in \partial\Omega, \tag{2.7.5}$$

where $\lambda \geq 0$ is a constant.

We denote by K the inverse operator of L_0 associated with the Dirichlet boundary condition (2.7.5). Note that, if $\lambda \neq 0$,

$$L_0[u] = \lambda u \qquad \Longleftrightarrow \qquad K[u] = \frac{1}{\lambda}u.$$

Since the operator L_0 is self-adjoint, its inverse operator K is also self-adjoint.

Suppose K is a compact operator from $L^2(\Omega)$ to $L^2(\Omega)$. Then, we consider two cases:

Case 1: $\mu := \frac{1}{\lambda}$ is not an eigenvalue of K.

For this case, the boundary value problem (2.7.4)–(2.7.5) has a solution for all $f(x) \in L^2(\Omega)$ by using the Fredholm Alternative. Namely, if λ is not an eigenvalue of L_0 associated with a Dirichlet boundary condition, then the boundary value problem (2.7.4)–(2.7.5) is solvable for all $f(x) \in L^2(\Omega)$.

Case 2: $\mu := \frac{1}{\lambda}$ is an eigenvalue of K (equivalently, $\lambda = \frac{1}{\mu}$ is an eigenvalue of L_0).

Let

$$N_\mu := \{u(x) \in L^2(\Omega) : K[u] = \mu u\}.$$

Then, for any $f(x) \in N_\mu^\perp$, the boundary value problem (2.7.4)–(2.7.5) has a solution $u(x)$. Namely, if λ is an eigenvalue of L_0 associated with a Dirichlet boundary condition, then the boundary value problem (2.7.4)–(2.7.5) is solvable if and only if $f(x) \in N_\mu^\perp$.

The compactness of the operator K relies on the regularity theory for the solution of the Laplace equation, which will be discussed in Chapter 6.

2.8 The Riesz representation and the Lax–Milgram theorem

In this section we state two important theorems that are used in the study of partial differential equations. Recall that a mapping from a Banach space to R^1 is called a functional. Define a new space

$$B' := \{\text{all bounded linear functionals from } B \text{ to } R^1\}.$$

Then, B' is called the dual space of B. B' is a Banach space equipped with the norm

$$||L|| = \sup_{u \in B, ||u||=1} |L[u]|.$$

Theorem 2.8.1. *(Riesz representation) Let H be a Hilbert space. Then, for every continuous linear functional $L \in H'$, there exists a unique vector $v \in H$ such that*

$$L[u] = (u, v), \qquad \forall u \in H.$$

Moreover,

$$||L|| = ||v||.$$

With the Riesz representation theorem, we can define a mapping

$$M : L \in H' \rightarrow v = M[L] \in H.$$

Then, M is linear and bijection. Hence, the dual space H' is isometric to H. We simply treat H and H' as identical. In this case, H is called reflexive.

Let H be a Hilbert space. Define a functional

$$B[u, v] : H \times H \rightarrow R^1.$$

$B[u, v]$ is said to be bilinear if $B[u, v]$ is linear with respect to u and v in H.

With the help of the Riesz representation theorem, we can state the following theorem that is essential in the existence theory for linear partial differential equations of elliptic type.

Theorem 2.8.2. *(Lax–Milgram theorem) Let H be a Hilbert space and $B[u, v]$ be a bilinear functional from $H \times H$ to R^1. If there exist two constants $\beta > 0$ and $M > 0$ such that*

$$(a) \quad B[u, u] \geq \beta ||u||^2, \qquad \forall u \in H,$$
$$(b) \quad |B[u, v]| \leq M ||u|| ||v||, \qquad \forall u, v \in H,$$

then, for every bounded linear functional $f \in H'$ there exists a unique u such that

$$B[u, v] = < f, v >, \qquad \forall v \in H.$$

The condition (a) is called the coercive condition, while the condition (b) is the boundedness condition. It will be seen in Chapter 6 that the existence of a solution for an elliptic equation is equivalent to the verification of the conditions in the Lax–Milgram theorem. We give a simple example to illustrate the basic idea.

Example 2.8.1. Let $f(x) \in L^2(\Omega)$. Consider the Laplace equation with a lower-order term:

$$-\Delta u + a(x)u = f(x), \; x \in \Omega,$$
$$u(x) = 0, \qquad x \in \partial\Omega.$$

For any function $u(x) \in H_0^1(\Omega)$, we define an inner production

$$< u, v >:= \int_\Omega [\nabla u \cdot \nabla v + uv] dx.$$

Define

$$B[u, v] = (\nabla u, \nabla v) + (a(x)u, v), \qquad \forall u, v \in H_0^1(\Omega).$$

Then, we see that $B[u, v]$ is bilinear. If $B[u, v]$ satisfies the Lax–Milgram conditions, then we obtain a solution in $H_0^1(\Omega)$. Suppose $a(x) \in L^\infty(\Omega)$ and $a(x) \geq a_0$ on Ω for some constant $a_0 > 0$. Then, one can easily verify that $B[u, v]$ satisfies the Lax–Milgram conditions, which implies the problem has a solution for every $f(x) \in L^2(\Omega)$. We will see more detail in Chapter 6.

2.9 Notes and remarks

Most materials in this chapter serve as an introduction or a review for students with or without the basic knowledge in functional analysis. Many concepts for Banach and Hilbert spaces are generalizations from the theory for a finite-dimensional space in linear algebra. Fourier series and function representations are interesting topics that have other applications in addition to the application in partial differential equations.

The materials from Sections 2.5 to 2.8 are some tools used to establish the existence theory for partial differential equations. Some of the tools can be used to deal with nonlinear equations. This part of the material is provided for students interested in theoretical analysis. For the beginners in PDEs, one may skip these sections without much difficulty for the rest of the book.

2.10 Exercises

1. In R^n, for each $x = (x_1, x_2, \cdots, x_n)$. Let $p \geq 1$. Define

$$||x||_p = \left(|x_1|^p + \cdots + |x_n|^p\right)^{\frac{1}{p}}, \qquad ||x||_\infty = \max_{1 \leq k \leq n} |x_k|.$$

(a) Prove both $||x||_p$ and $||x||_\infty$ are norms in R^n.

(b) Prove $||x||_\infty = \lim_{p \to \infty} ||x||_p$.

2. Let $\{v_n\}$ be a sequence in a Banach space V. Prove that each subsequence v_{n_k} of a Cauchy sequence $\{v_n\}$ converges to the same limit in V.

3. Let M be the vector space that consists of all $n \times n$ square matrices. For $A = (a_{ij})_{n \times n} \in M$, we define

$$||A|| = \max_{1 \leq i, j \leq n} |a_{ij}|.$$

Prove (a) $||A||$ is a norm for M.
(b) M is a Banach space.
(c) Define an inner product on M such that M is a Hilbert space.

4. Let H be a Hilbert space. Define an inner product for the product space $H \times H$ as follows: For any $u = (u_1, u_2)$, $v = (v_1, v_2)$ in $H \times H$,

$$< u, v > = < u_1, v_1 > + < u_2, v_2 > .$$

Verify that the above inner product satisfies the axioms of the inner product.

5. Let $w(x)$ be a continuous positive function in a bounded domain $\bar{\Omega}$ in R^n. Define

$$< f, g >_w := \int_B f(x)g(x)w(x)dx, \qquad f(x), g(x) \in L^2(\Omega).$$

Verify $< \cdot, \cdot >_w$ is an inner product on $L^2(\Omega)$. $w(x)$ is called a weight function

6. Prove the following generalized Hölder inequality: Let $p_1, \cdots, p_m \geq 1$ satisfy

$$\frac{1}{p_1} + \cdots + \frac{1}{p_m} = 1.$$

Prove

$$\int_\Omega |f_1(x) \cdots f_m(x)| dx \leq ||f_i||_{p_1} \cdots ||f_m||_{p_m},$$

where $f_i(x) \in L^{p_i}(\Omega)\ \forall 1 \leq i \leq m$.

7. Let Ω be a bounded domain. Suppose a sequence $f_n(x)$ is convergent to a function $f(x)$ in $L^p(\Omega)$. Then, $f_n(x)$ is convergent to $f(x)$ in $L^q(\Omega)$ provided that $q \leq p$. Is the conclusion true if $q > p$? (Hint: think about a sequence $f_n(x) = \sin(nx)$.)

8. Let $f_1(x) = 3$, $f_2(x) = 1 - x$ and $f_3(x) = x^2 + 2x - 1$.
(a) Prove the set $\{f_1(x), f_2(x), f_3(x)\}$ is linearly independent.
(b) Use the Gram–Schmidt process to find an orthogonal function set from $\{f_1, f_2, f_3\}$ over $[0, 1]$.

9. Determine whether or not the following is linearly independent. Justify your answer.
(a) $A := \{1, \sin x, \cos x\}$ in $[0, 2\pi]$.
(b) $B := \{\sin(nx), n = 1, 2, \cdots\}$ in $[0, \pi]$.
(c) $C := \{1, x, x^2, \sin x, \cos x\}$ in $[0, 1]$.

10. Let $\alpha, \beta \in (0, 1]$ with $\alpha < \beta$. Let Ω be a bounded domain in R^n. Prove

$$C^\beta(\bar{\Omega}) \subset C^\alpha(\bar{\Omega}).$$

Moreover, prove that the embedding operator $I : u(x) \in C^\beta(\bar{\Omega})$ into $I[u] = u(x) \in C^\alpha(\bar{\Omega})$ is compact.

11. Let Ω be a bounded domain in R^n. Prove

$$L^q(\Omega) \subset L^p(\Omega), \qquad \text{if } 1 \leq p < q.$$

Is the conclusion true if Ω is unbounded in R^n? Is the embedding operator I from $L^q(\Omega)$ into $L^p(\Omega)$ compact?

12. Let Ω be a bounded domain in R^n and $f(x) \in L^p(\Omega)$. Prove

$$\lim_{p\to\infty} ||f||_{L^p(\Omega)} = ess.\sup_{\bar{\Omega}} |f(x)|.$$

(Hint: Prove the result for $f(x) \in C(\bar{\Omega})$ as a first step.)

13. Let $f(x) = x(1-x)$.
(a) Find the Fourier series for $f(x)$ on $[-1, 1]$.
(b) Find the Fourier cosine series for $f(x)$ on $[0, 1]$.
(c) Find the Fourier sine series for $f(x)$ on $[0, 1]$.

14. Let $f(x)$ and $g(x)$ be Lipschitz continuous on a bounded domain $\bar{\Omega}$ in R^n. Prove $f(x) + g(x)$ and $f(x)g(x)$ are also Lipschitz continuous over $\bar{\Omega}$.

15. Let

$$M = \{\sin(nx) : n = 1, 3, 5, \cdots\}.$$

Can M form a basis for $L^2(0, \pi)$?

16. Let $Q(0, 2\pi)$ be the set of all irrational numbers in $(0, 2\pi)$.

$$M = \{\sin(p_i x) : p_i \in Q(0, 2\pi)\}.$$

Does M form a basis for $L^2(0, 1)$?

17. Let $L[u] = \frac{du}{dx}$ be an operator defined on $C^1[a, b]$. Prove that L is a linear operator, but not bounded on $C^1[a, b]$. Is the operator L invertible?

18. Let Ω be a bounded domain. Let $K(x, y)$ be a continuous function defined on $\bar{\Omega} \times \bar{\Omega}$. Define a mapping M from $C(\bar{\Omega})$ to $C(\bar{\Omega})$:

$$M : u(x) \in C(\bar{\Omega}) \to v = M[u] := \int_{\Omega} K(x, y)u(y)dy.$$

Prove the mapping M is a compact mapping. (Hint: Use Ascoli–Arzela's lemma.)

19. (a) Prove l^∞ is not separable.
(b) Prove $l_c^\infty \subset l^\infty$ is separable, where l_c^∞ consists of all convergent sequences in l^∞.

CHAPTER

Eigenvalue problems and eigenfunction expansions

3

3.1 The method of separation of variables

To find a solution for a PDE problem, one intends to use known methods and tools from the theory of ODEs. Hence, the separation of variables is introduced with the expectation that a function of one variable will satisfy an ODE in which its exact solution can be found from the ODE theory.

We use the heat equation as an example to illustrate the idea of the method. This equation will be investigated in detail in Chapter 4. Consider the heat equation with one-space dimension over interval $[0, L]$:

$$u_t = k^2 u_{xx}, \qquad 0 < x < L, t > 0, \tag{3.1.1}$$

where k is a positive constant.

Suppose that $u(x,t)$ can be expressed by the product of two functions with a single variable:

$$u(x,t) = h(t)\phi(x), \qquad 0 < x < L, t > 0,$$

where $h(t)$ and $\phi(x)$ are unknown functions.

From the heat equation (3.1.1), we see that

$$h'(t)\phi(x) = k^2 h(t)\phi''(x), \qquad 0 < x < L, t > 0.$$

It follows that

$$\frac{h'(t)}{k^2 h(t)} = \frac{\phi''(x)}{\phi(x)}, \qquad 0 < x < L, t > 0. \tag{3.1.2}$$

Note that the left-hand side of Eq. (3.1.2) is a function of t, while the right-hand side is a function of x. Hence, each side must be equal to a constant (unknown). Suppose the unknown constant is equal to $-\lambda$:

$$\frac{h'(t)}{k^2 h(t)} = \frac{\phi''(x)}{\phi(x)} = -\lambda, \qquad 0 < x < L, t > 0,$$

where the negative sign of λ is used for convenience.

Partial Differential Equations and Applications. https://doi.org/10.1016/B978-0-44-318705-6.00009-4

Consequently, we find that

$$h(t) = Ce^{-\lambda k^2 t}, \qquad t > 0,$$

where C is an arbitrary constant.

Also, $\phi(x)$ satisfies the following second-order ODE:

$$-\phi''(x) = \lambda\phi, \qquad 0 < x < L, \tag{3.1.3}$$

where the unknown constant λ and $\phi(x)$ need to be determined at the same time.

Eq. (3.1.3) with appropriate boundary conditions at $x = 0$ and $x = L$ forms an eigenvalue problem for the differential operator $L_o := \frac{d^2}{dx^2}$. We will see that the second-order ODE (3.1.3) with a boundary condition may not always have a nontrivial solution for every constant λ. When it does, λ is called an eigenvalue and the corresponding solution $\phi(x)$ is called an eigenfunction corresponding to λ.

An eigenvalue problem for a general ODE is the topic of the next section, which is called the Sturm–Liouville theory. When the space dimension is more than 1, it is the general eigenvalue problem for the Laplace (or an elliptic) operator.

3.2 The Sturm–Liouville theory

With the basic knowledge in the theory of ODEs, we can solve the eigenvalue problem completely when the space dimension is equal to one. This is often referred to as the Sturm–Liouville theory in the literature.

3.2.1 Eigenvalue problems in one-space dimension

Let $p(x)$ and $q(x)$ be continuous functions on $[a, b]$ with $p(x) > 0$ on (a, b). Let $\mathcal{L}$ be the following differential operator,

$$\mathcal{L}[\phi] := -\frac{d}{dx}\left[p(x)\frac{d\phi}{dx}\right] + q(x)\phi. \tag{3.2.1}$$

One can prove (see Exercise 3.1) that a general second-order linear ODE operator can always be expressed by the above form if one introduces a suitable integrating factor. To include some important applications, we consider an eigenvalue problem with a weight function.

Let $\rho(x)$ be a bounded, positive, and piecewise-continuous function in $[a, b]$.[1] All functions that are square-integrable with weight ρ are denoted by $L^2(a, b; \rho)$. That

[1] For advanced students, the positivity condition for the weight function $\rho(x)$ can be relaxed as $\rho(x)dx$ is a regular Lebesgue measure.

is,

$$L^2(a,b;\rho) = \{f(x) : \int_a^b f(x)^2 \rho(x) dx < \infty\}.$$

The inner product with the weight $\rho(x)$ is defined by

$$< f, g >_\rho = \int_a^b f(x) g(x) \rho(x) dx.$$

In most applications in this book, $\rho(x) = 1$ on $[a, b]$. It is easy to verify that $L^2(a, b; \rho)$ is a Hilbert space.

Consider the eigenvalue problem: Find a number λ and a nontrivial function ϕ such that

$$\mathcal{L}[\phi] = \lambda \rho(x) \phi, \qquad a < x < b, \tag{3.2.2}$$

subject to the following boundary conditions:

$$a_1 \phi(a) + a_2 p(a) \phi'(a) = 0, \tag{3.2.3}$$

$$b_1 \phi(a) + b_2 p(b) \phi(b) = 0, \tag{3.2.4}$$

where $a_1^2 + a_2^2 > 0$, $b_1^2 + b_2^2 > 0$.

In applications, there is another type of boundary conditions where the boundary conditions (3.2.3)–(3.2.4) are replaced by the following periodic type:

$$\phi(a) - \phi(b), \qquad p(a)\phi'(a) = p(b)\phi'(b). \tag{3.2.5}$$

The eigenvalue problem (3.2.2)–(3.2.4) (or (3.2.2) with (3.2.5)) is called regular if $[a, b]$ is a bounded interval and $p(x) \geq p_0 > 0$ on $[a, b]$ for some $p_0 > 0$. Otherwise, it is called singular.

Similar to eigenvalue problems in linear algebra, we are interested in only non-trivial solutions for an eigenvalue problem. We want to emphasize that a boundary value problem for an ODE may not have a solution.

Definition 3.2.1. We say λ is an eigenvalue of the problem (3.2.2)–(3.2.4) (or (3.2.2), (3.2.5)) if the problem (3.2.2)–(3.2.4) (or (3.2.2), (3.2.5)) has a nontrivial solution $\phi(x)$ corresponding to a real number λ. The solution $\phi(x)$ is called an eigenfunction and the number λ is called an eigenvalue.

Let

$$N_\lambda = Span\{\phi(x) \in H^1(a, b) : \mathcal{L}[\phi] = \lambda\phi \text{ associated with (3.2.3)–(3.2.4) or (3.2.5)}\}.$$

N_λ is called the eigenspace corresponding to the eigenvalue λ.

From the above definition, we see that any nonzero constant multiple of an eigenfunction is still an eigenfunction corresponding to the same eigenvalue. There are many interesting questions for an eigenvalue problem. For example, how many eigenvalues does an eigenvalue problem have? For each eigenvalue, how many linearly independent eigenfunctions are there? Are these eigenfunctions mutually orthogonal? We will answer these questions in the next section.

3.2.2 Some elementary examples

We first consider some simple examples. It will be seen that each function set in Example 2.2.1 in Chapter 2 are eigenfunctions for the differential operator $\frac{d^2}{dx^2}$ with different types of boundary conditions.

Example 3.2.1. Find all eigenvalues λ and the corresponding eigenfunctions $\phi(x)$ for the following eigenvalue problem:

$$-\phi''(x) = \lambda\phi, \qquad 0 < x < L, \tag{3.2.6}$$

$$\phi(0) = \phi(L) = 0. \tag{3.2.7}$$

Solution. Since λ is unknown, we need to find λ and the corresponding eigenfunction $\phi(x)$. We also keep in mind that we are only interested in nontrivial solutions.

We divide λ into three possible cases.

Case 1: Let $\lambda = 0$.

In this case, Eq. (3.2.6) becomes

$$\phi''(x) = 0, \qquad 0 < x < L.$$

The general solution of the ODE is equal to

$$\phi(x) = c_1 x + c_2,$$

where c_1 and c_2 are two arbitrary constants.

From the boundary condition (3.2.7), we see that

$$\phi(0) = c_2 = 0, \qquad \phi(L) = c_1 L = 0.$$

Hence, $c_1 = c_2 = 0$. Consequently, $\phi(x) = 0$ is the only trivial solution, which implies that $\lambda = 0$ is not an eigenvalue.

Case 2: Let $\lambda < 0$.

For convenience we set $\lambda = -k^2$ for some $k > 0$.

From the ODE (3.2.6):

$$\phi'' - k^2\phi = 0.$$

The characteristic equation is

$$r^2 - k^2 = 0,$$

which has two real roots $r_1 = k$ and $r_2 = -k$.

The general solution for the ODE of ϕ is equal to

$$\phi(x) = c_1 e^{kx} + c_2 e^{-kx}, \qquad 0 < x < L,$$

where c_1 and c_2 are arbitrary constants.

From the boundary condition (3.2.7), we see that

$$c_1 + c_2 = 0, \; c_1 e^{kL} + c_2 e^{-kL} = 0.$$

The linear system has only a trivial solution: $c_1 = c_2 = 0$.

It follows that $\phi(x) = 0$, $0 < x < L$, is the only trivial solution, which is not an eigenfunction. This implies that the eigenvalue λ cannot be negative.

Case 3: Let $\lambda > 0$.

We set $\lambda = k^2$ for some $k > 0$. From the ODE (3.2.6), we see that

$$\phi''(x) + k^2 \phi(x) = 0, \qquad 0 < x < L.$$

The characteristic equation is

$$r^2 + k^2 = 0,$$

which has two complex roots:

$$r = \pm ki.$$

The general real solution of the ODE (3.2.6) is equal to

$$\phi(x) = c_1 \cos(kx) + c_2 \sin(kx),$$

where c_1 and c_2 are arbitrary constants.

From the boundary condition $\phi(0) = 0$, we see that

$$\phi(0) = c_1 = 0.$$

It follows that

$$\phi(x) = c_2 \sin(kx).$$

From the boundary condition $\phi(L) = 0$, we see that

$$c_2 \sin(kL) = 0.$$

Since we want to find nontrivial solutions, we choose $c_2 \neq 0$. This implies that

$$\phi(x) = \sin(kL) = 0,$$

which yields

$$kL = n\pi, \qquad n = 1, 2, 3, \cdots.$$

Namely, for $k = \frac{n\pi}{L}$ we find the nontrivial solution $\phi(x) = \sin(\frac{n\pi x}{L})$. Hence, all eigenvalues and the corresponding eigenfunctions are

$$\lambda_n = \left(\frac{n\pi}{L}\right)^2, \qquad \phi_n(x) = \sin(\frac{n\pi x}{L}), \qquad n = 1, 2, 3, \cdots.$$

The set of all eigenfunctions for the problem (3.2.6)–(3.2.7) is the set M_3 in Chapter 2, which forms a basis for $L^2(0, L)$ (see Theorem 3.3.2 in Section 3.3). For any function $f(x) \in L^2(0, L)$, it can be expressed by the Fourier series:

$$f(x) = \sum_{n=1}^{\infty} c_n \sin(\frac{n\pi x}{L}),$$

where

$$c_n = \frac{< f, \sin(\frac{n\pi x}{L}) >}{< \sin(\frac{n\pi x}{L}), \sin(\frac{n\pi x}{L}) >} = \frac{2}{L}\int_0^L f(x)\sin(\frac{n\pi x}{L})dx, \qquad \forall n \geq 1.$$

This is the Fourier sine series.

Example 3.2.2. Find all the eigenvalues and the corresponding eigenfunctions for the following eigenvalue problem:

$$-\phi''(x) = \lambda\phi, \qquad 0 < x < L, \tag{3.2.8}$$

$$\phi'(0) = 0, \phi'(L) = 0. \tag{3.2.9}$$

Solution. By using the same steps, we easily obtain all the eigenvalues λ_n and the corresponding eigenfunctions are

$$\lambda_n = (\frac{n\pi}{L})^2, \qquad \phi_n(x) = \cos(\frac{n\pi x}{L}), \qquad n = 0, 1, 2, \cdots.$$

We will show that all the eigenfunctions also form a basis for $L^2(0, L)$. Note that $\lambda_0 = 0$ is an eigenvalue in this example, which is different from Example 3.2.1.

For any function $f(x) \in L^2(0, L)$, we can expand $f(x)$ as a series:

$$f(x) = \sum_{n=0}^{\infty} c_n \cos(\frac{n\pi x}{L}),$$

where

$$c_0 = \frac{1}{L}\int_0^L f(x)dx, c_n = \frac{< f, \cos(\frac{n\pi x}{L}) >}{< \cos(\frac{n\pi x}{L}), \cos(\frac{n\pi x}{L}) >}$$

$$= \frac{2}{L}\int_0^L f(x)\cos(\frac{n\pi x}{L})dx, \qquad \forall n \geq 1.$$

This is the Fourier cosine series.

Example 3.2.3. Find all the eigenvalues and eigenfunctions for the following eigenvalue problem:

$$-\phi''(x) = \lambda\phi, \qquad -L < x < L,$$
$$\phi(-L) = \phi(L), \ \phi'(-L) = \phi'(L).$$

Solution. We follow the same procedure as in Example 3.2.1 by dividing λ into three cases.

Case 1: Let $\lambda = 0$.

From the ODE (3.2.6), we know the general solution is

$$\phi(x) = c_1 x + c_2,$$

where c_1 and c_2 are arbitrary constants.

From the boundary conditions, we see $c_1 = 0$. It follows that

$$\phi(x) = c_2$$

is a solution for any constant c_2. We simply choose $\phi(x) = 1$.

Case 2: Let $\lambda < 0$, say, $\lambda = -k^2$ with $k > 0$.

By following exactly the same method as in Example 3.2.1, we find that $\phi(x) = 0$. This implies that there is no negative eigenvalue.

Case 3: Let $\lambda > 0$, say, $\lambda = k^2$ for some $k > 0$.

The general solution of the ODE is

$$\phi(x) = c_1 \cos(kx) + c_2 \sin(kx).$$

By using the periodic boundary conditions, we find the eigenvalues

$$\lambda_n = \left(\frac{n\pi}{L}\right)^2, \qquad n = 1, 2, \cdots,$$

and the corresponding eigenfunctions are

$$\phi_n(x) = \cos(\frac{n\pi x}{L}), \qquad \sin(\frac{n\pi x}{L}), \qquad n = 1, 2, \cdots.$$

Note that when $n \geq 1$, for each eigenvalue λ_n, there are two linearly independent eigenfunctions. This is due to the periodic boundary condition.

The set of all eigenfunctions is the set M_1 in Example 2.2.1 in Chapter 2. For any function $f(x) \in L^2(-L, L)$,

$$f(x) = c_0 + \sum_{n=1}^{\infty} \left[a_n \cos(\frac{n\pi x}{L}) + b_n \sin(\frac{n\pi x}{L})\right],$$

where

$$c_0 = \frac{2}{L}\int_{-L}^{L} f(x)dx,$$
$$a_n = \frac{1}{L}\int_{-L}^{L} f(x)\cos(\frac{n\pi x}{L})dx,$$
$$b_n = \frac{1}{L}\int_{-L}^{L} f(x)\sin(\frac{n\pi x}{L})dx.$$

In the next section we will show that the set of all linearly independent eigenfunctions for a regular eigenvalue problem forms a basis for $L^2(0, L)$. The representation by a Fourier series is a special case of eigenfunction expansion.

3.3 The main theorem in the Sturm–Liouville theory

In this section we prove the main theorem for the eigenvalue problem in one-space dimension.

3.3.1 Self-adjoint operators

Definition 3.3.1. A differential operator L defined by $H^2(a, b)$ associated with the boundary conditions (3.2.3)–(3.2.4) (or (3.2.5)) is said to be self-adjoint if for any differentiable function $\phi(x)$ and $\psi(x)$,

$$< L[\phi], \psi >=< \phi, L[\psi] >, \qquad \text{for any } \phi(x), \psi(x) \in H^2(a, b),$$

where $\phi(x)$, $\psi(x)$ satisfy the boundary conditions (3.2.3)–(3.2.4) or (3.2.5).

There are many interesting properties for a self-adjoint operator. The general theory for this type of operator requires advanced functional analysis. However, for one-space dimension we have a completed theory.

Theorem 3.3.1. *The differential operator (3.2.1) associated with boundary conditions (3.2.3)–(3.2.4) (or (3.2.5)) is self-adjoint in any bounded interval* $[a, b]$.

Proof. We first assume $u, v \in C^2[a, b]$;

$$u\frac{d}{dx}\left(p\frac{dv}{dx}\right) = \frac{d}{dx}[u\left(p\frac{dv}{dx}\right)] - p\frac{dv}{dx}\cdot\frac{du}{dx},$$
$$v\frac{d}{dx}\left(p\frac{du}{dx}\right) = \frac{d}{dx}[u\left(p\frac{du}{dx}\right)] - p\frac{du}{dx}\cdot\frac{dv}{dx}.$$

It follows that

$$\int_a^b \{uL[v] - vL[u]\}dx = \int_a^b \frac{d}{dx}[p\left(u\frac{dv}{dx} - v\frac{du}{dx}\right)]dx$$

$$= p(x)\left(u\frac{dv}{dx} - v\frac{du}{dx}\right)\Big|_{x=a}^{x=b}.$$

We only consider the case where $a_2 \neq 0, b_2 \neq 0$, other cases can be proved similarly. In this case,

$$p(a)u'(a) = -\frac{a_1}{a_2}u(a), \ p(a)v'(a) = -\frac{a_1}{a_2}v(a);$$
$$p(b)u'(b) = -\frac{b_1}{b_2}u(b), \ p(b)v'(b) = -\frac{b_1}{b_2}v(b).$$

Thus

$$p(a)\left[u(a)v'(a) - v(a)u'(a)\right] = 0,$$
$$p(b)\left[u(b)v'(b) - v(b)u'(b)\right] = 0,$$

which implies

$$< L[u], v > = < u, L[v] >, \qquad \forall u, v \in C^2[a, b].$$

By applying the approximation theorem, we see that the above equality holds for all $u(x), v(x) \in H^2(a, b)$. From the definition of a self-adjoint operator, we see that the operator L defined by (3.2.1) associated with the boundary conditions (3.2.3)–(3.2.4) or (3.2.5) is self-adjoint. □

3.3.2 The main theorem

Now, we can state the main result in the Sturm–Liouville theory.

Theorem 3.3.2. *For the eigenvalue problem (3.2.2) with boundary condition (3.2.3)–(3.2.4) or (3.2.5), the following statements hold:*

(a) All eigenvalues are real.

(b) There exists an infinite number of eigenvalues that can be ordered by

$$\lambda_1 < \lambda_2 < \cdots < \lambda_n \cdots,$$

and

$$\lim_{n \to \infty} \lambda_n = \infty.$$

(c) For each eigenvalue λ_n with (3.2.3)–(3.2.4), there exists a unique eigenfunction $\phi_n(x)$ in the sense of linearly independent functions. For the periodic boundary condition (3.2.5), there exist at most two linearly independent eigenfunctions.

(d) Eigenfunctions corresponding to different eigenvalues are orthogonal in $L^2(a, b; \rho)$. Moreover, all eigenfunctions $\{\phi_n(x)\}_{n=1}^{\infty}$ can be chosen such that the set of all linearly independent eigenfunctions forms an orthogonal basis of $L^2(a, b; \rho)$.

Namely, every function in $L^2(a,b;\rho)$ *can be expressed by an infinite series*

$$f(x)=\sum_{n=1}^{\infty} c_n\phi_n(x), \tag{3.3.1}$$

where

$$c_n=\frac{\int_a^b f(x)\phi_n\rho(x)dx}{\int_a^b \phi_n^2\rho(x)dx}, \qquad n=1,2,\cdots,$$

and the series converges to $f(x)$ *in the sense of a mean-square with weight* $\rho(x)$.

(e) Every eigenvalue λ *and the corresponding eigenfunction* $\phi(x)$ *satisfy the following identity (called the Rayleigh quotient):*

$$\lambda=\frac{-p\phi\phi'(x)|_a^b+\int_a^b[p\phi'^2+q\phi^2]\rho dx}{\int_a^b \phi^2\rho dx}. \tag{3.3.2}$$

Proof. The conclusion (a) is clear, since all coefficients of L and solution are real.

(b) The proof of this part requires a tedious discussion for all possible cases. We give a proof for a special case where $\rho(x)=1$, $q(x)=0$ with $\phi(a)=\phi(b)=0$. Note that the eigenvalue problem (3.2.2)–(3.2.4) (or (3.2.5)) is equivalent to the following eigenvalue problem:

$$K[\phi]=\frac{1}{\lambda}\phi,$$

where $K=L^{-1}$ along with the boundary condition (3.2.3)–(3.2.4) (or (3.2.5)). In the advanced PDE (see [5,8]) the operator K is compact from $L^2(a,b)$ into $L^2(a,b)$. Therefore we can apply the Fredholm Alternative for the operator K to obtain that all eigenvalues are distinct, bounded by $||K||$ and 0 is only the possible limit point of the eigenvalue set $\{\frac{1}{\lambda}\}$. Next, we show that $\lambda_n\to\infty$ as $n\to\infty$.

Let $H_0^1(a,b)$ be the Sobolev space defined in Chapter 2. As we indicated before, one may use a function sequence in $C^1[a,b]$ with zero value near $x=a$ and $x=b$ to approximate a function in $H_0^1(a,b)$. Define a functional $J[u]$ as follows:

$$J(u):=\frac{\int_a^b p(x)(u')^2dx}{\int_a^b u^2dx}, \qquad u\in H_0^1(a,b).$$

Since $J(u)$ is nonnegative, it has a minimum in $H_0^1(a,b)$, denoted by λ_1,

$$\lambda_1=\min_{u\in H_0^1(a,b)} J(u).$$

Suppose $\phi_1(x)$ is the minimum of $J(u)$ in $H_0^1(a,b)$.

Define

$$I(t):=J(\phi_1+tu)\geq J(\phi_1), \qquad \text{for all } u\in H_0^1(a,b),\ t\in R^1.$$

Then, $I(t)$ attains its minimum at $t = 0$ and $I'(0) = 0$. It follows for any $u \in H_0^1(a, b)$ that

$$\int_a^b p(x)u'\phi_1' dx = \lambda_1 \int_a^b \phi_1 u dx.$$

After performing integration by parts, we find that

$$\int_a^b \left\{ -(\frac{d}{dx}[(p(x)\phi_1'] + \lambda_1\phi_1)u \right\} dx = 0, \qquad \forall u \in H_0^1(a, b).$$

Since $u(x)$ is arbitrarily, we see that

$$L[\phi_1] = \lambda_1\phi_1(x),\ \phi_1(a) = \phi_1(b) = 0.$$

Let

$$N_1 = span\{\phi(x) \in H_0^1(a, b) : L[\phi_1] = \lambda_1\phi_1\}.$$

Recall the inner product in $H_0^1(a, b)$ is given by

$$< u, v > = \int_a^b [u'v' + uv]dx, \qquad u, v \subset H_0^1(a, b).$$

Since the dimension of N_1 is equal to 1 (see (c)), we can decompose $H_0^1(a, b)$ into two subspaces

$$H_0^1(a, b) = N_1 \oplus N_1^\perp.$$

Set

$$\lambda_2 = \min_{u \in N_1^\perp} J(u).$$

Again, we use the same argument to obtain λ_2 and corresponding minimum $\phi_2(x)$. We can continue this process to obtain a sequence

$$\{\lambda_n, \qquad n = 1, 2, 3, \cdots\}.$$

Moreover, since $N_1^\perp \subset H_0^1(a, b)$, we see $\lambda_1 < \lambda_2 < \cdots$. Since the dimension of $H_0^1(a, b)$ is infinite, we see that there exists an infinite number of eigenvalues λ_n. We will show $\lambda_n \to \infty$ as $n \to \infty$ below.

(c) Suppose there exist two linearly independent eigenfunctions ϕ and ψ in N_1. Then,

$$L[\phi] = \lambda\phi, \qquad L[\psi] = \lambda\psi.$$

It follows that

$$\frac{d}{dx}[p(x)\phi_x]\psi(x) - \frac{d}{dx}[p(x)\psi_x]\phi(x) = 0.$$

Consequently,

$$p(x)[\phi_x\psi - \phi\psi_x] = \text{constant}, \qquad 0 < x < L.$$

The boundary condition (3.2.3)–(3.2.4) implies that the constant must be equal to 0. Hence,

$$\frac{d}{dx}\left(\frac{\phi(x)}{\psi(x)}\right) = 0, \qquad 0 < x < L.$$

That is, $\phi(x)$ and $\psi(x)$ are linearly dependent on $[0, L]$.

For the periodic boundary condition (3.2.5), $\psi(x)$ and $\phi(x)$ may be linearly independent. However, for each eigenvalue, there exist at most two linearly independent eigenfunctions since a second-order ODE has at most two linearly independent solution.

(d) Suppose $n \neq m$, $\phi_n(x) \in N_n$, $\phi_m(x) \in N_m$. We see that

$$L[\phi_n] = \lambda_n\phi_n, \qquad L[\phi_m] = \lambda_m\phi_m.$$

Since L is a self-adjoint operator, we see that

$$< L[\phi_n], \phi_m > = < \phi_n, L[\phi_m] >,$$

i.e.,

$$< \lambda_n\phi_n, \phi_m > = < \phi_n, \lambda_m\phi_m > .$$

It follows that

$$(\lambda_n - \lambda_m) < \phi_n, \phi_m > = 0.$$

Consequently, since $\lambda_n \neq \lambda_m$,

$$< \phi_n, \phi_m > = 0,$$

i.e., ϕ_n and ϕ_m are orthogonal in $L^2(a, b)$.

Let

$$M = \{\phi_n(x) : n = 1, 2, \cdots\}.$$

For every $f(x) \in L^2(a, b)$, consider the boundary value problem:

$$L[u] = \mu u + f(x), a < x < b,$$
$$u(a) = u(b) = 0.$$

If μ is not an eigenvalue of the operator L, then the operator $L - \mu I$ is invertible. It follows that $(L - \mu I)^{-1}$ exists and is compact. By using the Fredholm Alternative, we see that M forms a mutual orthogonal basis for $L^2(a, b)$. The same conclusion holds for other types of boundary conditions. Next, we prove that $\{\lambda_n\}$ has no upper bound. Let $f(x) \in C^2[a,b]$. Then,

$$f(x) = \sum_{n=1}^{\infty} a_n \phi_n(x),$$

where

$$a_n = \frac{< f, \phi_n >}{< \phi_n, \phi_n >}, \qquad n = 1, 2, \cdots.$$

It follows that

$$L[f] = \sum_{n=1}^{\infty} a_n \lambda_n \phi_n(x).$$

Hence,

$$< L[f], \phi_m >= \lambda_m a_m < \phi_m, \phi_m >,$$

which yields

$$\lambda_m = \frac{< L[f], \phi_m >}{a_m < f, \phi_m >}, \text{ if } a_m \neq 0.$$

If we can construct $f_n(x)$ such that

$$L[f_n] \geq n f_n,$$

then we see that λ_m has no upper bound. The construction of $f_n(x)$ is easy, since $p(x) > 0$ on $[a,b]$. For example, for $p(x) = 1$, we can simply choose $f_n(x) = e^{nx}$. For a general $p(x)$, we set

$$f_n(x) = e^{n \int_a^x \frac{1}{p(s)} ds}.$$

(e) To prove the Rayleigh Quotient we take the inner product for the following equation with ϕ_n

$$L[\phi_n] = \lambda_n \phi_n.$$

It follows that

$$\lambda_n = \frac{< L[\phi_n], \phi_n >}{< \phi_n, \phi_n >}, \qquad n = 1, 2, \cdots. \qquad \square$$

Corollary 3.3.1. *Let M_1, M_2, M_3 be the sets of functions defined in Example 2.2.1. Then,*

(a) The orthogonal set M_1 forms a basis in $L^2(-L, L)$.
(b) Either M_2 or M_3 forms a basis for $L^2(0, L)$.

Proof. (a) The set M_1 consists of all eigenfunctions for eigenvalue problem (3.2.2) subject to periodic boundary conditions $\phi(-L)=\phi(L)$, $\phi'(-L)=\phi'(L)$.

(b) The set M_2 consists of all eigenfunctions for the eigenvalue problem (3.2.2) subject to $\phi(0)=\phi(L)=0$. The set M_3 consists of all eigenfunctions for the eigenvalue problem (3.2.2) subject to the boundary conditions: $\phi'(0)=\phi'(L)=0$. □

3.4 Eigenvalues problems in several space dimensions

The Sturm–Liouville theory can be extended into higher-space dimension. We only state the main results. The reader may find the proof in an advanced PDE book, such as [5,8].

Let Ω be a bounded domain in R^n with C^2-boundary $\partial\Omega$. Let $p(x)\geq p_0>0$ on $\bar{\Omega}$ and $p(x), q(x)\in L^\infty(\Omega)$.

Define an operator

$$L[u]:=-\nabla[p(x)\nabla u]+q(x)u, \qquad x\in\Omega.$$

We consider the eigenvalue problem:

$$L[u]=\lambda\rho(x)u, \qquad x\in\Omega, \tag{3.4.1}$$
$$a(x)p(x)\nabla_\nu u+b(x)u(x)=0, \qquad x\in\partial\Omega, \tag{3.4.2}$$

where the weight function $\rho(x)\in C(\bar{\Omega})$ and $0<\rho(x)<\rho_0$ in Ω, ν represents the outward unit normal on $\partial\Omega$. $a^2+b^2>0$ on $\partial\Omega$.

The boundary condition (3.4.2) contains several interesting cases that will be seen in later chapters. When $a(x)\neq 0$, we may rewrite (3.4.2) as follows:

$$p(x)\nabla_\nu u(x)+c(x)u=0, \qquad x\in\partial\Omega, \tag{3.4.3}$$

where $c(x)=\frac{b(x)}{a(x)}$.

Proposition 3.4.1. *The operator L associated with the boundary condition (3.4.2) is self-adjoint*

$$<L[u], v>=<u, L[v]>, \qquad u, v\in H^2(\Omega).$$

Proof. The proof is based on the following Green's identity in Proposition 1.5.1: for any $u, v\in H^2(\Omega)$,

$$\int_\Omega\big[v\nabla(p(x)\nabla u)-u(\nabla(p(x)\nabla v)\big]dx=\int_{\partial\Omega}p(x)(v\nabla_\nu u-u\nabla_\nu v)ds. \tag{3.4.4}$$

Suppose $u(x)$ and $v(x)$ are solutions of Eqs. (3.4.1)–(3.4.2). For the Dirichlet boundary condition or Neumann boundary condition, we see by Green's identity (1.5.4) in

Chapter 1 that:

$$< L[u], v > - < u, L[v] >= \int_{\partial\Omega} p(x)(v\nabla_\nu u - u\nabla_\nu v)ds = 0.$$

For the Robin type of boundary condition with $a(x) \neq 0$, we have

$$p(x)\nabla_\nu u = -\frac{b(x)}{a(x)}u(x), \qquad p(x)\nabla_\nu v = -\frac{b(x)}{a(x)}v(x), x \in \partial\Omega.$$

Again, Green's identity in Proposition 1.5.1 yields

$$< L[u], v >=< u, L[v] > . \qquad \square$$

Similar to the Sturm–Liouville theory, we have the following theorem for the eigenvalue problem (3.4.1)–(3.4.2). For convenience, we only consider the first type or second type of the boundary condition (3.4.2). For the condition (3.4.3) we need to impose an additional sign condition for $c(x)$ (see Chapter 6). Again, the eigenvalue problem (3.4.1)–(3.4.2) is equivalent to

$$K[\psi] = \frac{1}{\lambda}\psi,$$

where $K := L^{-1}$ associated with the boundary condition (3.4.2). The operator K is compact from $L^2(\Omega)$ into $L^2(\Omega)$. Therefore the Fredholm Alternative for the operator K holds. The rigorous proof can be found in an advanced PDE book, such as [8].

Theorem 3.4.1. *Let $p(x), q(x) \in L^\infty(\Omega)$ and $p(x) \geq p_0 > 0$ for some $p_0 > 0$ on $\bar{\Omega}$. For the eigenvalue problem (3.4.1)–(3.4.2), then the following statements hold:*

(a) All eigenvalues are real.

(b) There exist an infinite number of eigenvalues that can be ordered

$$\lambda_1 < \lambda_2 < \cdots < \lambda_n \cdots,$$

and

$$\lim_{n\to\infty} \lambda_n = \infty.$$

(c) For each eigenvalue λ_n, there exists a set of mutually orthogonal eigenfunctions in N_{λ_n} and the dimension of the eigenspace N_{λ_n} is finite.

(d) Eigenfunctions corresponding to different eigenvalues are orthogonal in $L^2(\Omega; \rho)$. Moreover, all eigenfunctions $\{\phi_n(x)\}_{n=1}^\infty$ form an orthogonal basis of $L^2(\Omega; \rho)$. Namely, every function in $L^2(\Omega; \rho)$ can be expressed by an infinite series

$$f(x) = \sum_{n=1}^\infty c_n\phi_n(x), \qquad x \in \Omega,$$

where

$$c_n = \frac{\int_\Omega f(x)\phi_n \rho(x)dx}{\int_\Omega \phi_n^2 \rho(x)dx}, \qquad n = 1, 2, \cdots,$$

and the series converges to $f(x)$ in the sense of a mean-square with weight $\rho(x)$.

(e) Every eigenvalue λ and the corresponding eigenfunction $\phi(x)$ satisfy the following identity (the Rayleigh quotient):

$$\lambda = \frac{-p\phi\nabla_\nu\phi(x)|_{\partial\Omega} + \int_\Omega [p|\nabla\phi|^2 + q\phi^2]\rho(x)dx}{\int_\Omega \phi^2 \rho dx}. \quad \square \tag{3.4.5}$$

When the space dimension is greater than 1, the dimension of an eigenfunction space N_λ is usually greater than 1. Finding eigenvalues and eigenfunctions for (3.4.1)–(3.4.2) in higher-space dimension is much more complicated than that in one-space dimension. Here we give two examples. The reader will see more examples in Chapter 6.

Example 3.4.1. Let $R = [0, L] \times [0, H]$. Consider the eigenvalue problem:

$$-\Delta u = \lambda u, \ (x, y) \in R, \tag{3.4.6}$$

$$u(x, y) = 0, \ (x, y) \in \partial R. \tag{3.4.7}$$

Find all eigenvalues and the corresponding eigenfunctions.

Solution. We use the method of separation of variables. Let

$$u(x, y) = \phi(x)\psi(y), \qquad (x, y) \in R.$$

Then, we see that

$$-[\phi''(x)\psi(y) + \phi(x)\psi''(y)] = \lambda\phi(x)\psi(y), \qquad (x, y) \in R.$$

The boundary conditions for $\phi(x)$ and $\psi(y)$ are

$$\phi(0) = \phi(L) = 0; \ \psi(0) = \psi(H) = 0.$$

It follows that

$$\frac{\phi''(x)}{\phi(x)} + \frac{\psi''(y)}{\psi(y)} = -\lambda.$$

Let

$$\frac{\phi''(x)}{\phi(x)} = -\mu, \qquad 0 < x < L.$$

The boundary condition implies

$$\phi(0) = \phi(L) = 0.$$

It follows that the eigenvalues are

$$\mu_n = \left(\frac{n\pi}{L}\right)^2, \qquad n = 1, 2, \cdots .$$

The corresponding eigenfunctions are

$$\phi_n(x) = \sin(\frac{n\pi x}{L}), \qquad n = 1, 2, \cdots .$$

On the other hand, from

$$\frac{\psi''(y)}{\psi(y)} = \mu_n - \lambda,$$

subject to the boundary conditions,

$$\psi(0) = \psi(H) = 0.$$

We see the eigenvalues from Example 3.2.1 are

$$\mu_n - \lambda_{nm} = -\left(\frac{m\pi x}{H}\right)^2, m = 1, 2, \cdots .$$

The corresponding eigenfunctions are

$$\psi_m(y) = \sin(\frac{m\pi y}{H}).$$

Hence, all the eigenvalues are

$$\sigma = \left\{\lambda_{nm} = \left(\frac{n\pi}{L}\right)^2 + \left(\frac{m\pi}{H}\right)^2; \; n, m = 1, 2, \cdots\right\}.$$

The corresponding eigenfunctions are

$$\{\phi_n(x)\psi_m(y) = \sin(\frac{n\pi x}{L})\sin(\frac{m\pi y}{H}); \; n, m = 1, 2, \cdots\}.$$

Example 3.4.2. Let $\Omega = \{(x, y, z) : 0 < x < L, 0 < y < H, 0 < z < K\}$. Find all eigenvalues and eigenfunctions for the following eigenvalue problem in Ω:

$$-\Delta u = \lambda u, (x, y) \in \Omega, \tag{3.4.8}$$

$$u(x, y, z) = 0, (x, y, z) \in \partial\Omega. \tag{3.4.9}$$

Solution. We follow the same idea as in Example 3.4.1. Set

$$u(x, y, z) = \phi(x)\psi(y)h(z).$$

Then,

$$\frac{\phi''(x)}{\phi(x)} + \frac{\psi''(y)}{\psi(y)} + \frac{h''(z)}{h(z)} = -\lambda.$$

Let

$$\frac{\phi''(x)}{\phi(x)} = -\alpha, \qquad \phi(0) = \phi(L) = 0;$$
$$\frac{\psi''(y)}{\psi(y)} = -\beta, \qquad \psi(0) = \psi(H) = 0;$$
$$\frac{h''(z)}{h(z)} = -\gamma, \qquad h(0) = h(K) = 0.$$

Then, from Section 3.1, we know that the eigenvalues and the corresponding eigenfunctions for each eigenvalue problem are:

$$\alpha_n = (\frac{n\pi}{L})^2, \qquad \phi_n(x) = \sin(\frac{n\pi x}{L});$$
$$\beta_m = (\frac{m\pi}{H})^2, \qquad \psi_m(y) = \sin(\frac{m\pi y}{H});$$
$$\gamma_k = (\frac{k\pi}{K})^2, \qquad h_k(z) = \sin(\frac{k\pi z}{K}).$$

Therefore the eigenvalues for the problem (3.4.7)–(3.4.8) are

$$\lambda_{nmk} = (\frac{n\pi}{L})^2 + (\frac{m\pi}{H})^2 + (\frac{k\pi}{K})^2, \qquad n, m, k = 1, 2, \cdots.$$

The corresponding eigenfunctions are

$$u_{nmk}(x, y, z) = \sin(\frac{n\pi x}{L}) \sin(\frac{m\pi y}{H}) \sin(\frac{k\pi z}{K}), \qquad n, m, k = 1, 2, \cdots.$$

We will see more examples in Chapter 6 with different domains.

3.5 Boundary eigenvalue problems

When the space dimension is higher than one, there is a different type of eigenvalue problem in which the eigenvalue occurs on the boundary.

Let $\Omega \subset R^n$ be a bounded domain with C^2-boundary. Let the operator L be defined the same as in Section 3.4 with $p(x) \geq p_0 > 0$, $p(x), q(x) \in L^\infty(\Omega)$ and $q(x) \geq 0$ over Ω. Consider the following eigenvalue problem: Find $u(x)$ and μ such that

$$L[u] = 0, \qquad x \in \Omega, \tag{3.5.1}$$
$$p(x)\nabla_\nu u + c(x)u(x) = \mu u, \qquad x \in \partial\Omega, \tag{3.5.2}$$

where $c(x) \in L^\infty(\partial\Omega)$ with $c(x) \geq 0$ over $\partial\Omega$.

It will be seen that the eigenvalue problem on the boundary is quite different from the one in a bounded domain.

Example 3.5.1. Consider the boundary eigenvalue problem (3.5.1)–(3.5.2) where the dimension is equal to 1. Find μ and $\psi(x)$ such that

$$\psi''(x) = 0, \qquad 0 < x < L, \tag{3.5.3}$$

$$-\psi'(0) = \mu\psi(0), \qquad \psi'(L) = \mu\psi(L). \tag{3.5.4}$$

Solution. The general solution of (3.5.3) is equal to

$$\psi(x) = c_1 x + c_2, \qquad 0 < x < L.$$

From the boundary condition (3.5.4), we have

$$-c_1 = \mu c_2, \qquad c_1 = \mu(c_1 L + c_2).$$

As $\mu \neq 0$, we find

$$\mu = \frac{2}{L}, \qquad \psi(x) = (x - \frac{L}{2}).$$

Namely, there exists one eigenvalue $\mu = \frac{2}{L}$ and the corresponding eigenfunction $\psi(x) = x - \frac{2}{L}$.

Example 3.5.2. Let $R = \{(x, y) : 0 < x < L, 0 < y < H\}$ be a rectangle in R^2. Find all eigenvalues and corresponding eigenfunctions for the following boundary eigenvalue problem:

$$\Delta\psi = 0, \qquad (x, y) \in R, \tag{3.5.5}$$

$$\nabla_\nu \psi = \mu\psi, \qquad (x, y) \in \partial R. \tag{3.5.6}$$

Solution. Suppose

$$\psi(x, y) = v(x)w(y), \qquad (x, y) \in R^2.$$

Then, Eq. (3.5.5) is equivalent to

$$\frac{v''(x)}{v(x)} + \frac{w''(y)}{w(y)} = 0.$$

The boundary condition (3.5.6) is equivalent to

$$-v'(0) = \mu v(0),\ v'(L) = \mu v(L), \qquad -w'(0) = \mu w(0),\ w'(H) = \mu w(H).$$

Set

$$\frac{v''(x)}{v(x)} = -\lambda, \qquad 0 < x < L.$$

If $\lambda = 0$, we obtain from Example 3.2.3 that there exists a unique μ_1 with

$$\mu_1 = \frac{2}{L}, \qquad v(x) = x - \frac{L}{2}.$$

Similarly, for $w(y)$ we obtain

$$\mu_2 = \frac{2}{H}, \qquad w(y) = y - \frac{H}{2}.$$

Since $\mu_1 \neq \mu_2$ unless $L = H$, there is no nontrivial solution for the eigenvalue problem. On the other hand, when $L = H$, there exists a nontrivial solution with

$$\mu = \frac{2}{L}, \qquad \psi(x, y) = (x - \frac{L}{2})(y - \frac{L}{2}).$$

We assume $\lambda > 0$ (the case for $\lambda < 0$ is the same if we solve the eigenvalue problem for $w(y)$). Then, the general solution for $v(x)$ is equal to

$$v(x) = c_1 \cos(\sqrt{\lambda} x) + c_2 \sin(\sqrt{\lambda} x), \qquad 0 < x < L.$$

From the boundary condition (3.5.6), we find that

$$-c_2\sqrt{\lambda} = \mu c_1, \qquad (c_2\sqrt{\lambda} - \mu c_1)\cos(\sqrt{\lambda}L) = (\mu c_2 + c_1\sqrt{\lambda}\sin(\sqrt{\lambda}L).$$

Let $\gamma = -\frac{c_2}{c_1}$. Then, we have

$$\mu = \gamma\sqrt{\lambda}, \qquad \tan(\sqrt{\lambda}L) = \frac{2\gamma}{\gamma^2 - 1}.$$

It follows that for any fixed $\gamma > 1$ there exists a sequence of λ_n such that

$$\lambda_n = n\pi + \frac{1}{L}\tan^{-1}\left(\frac{2\gamma}{1-\gamma^2}\right), n = 1, 2 \cdots.$$

Consequently, we obtain

$$\mu_n = \gamma\sqrt{\lambda_n}, \qquad n = 1, 2, \cdots.$$

Once λ_n is found, we can solve for $w(y)$:

$$\begin{aligned} w''(y) &= \lambda_n w(y), & 0 < y < H, \\ -w'(0) &= \mu_n w(0), & w'(H) = \mu_n w(H). \end{aligned}$$

We find that

$$w(y) = c_1 e^{\lambda_n y} + c_2 e^{-\lambda_n y}, \qquad 0 < y < H,$$

where c_1 and c_2 are arbitrary constants.

To satisfy the boundary condition (3.5.6), we need to choose c_1 and c_2 such that

$$\begin{aligned} &-(c_1 - c_2)\sqrt{\lambda_n} = \mu(c_1 + c_2), \\ &(c_1 e^{\sqrt{\lambda_n}H} - c_2 e^{-\lambda_n H})\sqrt{\lambda_n} = \mu\big(c_1 e^{\sqrt{\lambda_n}} + c_2 e^{-\sqrt{\lambda_n}H}\big). \end{aligned}$$

If we choose $\gamma > 1$ such that

$$\frac{(\gamma+1)^2}{\gamma^2-1} = e^{2\sqrt{\lambda_n}H},$$

then there exists a nontrivial solution c_1 and c_2. This implies that we obtain a sequence of eigenvalues $\{\mu_n\}$ that are positive and $\mu_n \to \infty$ as $n \to \infty$.

Theorem 3.5.1. *Let $p(x), q(x) \in L^\infty(\Omega)$ and $p(x) \geq p_0 > 0$ for some $p_0 > 0$, $q(x) \geq 0$ on $\bar{\Omega}$. Moreover, $c(x) \in L^\infty(\partial\Omega)$ with $c(x) \geq 0$. For the boundary eigenvalue problem (3.5.1)–(3.5.2) with $n \geq 2$, then the following statements hold:*

(a) All eigenvalues are real and nonnegative.

(b) There exists an infinite number of eigenvalues that can be ordered

$$0 \leq \mu_1 < \mu_2 < \cdots < \mu_n \cdots,$$

and

$$\lim_{n\to\infty} \mu_n = \infty.$$

(c) For each eigenvalue μ_n, there exists a set of mutually orthogonal eigenfunctions in N_n and the dimension of the eigenspace N_n is finite.

(d) Eigenfunctions corresponding to different eigenvalues are orthogonal in $L^2(\partial\Omega)$. Moreover, all eigenfunctions $\{\phi_n(x)\}_{n=1}^\infty$ form an orthogonal basis of $L^2(\partial\Omega)$. Namely, every function in $L^2(\partial\Omega)$ can be expressed by an infinite series

$$f(x) = \sum_{n=1}^\infty c_n\phi_n(x), \qquad x \in \Omega,$$

where

$$c_n = \frac{\int_{\partial\Omega} f(x)\phi_n\rho(x)ds}{\int_{\partial\Omega}\phi_n^2\rho(x)ds}, \qquad n = 1, 2, \cdots,$$

and the series converges to $f(x)$ in the sense of a mean-square in $L^2(\partial\Omega)$.

(e) Every eigenvalue μ and the corresponding eigenfunction $\phi(x)$ satisfy the following identity (the Rayleigh quotient):

$$\mu = \frac{\int_\Omega [p|\nabla\phi|^2 + q\phi^2]\rho(x)dx + \int_{\partial\Omega} c(x)\phi^2\rho(x)ds}{\int_{\partial\Omega}\phi^2\rho(x)ds}. \tag{3.5.7}$$

The proof is very much similar to the case for the eigenvalue problem in Ω. We omit it here. □

3.6 Convergence of function series

Let $f(x) \in C^1[0, L]$ and $f(x)$ be expressed by the sine series

$$f(x) = \sum_{n=1}^{\infty} c_n \sin(\frac{n\pi x}{L}).$$

From calculus we can easily show that

$$f'(x) = \sum_{n=1}^{\infty} \frac{n\pi c_n}{L} \cos(\frac{n\pi x}{L}).$$

A simple approximation shows that the direct derivative for the sine series of $f(x)$ holds if $f'(x) \in L^2(0, L)$, where the convergence is under the $L^2(0, L)$-norm. It turns out that the direct derivative of the eigenfunction series also holds as long as the eigenfunction series has certain required regularity. We state a theorem without giving a rigorous proof (see [23]), since it needs the measure theory and the theory of Sobolev spaces.

Theorem 3.6.1. *Suppose $\{\phi_n(x)\}$ is a basis for $L^2(\Omega)$ and*

$$f(x) = \sum_{n=1}^{\infty} c_n \phi_n(x), \qquad x \in \Omega.$$

Let $\phi_n(x) \in H^1(\Omega)$ for all n and $f(x) \in H^1(\Omega)$, then

$$f(x)_{x_j} = \sum_{n=1}^{\infty} c_n \phi_n(x)_{x_j}, \qquad x \in \Omega, j = 1, 2, \cdots,$$

where the convergence holds under the $L^2(\Omega)$-norm. □

Theorem 3.6.1 justifies the method by using a series solution in terms of eigenfunctions for a PDE problem. We will see many examples in later chapters.

3.7 Notes and remarks

The classical Sturm–Liouville theory is well understood for students with a basic knowledge of the elementary theory of ODEs. This theory reveals many distinguishing features for eigenvalue problems in infinite-dimensional space. It also shows that the classical Fourier series is a special case for representing a function in terms of eigenfunctions. Students may find that many results for eigenvalue problems in linear algebra are still valid for the eigenvalue problems associated with self-adjoint differential operators.

For a general domain in R^n with $n > 1$ it is difficult to find explicit eigenvalues and the corresponding eigenfunctions even for the Laplace operator associated with the Dirichlet boundary condition. Nevertheless, one can see that most results in the Sturm–Liouville theory hold when the dimension is greater than one. In Chapter 6 we will see the connection between the Fredholm Alternative and solvability for an elliptic boundary value problem. For a general elliptic operator that may not be self-adjoint, there is a similar theory about the eigenvalue problem. The interested reader may find these advanced materials about the general elliptic equations in, for example, [8,12,20].

3.8 Exercises

1. Let a differential operator

$$\mathcal{L}[u] := a(x)u'' + b(x)u' + c(x)u = f(x), \qquad \alpha < x < \beta,$$

where $a(x) > 0$, $b(x)$, and $c(x)$ are smooth functions on $[a, b]$. Introduce a suitable integrating factor to rewrite operator $\mathcal{L}$ as $\hat{\mathcal{L}}$ such that $\hat{\mathcal{L}}$ is a self-adjoint operator from (3.2.1) and

$$\hat{\mathcal{L}}[u] = \hat{f}, \qquad \alpha < x < \beta.$$

2. Prove that the differential operator (3.2.1) along with periodic boundary conditions

$$\phi(a) = \phi(b),\ \phi'(a) = \phi'(b)$$

is self-adjoint.
3. Find the eigenvalues and the corresponding eigenfunctions for the following eigenvalue problem

$$-u'' = \lambda u, \qquad 0 < x < L,$$
$$u(0) = 0,\ u'(L) = 0.$$

4. Let the differential operator L be defined as in (3.2.1). Assume the boundary conditions are given as follows:

$$\phi(a) = 0,\ \phi'(b) = c\phi(b),$$

where c is a positive constant.
Prove that the dimension of every eigenspace is equal to 1. Namely, there exists only one linearly independent eigenfunction for each eigenvalue.
5. Let $p(x) \in C^2[0, L]$ with $\geq p_0 > 0$ on $[0, L]$. Consider the eigenvalue problem:

$$L[u] := -\frac{d^2}{dx^2}\left[p(x)\frac{d^2u}{dx^2}\right] = \lambda u, 0 < x < L$$

subject to the following boundary conditions

$$u(0) = u''(0) = 0, u(L) = u''(L) = 0.$$

(a) Prove the Sturm–Liouville theorem holds.
(b) Find all eigenvalues and corresponding eigenfunctions for $p(x) = 1$.

6. Let $R = [0, L] \times [0, H]$. Find all eigenvalues and corresponding eigenfunctions for the following two-dimensional Laplace operator:

$$-\Delta u = \lambda u, (x, y) \in R,$$

subject to the following boundary conditions:

$$u(0, y) = u(L, y) = 0, u_y(x, 0) = u_y(x, H) = 0, (x, y) \in \partial R.$$

7. Let $f(x, y) = 3 + xy$. Find the eigenfunction representation to express $f(x, y)$ in terms the eigenfunctions obtained in Exercise 6.

8. Let $B_a(0)$ be a sphere centered at 0 with radius a in R^3:

$$B_a(0) = \{(x, y, z) : x^2 + y^2 + z^2 < a^2\}.$$

Rewrite the Laplace equation

$$-\Delta u = 0$$

in spherical coordinate form.

9. Let

$$B_a(0) = \{(x, y, z) : x^2 + y^2 + z^2 < a^2\}.$$

Find all eigenvalues and the corresponding eigenvector for the following eigenvalue problem:

$$-\Delta u = \lambda u, (x, y, z) \in B_a(0),$$
$$u(x, y, z) = 0, \text{on}\, r = a,$$

where $r = \sqrt{x^2 + y^2 + z^2}$. (Hint: use the spherical coordinates form.)

10. Let $q(x) \geq 0$ on Ω. Prove that the principal eigenvalue for the eigenvalue problem (3.4.1)–(3.4.2) is positive when $a = 0$ and $b = 1$.

11. Let Ω be a bounded domain with C^1-boundary $\partial\Omega$. Suppose the boundary $\partial\Omega$ is divided into two parts Γ_1 and Γ_2 with $|\Gamma_i| > 0$ for $i = 1, 2$. Let

$$u(x) = 0, \text{ on } \Gamma_1, \qquad \nabla_\nu u(x) = 0, \text{ on } \Gamma_2.$$

Prove all eigenvalues associated with the operator L defined in (3.4.1) are positive if $q(x) \geq 0$ on Ω.

12. Let $a(x)$ be continuous on $\partial\Omega$ and $b(x) \in L^\infty(\Omega)$. Suppose the boundary condition (3.4.2) is given by

$$a(x)\nabla_\nu u + b(x)u = 0, a(x)^2 + b(x)^2 > 0, \qquad x \in \partial\Omega.$$

Prove that the elliptic operator L defined by (3.4.1) associated with the above boundary condition is self-adjoint.

13. Let $q(x) \in C[0, L]$ be nonnegative. Consider the following eigenvalue problem:

$$\begin{aligned} &-\phi''(x) + q(x)\phi(x) = \lambda\phi(x), \qquad 0 < x < L, \\ &\phi(0) = \phi(L) = 0. \end{aligned}$$

Prove all eigenvalues must be positive if there exists an interval $(a, b) \subset [0, L]$ in which $q(x) > 0$ on $[a, b]$.

14. Let λ_1 be the smallest eigenvalue (principal eigenvalue) for the following eigenvalue problem:

$$\begin{aligned} &-\Delta\phi = \lambda\phi, \qquad x \in \Omega, \\ &\phi(x) = 0, \qquad x \in \partial\Omega. \end{aligned}$$

Prove the smallest eigenvalue for the following eigenvalue problem:

$$\begin{aligned} &-\Delta\phi + q(x)\phi = \lambda\phi, \qquad x \in \Omega, \\ &\phi(x) = 0, \qquad x \in \partial\Omega, \end{aligned}$$

is positive if $||q||_{L^\infty(\Omega)} < \lambda_1$, where λ_1 is the principal eigenvalue of the Laplace operator associated with the homogeneous Dirichlet boundary condition.

CHAPTER

The heat equation

4

4.1 The mathematical model of heat conduction

The mathematical model for heat conduction is based on two fundamental laws in physics: the energy conservation principle and the experimental Fourier's law. We will demonstrate step by step the process of how we use the basic laws to establish the mathematical model for heat conduction in media. The model can be used for describing a chemical reaction–diffusion process in which the physical laws are based on mass conservation and Fick's law.

4.1.1 Derivation of the heat equation

We begin with a physical description. Consider an object occupying a domain Ω in R^n. Suppose there is a heat generator (heat source) or a cooling system (heat sink) running on the inside of the object. Suppose we can make a measurement of the temperature at the surface $\partial\Omega$ and also know the temperature distribution in Ω at an initial moment. How can we find the temperature distribution for the object in future time?

To answer the above question, we first introduce some physical variables. Let $u(x,t)$ be the temperature at location $x \in \Omega$ and time $t > 0$. In physics, the heat-energy density at (x,t) is defined by

$$E(x,t) := \rho_0 c_0 u(x,t),$$

where c_0 is the specific heat and ρ_0 is the density of the object.

Suppose D is an arbitrary subregion of Ω and $[t_1, t_2] \subset [0, \infty)$.

$$\text{The total heat energy in } D \text{ at time } t = \int_D E(x,t)dx.$$

We also assume that the density of the heat generator (or sink) is equal to $f_0(x,t)$. Then, the total heat energy generated in D during a time period $[t_1, t_2]$ is equal to

$$\int_{t_1}^{t_2} \int_D f_0(x,t)dxdt.$$

The energy conservation states that

Partial Differential Equations and Applications. https://doi.org/10.1016/B978-0-44-318705-6.00010-0

the total energy change over D from $t = t_1$ to $t = t_2$
$= -$total energy flowing out across the boundary of D in a time period $[t_1, t_2]$
$+$ the total energy generated (or lost) in D over the time period $[t_1, t_2]$,

where the negative sign means that the energy is flowing out across ∂D.

Suppose the heat flux is denoted by $\vec{q}(x,t)$. Then, the total energy flows out across the boundary of D in the time period $[t_1, t_2]$ is equal to

$$\int_{t_1}^{t_2} \oint_{\partial D} \vec{q} \cdot \nu ds,$$

where ν is the outward unit normal on ∂D.

Fourier's law states that the heat flux is proportional to the gradient of the temperature:

$$\vec{q}(x,t) = -k_0 \nabla u(x,t),$$

where $k_0 > 0$ represents the heat conductivity and the negative sign means the temperature flows from higher to lower values.

From Fourier's law, we see that the total energy flowing out across the boundary ∂D is equal to

$$\int_{t_1}^{t_2} \oint_{\partial D} \vec{q} \cdot \nu ds = -\int_{t_1}^{t_2} \oint_{\partial D} k_0 \nabla u \cdot \nu ds.$$

It follows from the energy conservation that

$$\int_D c_0 \rho_0 (u(x,t_2) - u(x,t_1)) dx = \int_{t_1}^{t_2} \oint_{\partial D} k_0 \nabla_\nu u ds dt + \int_{t_1}^{t_2} \int_D f_0(x,t) dx dt.$$

Gauss's divergence theorem states that for any vector field $\vec{V}$:

$$\int_D \nabla \cdot \vec{V} dx = \oint_{\partial D} (\vec{V} \cdot \nu) ds.$$

We replace t_1 by t and set $\Delta t = t_2 - t_1 \to 0$ to obtain

$$\int_D c_0 \rho_0 u_t(x,t) dx = \int_D \nabla [k_0 \nabla u] dx + \int_D f_0(x,t) dx.$$

Since $D \subset \Omega$ is arbitrary, we see that $u(x,t)$ satisfies

$$c_0 \rho_0 u_t = k_0 \Delta u + f_0(x,t), \qquad x \in \Omega, t > 0,$$

where k_0 is assumed to be a constant.

Finally, we arrive at the following heat equation:

$$u_t = k^2 \Delta u + f(x,t), \qquad x \in \Omega, t > 0, \tag{4.1.1}$$

where

$$k^2 = \frac{k_0}{c_0 \rho_0}, \qquad f(x,t) = \frac{f_0(x,t)}{c_0 \rho_0}.$$

4.1.2 Initial and boundary conditions

From the physical model, we need to know the temperature at the beginning $t = 0$ (called initial condition) and the information on the boundary $\partial\Omega$ (called the boundary condition).

Let $u_0(x)$ be the initial temperature distribution in Ω:

$$u(x,0) = u_0(x), \qquad x \in \Omega. \tag{4.1.2}$$

Again from the physical model, we may have the following types of boundary condition.

Case 1: The first type

This condition simply means that the temperature at the surface of the object is known, denoted by $g(x,t)$:

$$u(x,t) = g(x,t), \qquad x \in \partial\Omega, t > 0. \tag{4.1.3}$$

This type of boundary condition is also called *the Dirichlet type* in the literature.

Case 2: The second type

If we know that the heat flows out or flows in across the boundary of Ω, then the heat flux is known on the boundary of Ω:

$$k_0 \nabla_\nu u(x,t) = g(x,t), \qquad x \in \partial\Omega, t > 0, \tag{4.1.4}$$

where ν represents the outward unit normal on $\partial\Omega$.

For example, if we know that the surface of the object is insulated, i.e., no heat can flow in or out across the boundary, then the heat flux on the boundary is equal to zero.

This type of boundary condition is also called *the Neumann type*.

Case 3: The third type

In addition to the above two types of boundary conditions, some experiments show that there is a certain relation between the heat flux and the temperature on the boundary. A notable example is Newton's cooling law that states that the heat flux across the boundary is proportional to the difference of the temperature between the surface temperature and the surrounding temperature. Hence, we obtain the following type of boundary condition:

$$k_0 \nabla_\nu u + a(x,t) u(x,t) = g(x,t), \qquad x \in \partial\Omega, t > 0, \tag{4.1.5}$$

where $a(x,t)$ and $g(x,t)$ are known functions.

This type of boundary condition is also called *the Robin type*. This condition can also be considered as the radiation effect for the heat flux (see Remark 4.1.1 below).

Case 4: The mixed type

This boundary condition is simply to give one type of boundary condition on one part of the boundary and another type of boundary condition on the rest of the boundary.

In the research literature, there are other types of boundary conditions. It is often deduced from a concrete physical model and experiments.

We summarize the above discussion to obtain a completed mathematical model as follows: Find $u(x,t)$ such that

$$u_t = k^2 \Delta u + f(x,t), \qquad x \in \Omega, t > 0, \tag{4.1.6}$$

subject to an initial condition (4.1.2) and one of these boundary conditions, (4.1.3), (4.1.4) or (4.1.5). For brevity, sometimes we call the problem an initial-boundary value problem or IBVP.

Case 5: The Cauchy problem

When a domain is the whole space R^n, there is no boundary condition needed except that there is a certain restriction for the solution as $|x| \to \infty$. We only need an initial value. This type of problem is called a Cauchy problem (or called IVP).

When a mathematical model is established, the next task is to justify if the model is suitable, which means we need to study the well-posedness of the model. Another task is to find the solution. In the next two sections, we will show how to obtain the series solution by using the method of separation of variables.

Remark 4.1.1. The classical Fourier's law does not include the radiation of the heat in the conduction process. With the radiation effect, the heat flux is equal to

$$\vec{q}(x,t) = -k_0 \nabla u(x,t) + cu(x,t),$$

where c is the radiation coefficient.

4.2 Solution of the heat equation in one-space dimension

In this section we will use the results from the Sturm–Liouville theory in Chapter 3 to find the explicit series solution for the heat equation in one-space dimension.

4.2.1 Homogeneous problems

Let $f(x) \in L^2(0,L)$ and $k > 0$. Consider the following initial-boundary value problem in one-space dimension. Let $Q = (0,L) \times (0,\infty)$;

$$u_t = k^2 u_{xx}, \qquad (x,t) \in Q, \tag{4.2.1}$$

$$a_1 u_x(0,t) + a_2 u(0,t) = 0, \qquad t > 0, \tag{4.2.2}$$
$$b_1 u_x(L,t) + b_2 u(L,t) = 0, \qquad t > 0, \tag{4.2.3}$$
$$u(x,0) = f(x), \qquad 0 < x < L, \tag{4.2.4}$$

where $a_1^2 + a_2^2 > 0$, $b_1^2 + b_2^2 > 0$.

The boundary conditions (4.2.2)–(4.2.3) include all cases discussed in Section 4.1 with different choices of coefficients.

We use the method of separation of variables to find the solution. Suppose

$$u(x,t) = \psi(x)v(t), \qquad 0 < x < L, t > 0.$$

Then, Eq. (4.2.1) is equivalent to the following:

$$v'(t)\psi(x) = k^2 v(t)\psi''(x), \qquad 0 < x < L, t > 0,$$

which is the same as:

$$\frac{v'(t)}{k^2 v(t)} = \frac{\psi''(x)}{\psi(x)}, \qquad 0 < x < L, t > 0. \tag{4.2.5}$$

Since the left-hand side of Eq. (4.2.5) is a function of t while the right-hand side is a function of x, it must be equal to a constant, denoted by $-\lambda$, which is unknown. Here, the negative sign is for convenience.

Now, we have from Eq. (4.2.5) that

$$v'(t) = -\lambda k^2 v(t), \qquad \psi''(x) + \lambda\psi(x) = 0, \qquad 0 < x < L, t > 0.$$

Thus,

$$v(t) = Ce^{-\lambda k^2 t},$$

where C is an arbitrary constant and λ will be determined later.

From the boundary conditions (4.2.2)–(4.2.3), we see that $\psi(x)$ is the solution of the following problem:

$$\psi''(x) + \lambda\psi(x) = 0, \qquad 0 < x < L, \tag{4.2.6}$$
$$a_1\psi'(0) + a_2\psi(0) = 0, \tag{4.2.7}$$
$$b_1\psi'(L) + b_2\psi(L) = 0. \tag{4.2.8}$$

By the Sturm–Liouville Theorem, we see that for the above eigenvalue problem there exist eigenvalues

$$\lambda_1 < \lambda_2 < \cdots < \lambda_n < \cdots,$$

with $\lambda_n \to \infty$, and the corresponding eigenfunctions $\psi_1(x), \psi_2(x), \cdots, \psi_n(x), \cdots$. Once λ_n is determined, we have

$$v_n(t) = e^{-\lambda_n k^2 t}, \qquad n = 1, 2, 3, \cdots.$$

It is clear that for any $n \geq 1$

$$u_n(x,t) = v_n(t)\psi_n(x)$$

satisfies Eq. (4.2.1) and the boundary conditions (4.2.2)–(4.2.3). By the superposition principle, we see that for any constants $c_1, \cdots, c_m$,

$$u(x,t) = \sum_{n=1}^{m} c_n v_n(t)\psi_n(x)$$

also satisfies the heat equation (4.2.1) and boundary conditions (4.2.2)–(4.2.3).

Naturally, we expect that the above form of series solution holds for $m \to \infty$. With the help of advanced functional analysis, this step can be proved rigorously (see [5]).

Finally, we need to choose coefficients $\{c_n\}$ such that the series solution satisfies the initial condition (4.2.4). Since $\{\psi_n(x)\}$ is an orthogonal basis for $L^2(0, L)$, we can express $f(x)$ as a series

$$f(x) = \sum_{n=1}^{\infty} a_n \psi_n(x), \qquad 0 < x < L,$$

where a_n are the Fourier coefficients:

$$a_n = \frac{< f, \psi_n >}{< \psi_n, \psi_n >}, \qquad n = 1, 2, \cdots.$$

Hence, if we choose $c_n = a_n$ for $\forall n \geq 1$, then

$$u(x,t) = \sum_{n=1}^{\infty} a_n v_n(t)\psi_n(x)$$

is a solution of the problem (4.2.1)–(4.2.4).

Example 4.2.1. Solve the problem (4.1.1)–(4.1.4) where the boundary conditions are of the Dirichlet type:

$$u_t = k^2 u_{xx}, \qquad 0 < x < L, t > 0, \tag{4.2.9}$$

$$u(0,t) = u(L,t) = 0, \qquad t > 0, \tag{4.2.10}$$

$$u(x,0) = f(x), \qquad 0 < x < L, \tag{4.2.11}$$

where $f(x) \in L^2(0, L)$.

Solution. We follow the same procedure to have the following eigenvalue problem:

$$-\psi''(x) = \lambda\psi, \qquad 0 < x < L,$$

$$\psi(0) = \psi(L) = 0.$$

From the previous example in Chapter 3, the eigenvalues and corresponding eigenfunctions are:

$$\lambda_n = \left(\frac{n\pi}{L}\right)^2, \qquad \psi_n(x) = \sin(\frac{n\pi x}{L}), \qquad n = 1, 2, \cdots.$$

Hence, the solution is equal to

$$u(x,t) = \sum_{n=1}^{\infty} a_n e^{-\lambda_n k^2 t} \sin(\frac{n\pi x}{L}),$$

where

$$a_n = \frac{< f, \psi_n >}{< \psi_n, \psi_n >} = \frac{2}{L}\int_0^L f(x) \sin\left(\frac{n\pi x}{L}\right) dx, \qquad n = 1, 2, \cdots.$$

Consider a concrete case where

$$f(x) = 1, \qquad 0 < x < L.$$

The Fourier coefficients are

$$a_n = \frac{1 + (-1)^2}{L}, \qquad n = 1, 2, \cdots.$$

There is an interesting phenomenon when one approximates the solution near $t = 0$ and near $x = 0$ and $x = L$ since the solution $u(x, t)$ is discontinuous at the corner points $(0, 0)$ and $(0, L)$.

Example 4.2.2. Solve the problem (4.1.1)–(4.1.4) where the boundary conditions are of the Neumann type:

$$u_t = k^2 u_{xx}, \qquad 0 < x < L, t > 0, \tag{4.2.12}$$

$$u_x(0,t) = u_x(L,t) = 0, \qquad t > 0, \tag{4.2.13}$$

$$u(x,0) = f(x), \qquad 0 < x < L. \tag{4.2.14}$$

The eigenvalues and corresponding eigenfunctions in this case are:

$$\lambda_n = \left(\frac{n\pi}{L}\right)^2, \qquad \psi_n(x) = \cos(\frac{n\pi x}{L}), \qquad n = 0, 1, 2, \cdots.$$

The solution is equal to

$$u(x,t) = \sum_{n=0}^{\infty} a_n e^{-\lambda_n t} \cos(\frac{n\pi x}{L}),$$

where

$$a_0 = \frac{1}{L}\int_0^L f(x)dx,$$
$$a_n = \frac{< f, \psi_n >}{< \psi_n, \psi_n >} = \frac{2}{L}\int_0^L f(x)\cos\left(\frac{n\pi x}{L}\right)dx, \qquad n = 1, 2, \cdots.$$

Note that $\lambda_0 = 0$ is an eigenvalue for the corresponding eigenvalue problem, the solution does not decay as $t \to \infty$.

The next example is different from the boundary conditions (4.2.2)–(4.2.3). We consider a periodic boundary condition. For convenience, we consider the heat equation in $[-L, L]$.

Example 4.2.3. Solve the problem (4.2.1), (4.2.4) where the boundary conditions are of the periodic type:

$$u_t = k^2 u_{xx}, \qquad -L < x < L, t > 0, \tag{4.2.15}$$
$$u(-L, t) = u(L, t), u_x(-L, t) = u_x(L, t), \qquad t > 0, \tag{4.2.16}$$
$$u(x, 0) = f(x), \qquad -L < x < L, \tag{4.2.17}$$

where $f(x) \in L^2(0, L)$.

The eigenvalues are:

$$\lambda_n = \left(\frac{n\pi}{L}\right)^2, \qquad n = 0, 1, 2, \cdots.$$

The corresponding eigenfunctions are

$$\psi_0(x) = 1, \psi_n(x) = \left\{\cos\left(\frac{n\pi x}{L}\right), \sin\left(\frac{n\pi x}{L}\right)\right\} \qquad n = 1, 2, \cdots.$$

Note that there are two linearly independent eigenfunctions for each eigenvalue λ_n when $n \geq 1$.

We assume that the initial function $f(x)$ can be expressed as a Fourier series:

$$f(x) = a_0 + \sum_{n=1}^{n}[a_n \cos\left(\frac{n\pi x}{L}\right) + b_n \sin\left(\frac{n\pi x}{L}\right)], \qquad -L < x < L,$$

where the Fourier coefficients are given by

$$a_0 = \frac{2}{L}\int_{-L}^{L} f(x)dx,$$
$$a_n = \frac{1}{L}\int_{-L}^{L} f(x)\cos\left(\frac{n\pi x}{L}\right)dx, \qquad n = 1, 2, 3, \cdots,$$

$$b_n = \frac{1}{L}\int_{-L}^{L} f(x)\sin(\frac{n\pi x}{L})dx, \qquad n = 1, 2, 3, \cdots.$$

Then, the solution to the problem (4.2.15)–(4.2.17) is equal to

$$u(x,t) = a_0 + \sum_{n=1}^{\infty} e^{-\lambda_n k^2 t}[a_n \cos(\frac{n\pi x}{L}) + b_n \cos(\frac{n\pi x}{L})].$$

It is interesting to see that the solution $u(x,t)$ of the problem (4.2.15)–(4.2.17) may not decay as $t \to \infty$.

4.2.2 Nonhomogeneous problems

In this subsection we consider the model problem in which there is a heat source in the model. Moreover, the boundary conditions may not be homogeneous.

We begin with the nonhomogeneous equation subject to homogeneous boundary conditions. Let $f(x) \in L^2(0, L)$ and $g(x,t) \in L^2(R^1 \times R^1_+)$. Consider the following problem:

$$u_t = k^2 u_{xx} + g(x,t), \qquad 0 < x < L, t > 0, \qquad (4.2.18)$$

$$a_1 u_x(0,t) + a_2 u(0,t) = 0, \qquad t > 0, \qquad (4.2.19)$$

$$b_1 u_x(L,t) + b_2 u(L,t) = 0, \qquad t > 0, \qquad (4.2.20)$$

$$u(x,0) = f(x), \qquad 0 < x < L, \qquad (4.2.21)$$

where $a_1^2 + a_2^2 > 0$, $b_1^2 + b_2^2 > 0$.

The method is based on choosing the same eigenfunctions for the homogeneous problem as the basis and then using the superposition property.

Suppose the eigenvalues and corresponding eigenfunctions for the homogeneous problem are:

$$\{\lambda_n\}_{n=1}^{\infty}, \qquad \{\phi_n(x)\}_{n=1}^{\infty}.$$

Since $\{\phi_n(x)\}_{n=1}^{\infty}$ forms a basis for $L^2(0, L)$, we can expand $g(x,t)$ as a series:

$$g(x,t) = \sum_{n=1}^{\infty} q_n(t)\phi_n(x),$$

where the Fourier coefficients are

$$q_n(t) = \frac{< g, \phi_n >}{< \phi_n, \phi_n >}, \qquad n = 1, 2, \cdots.$$

Assume a solution of (4.2.18)–(4.2.21) is in the following form

$$u(x,t) = \sum_{n=1}^{\infty} a_n(t)\phi_n(x).$$

Then, we have

$$u_t = \sum_{n=1}^{\infty} a_n'(t)\phi_n(x), \; u_{xx} = \sum_{n=1}^{\infty} a_n(t)\phi_n(x)''.$$

Since $\phi_n(x)$ is an eigenfunction corresponding to eigenvalue λ_n, we know that

$$-\phi_n'' = \lambda_n \phi_n, n = 1, 2, \cdots.$$

It follows that Eq. (4.2.18) is equivalent to

$$\sum_{n=1}^{\infty} [a_n'(t) + \lambda_n a_n(t)]\phi_n(x) = \sum_{n=1}^{\infty} q_n(t)\phi_n(x).$$

Hence, $a_n(t)$ must satisfy

$$a_n'(t) + \lambda_n a_n(t) = q_n(t), \qquad t > 0, n = 1, 2, \cdots.$$

We solve the ODE to obtain

$$a_n(t) = a_n(0)e^{-\lambda_n t} + \int_0^t q_n(s)e^{-\lambda_n(t-s)}ds, \qquad t \geq 0, \; n = 1, 2, \cdots,$$

where $a_n(0)$ is the initial value to be determined.

Finally, from the initial condition (4.2.21), we see that

$$u(x, 0) = \sum_{n=1}^{\infty} a_n(0)\phi_n(x) = f(x).$$

Consequently, if we choose $a_n(0)$ to be the Fourier coefficient of $f(x)$:

$$a_n(0) = \frac{< f, \phi_n >}{< \phi_n, \phi_n >}, \qquad \forall n \geq 1,$$

then $u(x, t)$ is the solution of the problem (4.2.18)–(4.2.21).

We summarize the above derivation to obtain the following theorem.

Theorem 4.2.1. *Let $\{\lambda_n\}_{n=1}^{\infty}$ be the eigenvalues and $\{\phi_n(x)\}_{n=1}^{\infty}$ be the corresponding eigenfunctions for the eigenvalue problem (4.2.6)–(4.2.8). Let*

$$g(x, t) = \sum_{n=1}^{\infty} q_n(t)\phi_n(x), \qquad (x, t) \in Q,$$

where

$$q_n(t) = \frac{< g, \phi_n >}{< \phi_n, \phi_n >}, \qquad n = 1, 2, \cdots.$$

Then, the solution to the problem (4.2.18)–(4.2.21) is equal to

$$u(x,t) = \sum_{n=1}^{\infty} a_n(t)\phi_n(x), \qquad (x,t) \in Q,$$

where

$$a_n(t) = a_n(0)e^{-\lambda_n t} + \int_0^t q_n(s)e^{-\lambda_n(t-s)}ds, \qquad t \geq 0,\ n = 1, 2, \cdots,$$

$$a_n(0) = \frac{< f, \phi_n >}{< \phi_n, \phi_n >}, \qquad \forall n \geq 1. \quad \square$$

Next, we consider a nonhomogeneous problem with nonhomogeneous boundary conditions. Namely, the boundary conditions (4.2.19)–(4.2.20) are given by the following nonhomogeneous conditions:

$$a_1 u_x(0,t) + a_2 u(0,t) = h_1(t), \qquad t > 0, \tag{4.2.22}$$

$$b_1 u_x(L,t) + b_2 u(L,t) = h_2(t), \qquad t > 0. \tag{4.2.23}$$

To deal with nonhomogeneous boundary conditions, we introduce a new function $w(x,t)$ such that

$$a_1 w_x(0,t) + a_2 w(0,t) = h_1(t), \qquad t > 0, \tag{4.2.24}$$

$$b_1 w_x(L,t) + b_2 w(L,t) = h_2(t), \qquad t > 0. \tag{4.2.25}$$

The direct construction of such a function depends on the coefficients a_1, a_2 and b_1, b_2.

For example, when $a_1 = b_1 = 0$, $a_2 = b_2 = 1$, we can easily construct such a function:

$$w(x,t) = \frac{L-x}{L}h_1(t) + \frac{x}{L}h_2(t).$$

After $w(x,t)$ is constructed, we set $v(x,t) = u(x,t) - w(x,t)$. Then, $v(x,t)$ will satisfy a nonhomogeneous heat equation subject to homogeneous boundary conditions. Consequently, we can follow the same procedure as for the problem (4.2.18)–(4.2.21) to find a solution for the nonhomogeneous problem.

4.3 Solution of the heat equation in higher-space dimension

In this section we extend the method in the previous section to a domain in several space dimensions. We use the Dirichlet boundary condition as an example.

4.3.1 Homogeneous problems and examples

Let Ω be a bounded domain in R^n. A point in R^n is still denoted by $x = (x_1, x_2, \cdots x_n)$. From the context, the reader should have no confusion when x represents a single variable or several variables.

Let $f(x) \in L^2(\Omega)$. Consider the heat equation in $Q = \Omega \times (0, \infty)$:

$$u_t = k^2 \Delta u, \qquad x \in \Omega, t > 0, \tag{4.3.1}$$

$$u(x,t) = 0, \qquad x \in \partial\Omega, t > 0, \tag{4.3.2}$$

$$u(x,0) = f(x), \qquad x \in \Omega. \tag{4.3.3}$$

Set

$$u(x,t) = \phi(x)v(t).$$

Eq. (4.3.1) is equivalent to

$$\frac{v'(t)}{k^2 v(t)} = \frac{\Delta\phi}{\phi} = -\lambda, \qquad x \in \Omega, \tag{4.3.4}$$

where λ is an unknown constant.

It follows that

$$v(t) = Ce^{-\lambda k^2 t}.$$

For $\phi(x)$, we use the boundary condition (4.3.2) to obtain the following eigenvalue problem:

$$-\Delta\phi = \lambda\phi, \qquad x \in \Omega, \tag{4.3.5}$$

$$\phi(x) = 0, \qquad x \in \partial\Omega. \tag{4.3.6}$$

Suppose λ is an eigenvalue of the problem (4.3.5)–(4.3.6) and N_λ is the corresponding eigenspace. Then, by the Fredholm Alternative, the dimension of N_λ is finite. Moreover, we can use the Gram–Schmidt process to find an orthogonal basis for N_λ. We always choose an orthogonal basis of N_λ as the set of independent eigenfunctions corresponding to λ.

Suppose we can find all eigenvalues $\{\lambda_n\}_{n=1}^{\infty}$ and also all corresponding eigenfunctions $\{\phi_m(x)\}_{m=1}^{\infty}$, then we can follow the same steps as for one-space dimension to find a series solution for the problem (4.3.1)–(4.3.3). Here is just one example to illustrate the procedure.

Example 4.3.1. Let $D = [0, L] \times [0, H]$ be a rectangular domain in R^2. Solve the problem (4.3.1)–(4.3.3).

Solution. In this example, we use $(x, y) \in D$ instead of $x = (x_1, x_2)$.

Note that the eigenvalue problem in D is:

$$-\Delta\phi = \lambda\phi, \qquad (x,y) \in D, \tag{4.3.7}$$

$$\phi(x, y) = 0, \qquad (x, y) \in \partial D. \tag{4.3.8}$$

Recall Example 3.4.1 in Section 3.4 of Chapter 3, we see that all eigenvalues are

$$\lambda_{nm} = \left(\frac{n\pi}{L}\right)^2 + \left(\frac{m\pi}{H}\right)^2, \qquad n, m = 1, 2, \cdots.$$

All corresponding eigenfunctions are

$$\phi_{nm}(x, y) = \sin(\frac{n\pi x}{L})\sin(\frac{m\pi y}{H}), \qquad n, m = 1, 2, \cdots.$$

It follows that the series solution is

$$u(x, y, t) = \sum_{n=1}^{\infty}\sum_{m=1}^{\infty} a_{nm} e^{-\lambda_{nm}k^2 t}\phi_{nm}(x, y),$$

where

$$a_{nm} = \frac{< f, \phi >}{< \phi_{nm}, \phi_{nm} >} = \frac{4}{HL}\int\int_R f(x, y)\sin(\frac{n\pi x}{L})\sin(\frac{m\pi y}{H})dxdy, \forall n, m \geq 1.$$

4.3.2 Nonhomogeneous problems

When there is a heat-source generator in the system, Eq. (4.3.1) becomes nonhomogeneous. Also, the temperature on the boundary may not be equal to 0. However, we can follow the same idea as for one-space dimension to find the series solution as in the case for one-space dimension.

Consider the following nonhomogeneous problem:

$$u_t = k^2\Delta u + g(x, t), \qquad (x, t) \in Q, \tag{4.3.9}$$

$$u(x, t) = h(x, t), \qquad x \in \partial\Omega, t > 0, \tag{4.3.10}$$

$$u(x, 0) = f(x), \qquad x \in \Omega, \tag{4.3.11}$$

where $\Omega \in R^n$.

To find the series solution, we first construct a smooth function $w(x, t)$ such that

$$w(x, t) = h(x, t), \qquad x \in \partial\Omega, t \geq 0.$$

There are many different ways to construct such a function. When $\partial\Omega$ is smooth, we can use the distance function

$$d(x) = dist\{x, \partial\Omega\}, \qquad x \in \Omega.$$

Set

$$w(x, t) = (1 - d(x))h(x, t).$$

Introduce

$$v(x,t) = u(x,t) - w(x,t), \qquad (x,t) \in \Omega \times (0,\infty).$$

Then, $v(x,t)$ satisfies a nonhomogeneous equation subject to a homogeneous boundary condition. Next, we follow the same steps as the case for one-space dimension to construct a series solution. We skip the details here.

4.4 The well-posedness and energy estimates

In this section we study some theoretical questions for the heat equation. There are different ways to establish the well-posedness for the problem (4.3.1)–(4.3.3). Here, we introduce a method that can deal with a general parabolic equation (see [8,19,21]).

4.4.1 Definition of a weak solution

Let $T > 0$ be arbitrary and $Q_T = \Omega \times (0,T]$ and $S_T := \partial\Omega \times (0,T]$. When $T = \infty$, we use $Q = \Omega \times (0,\infty)$.

Consider the general heat equation with a source:

$$u_t + L[u] = g(x,t), \qquad (x,t) \in Q_T, \tag{4.4.1}$$

$$B[u](x,t) = 0, \qquad x \in \partial\Omega \times (0,T], \tag{4.4.2}$$

$$u(x,0) = f(x), \qquad x \in \Omega, \tag{4.4.3}$$

where the operators L and B are defined as before:

$$L[u] := -\nabla[p(x,t)\nabla u] + q(x,t)u,$$

$$B[u] := u, \qquad \text{or} \qquad B[u] := p\nabla_\nu u + \beta(x,t)u, \qquad (x,t) \in S_T.$$

The following basic assumptions are assumed throughout this section.

H(4.1) Let $p(x,t), q(x,t) \in L^\infty(Q_T)$ and $p(x,t) \geq p_0 > 0$ on $\bar{Q}$ for some constant $p_0 > 0$.

H(4.2) Let $g(x,t) \in L^2(Q_T)$ and $f(x) \in L^2(\Omega)$. Let $\beta(x,t) \in L^\infty(S_T)$.

Since we only assume that $p(x,t) \in L^\infty(Q_T)$, we need to extend our solution in a broader class, which we call a weak solution. The regularity theory in advanced PDEs shows that the weak solution is also classical if $p(x,t)$ and $q(x,t)$ are smooth (see [8,21]).

Define a space $V_2(Q_T) := C([0,T]; H_0^1(\Omega))$ equipped with the following norm:

$$||u||_{V_2(Q_T)} := \max_{0\leq t\leq T} ||u||_{L^2(\Omega)} + ||\nabla u||_{L^2(Q_T)}.$$

For the Dirichlet boundary condition we define a weak solution for (4.4.1)–(4.4.3) as follows.

Definition 4.4.1. We call $u(x,t)$ a weak solution of the problem (4.4.1)–(4.4.3) if $u(x,t) \in C([0,T]; H_0^1(\Omega))$ such that the following integral identity holds for all

$$v \in L^2([0,T], H_0^1(\Omega)) \bigcap H^1(0,T; L^2(\Omega)) \text{ with } v(x,T) = 0 \text{ in } \Omega:$$

$$\int_0^T \int_\Omega [-uv_t + p\nabla u \cdot \nabla v + quv]\,dxdt = \int_\Omega f(x)v(x,0)dx + \int_0^T \int_\Omega g(x,t)v(x,t)dxdt. \tag{4.4.4}$$

For the Neumann or Robin type of boundary condition, we modify Definition 4.4.1 as follows:

Definition 4.4.2. We call $u(x,t)$ a weak solution of the problem (4.4.1)–(4.4.3) with

$$B[u] = p\nabla_\nu u + \beta(x,t)u = 0$$

if $u(x,t) \in C([0,T]; H^1(\Omega))$ such that the following integral identity holds for all $x \subset \Omega$ and

$$v \in C([0,T], H^1(\Omega)) \bigcap H^1(0,T; L^2(\Omega)) \text{ with } v(x,T) = 0:$$

$$\int_0^T \int_\Omega [-uv_t + p\nabla u \cdot \nabla v + quv]\,dxdt + \int_0^T \int_{\partial\Omega} \beta uv dsdt = \int_\Omega f(x)v(x,0)dx + \int_0^T \int_\Omega gvdxdt. \tag{4.4.5}$$

One can easily verify that a classical solution must be a weak solution.

4.4.2 The energy estimate and uniqueness

The uniqueness and continuous dependence can be proved by using an elementary argument called the energy method. We demonstrate the method in detail here.

Suppose there exist two solutions $u_1(x,t)$ and $u_2(x,t)$ corresponding to two sets of known data $(g_1(x,t), f_1(x))$ and $(g_2(x,t), f_2(x))$.

Set

$$g(x,t) = g_1(x,t) - g_2(x,t),\ f(x) = f_1(x) - f_2(x).$$

Then, $u(x,t) = u_1(x,t) - u_2(x,t)$ is a solution in the weak sense for the following problem:

$$u_t + L[u] = g(x,t), \qquad x \in \Omega, t > 0, \tag{4.4.6}$$

$$B[u](x,t) = 0, \qquad x \in \partial\Omega, t > 0, \tag{4.4.7}$$

$$u(x,0) = f(x), \qquad x \in \Omega. \tag{4.4.8}$$

Define the energy function

$$E(t) = \frac{1}{2}\int_{\Omega} u^2 dx.$$

We assume that $u(x,t)$ is suitably smooth in Q_T for simplicity (otherwise, we can use the Steklov approximation [21]). The following calculation can be carried out for the weak solution without this condition.

Then,

$$\begin{aligned} E'(t) &= \int_{\Omega} u u_t dx \\ &= \int_{\Omega} (-uL[u] + gu) dx \\ &= -\int_{\Omega} p(x,t)|\nabla u|^2 dx - \int_{\Omega} q u^2 dx + \int_{\Omega} ug dx + \int_{\partial\Omega} up(x,t)(\nabla_{\nu} u) ds \\ &:= I_1 + I_2 + I_3 + I_4. \end{aligned}$$

We first consider the case with a Dirichlet or Neumann boundary condition and $q(x,t) \geq 0$ on Q. In this case, $I_2 \leq 0$ and $I_4 = 0$.

Next, we use Cauchy–Schwarz's inequality with a parameter $\varepsilon > 0$ to obtain

$$\begin{aligned} |I_3| &= |\int_{\Omega} ug dx| \\ &\leq \varepsilon \int_{\Omega} u^2 dx + \frac{4}{\varepsilon}\int_{\Omega} g^2 dx \\ &:= J_1 + J_2. \end{aligned}$$

By using Poincare's inequality, we obtain

$$|J_1| \leq c_0 \int_{\Omega} |\nabla u|^2 dx, \qquad t \geq 0.$$

If we choose $\varepsilon = \frac{p_0}{2c_0}$, then we obtain the following energy estimate:

$$E'(t) + a_0 E(t) \leq A_0 ||g||^2_{L^2(\Omega)},$$

where $a_0 > 0$ and A_0 are two constants.

We use an integrating factor $e^{a_0 t}$ to rewrite the above energy inequality in the following form:

$$\frac{d}{dt}\left(e^{a_0 t} E(t)\right) \leq e^{a_0 t} ||g||^2_{L^2(\Omega)}.$$

Consequently, we have

$$E(t) \leq E(0)e^{-a_0 t} + A_0 \int_0^t \int_{\Omega} e^{-a_0(t-s)} g(x,s)^2 dx ds, \qquad t \geq 0.$$

When the boundary condition (4.4.2) is Robin type and the sign condition on $q(x,t)$ is dropped, then the proof is more complicated, since we have to deal with the term involving a boundary integral I_4. One needs to use a trace type of estimate in order to derive a similar result.

Since $q(x,t) \in L^\infty(Q)$, we see that

$$|I_2| \le CE(t), \qquad t \ge 0,$$

where C is the upper bound of $q(x,t)$ in Q.

For the Robin type of boundary condition (4.4.2), we have

$$|I_4| \le C \int_{\partial\Omega} u^2 ds, \qquad t \ge 0,$$

where C depends only on known data.

Now, we use the trace estimate in Chapter 1 with a small parameter $\varepsilon > 0$:

$$\int_{\partial\Omega} u^2 ds \le \varepsilon \int_\Omega |\nabla u|^2 dx + C(\varepsilon) \int_\Omega u^2 dx,$$

where $C(\varepsilon)$ depends only on Ω.

We choose ε sufficiently small to obtain

$$E'(t) \le C_1 \int_\Omega g^2 dx + C_2 E(t), \qquad t \ge 0.$$

We use Gronwall's inequality to obtain

$$||u||_{L^2(Q)} \le C_1 e^{C_2 t}[||g||_{L^2(Q_T)} + ||f||_{L^2(\Omega)}],$$

where C_1 and C_2 depend only on known data.

We summarize the above analysis to obtain the following result.

Theorem 4.4.1. *Under the assumptions H(4.1)–H(4.2) the solution of the problem (4.4.1)–(4.4.3) is unique and continuously depends on the known data in L^2-sense. Moreover,*

$$||u||_{L^2(Q_T)} \le C_1 e^{C_2 t}\left[||g||_{L^2(Q_T)} + ||f||_{L^2(\Omega)}\right],$$

where C_1 and C_2 depend only on known data. □

The reader will see a different method for the uniqueness proof by using the maximum principle for a smooth solution in the next section.

4.4.3 The existence and regularity

There are different ways to establish the global existence of a weak solution to Problem (4.4.1)–(4.4.3). Here, we use a finite-element method (Galerkin method). The idea of the method is based on the fact that an infinite-dimensional function space can be approximated by a finite-dimensional space.

Theorem 4.4.2. *Under the assumptions H(4.1)–(4.2), the problem (4.4.1)–(4.4.3) has a unique weak solution $u(x,t)$ and the weak solution is smooth in Q if $p(x,t), q(x,t)$ and $g(x,t), f(x)$ are smooth in Q.*

Proof. We use the Dirichlet boundary condition as an example to illustrate the basic idea of the proof. Choose a basis $\{\psi_k(x)\}_{k=1}^{\infty}$ for the space $H_0^1(\Omega)$ and orthogonal in $L^2(\Omega)$. For example, we choose all eigenfunctions for the following eigenvalue problem:

$$\begin{aligned} -\Delta u &= \lambda u, & x &\in \Omega, \\ u(x) &= 0, & x &\in \partial\Omega. \end{aligned}$$

Then, we can construct an approximate solution $u_N(x,t)$ as follows:

$$u_N(x,t) = \sum_{k=1}^{N} d_k^N(t)\psi_k(x), \ N = 1, 2, \cdots,$$

where $d_k^N(t)$ is the solution of the ODE system:

$$\frac{d}{dt}d_k^N(t) + \sum_{j=1}^{N} a_{kj}(t)d_j^N(t) + d_k^N(t)q_k(t) = g_k^N(t), \tag{4.4.9}$$

$$d_k^N(0) =< f, \psi_k >, \qquad k = 1, \cdots, N, \tag{4.4.10}$$

where

$$a_{kj}(t) =< p\nabla\psi_k, \nabla\psi_j >, \qquad q_k(t) =< q, \psi_k >, \qquad g_k =< g, \psi_k > .$$

From the ODE theory, we know that the system (4.4.9)–(4.4.10) has a unique solution $\{d_k^N(t)\}$ for all $N \geq 1$.

From the construction of $u_N(x,t)$, we have

$$\begin{aligned} &\frac{1}{2}\frac{d}{dt} < u_N, u_N > + < p\nabla u_N, \nabla u_N > + < qu_N, u_N >=< g_N, u_N >, \qquad t \geq 0, \\ &u_N(x,0) = f_N(x), \qquad x \in \Omega, \end{aligned}$$

where

$$g_N(x,t) = \sum_{k=1}^{N} g_k(t)\psi_k(x), \qquad f_N(x) = \sum_{k=1}^{N} f_k\psi_k(x).$$

Now, from the construction of $u_N(x,t)$ we find the following uniform estimate:

$$\sup_{0\leq t\leq T} ||u_N||_{L^2(\Omega)} + ||\nabla u_N||_{L^2(Q_T)} \leq C,$$

where C depends only on the known constants and an upper bound of T.

Moreover, from the construction of $u_N(x,t)$, we see that

$$u_N(x,t)_t \in L^2((0,T]; H^{-1}(\Omega)).$$

It follows by Aubin–Lions' compactness theorem ([28]) that we can extract a subsequence from $u^N(x,t)$ that converges to a limit function $u(x,t)$ in $V_2(Q_T)$. The subsequence is still denoted by $u_N(x,t)$. Hence,

$$\begin{aligned} \nabla u_N(x,t) \to \nabla u(x,t) \qquad & \text{weakly in } L^2(Q_T); \\ u_N(x,t) \to u(x,t) \qquad & \text{strongly in } L^2(Q_T). \end{aligned}$$

Finally, we assume that $v(x,t) \in L^2(0,T); H_0^1(\Omega)) \bigcap H^1(0,T); L^2(\Omega))$ is expressed in the following form:

$$v(x,t) = \sum_{k=1}^{\infty} v_k(t)\psi_k(x),$$

where $v_k(T) = 0$.

From the equation of $d_k^N(t)$ we have

$$\begin{aligned} &\int_0^T \int_\Omega [-u_N v_t + p\nabla u_N \nabla v + q u_N v]\, dxdt \\ &= \int_\Omega f_N v(x,0) dx + \int_0^T \int_\Omega g_N v dxdt. \end{aligned}$$

After taking the limit as $N \to \infty$, we see that $u(x,t)$ satisfies the integral equation. From the integral equation (4.4.4), we see that $u_t \in L^2(0,T); H^{-1}(\Omega))$, which implies $u(x,t) \in C([0,T]; H_0^1(\Omega))$ (see [8,19,21]). Consequently, $u(x,t)$ is a weak solution.

Moreover, the regularity theory ([19,21]) implies that the solution is actually in $C^{2+\alpha,1+\frac{\alpha}{2}}(Q_T)$ if the coefficients $p(x,t) \in C^{1+\alpha,\frac{\alpha}{2}}(Q_T)$, $q(x,t) \in C^{\alpha,\frac{\alpha}{2}}(Q_T)$ and the nonhomogeneous term $g(x,t) \in C^{\alpha,\frac{\alpha}{2}}(Q_T)$. The reader can find its rigorous proof in advanced PDE books, such as [8,19,21]. □

Remark 4.4.1. In advanced PDEs, the regularity of a weak solution to a PDE problem is one of the major research topics.

4.5 A qualitative property: The maximum principle

The maximum principle is the most important property for the heat equation and the Laplace equation. It is the fundamental difference between the wave equation and the other two types of equations. Physically, it shows that the propagation speed of heat is infinite, while the propagation speed of sound is finite. We will see more precise explanations in later chapters. Throughout this section, a solution for a PDE problem always means in the classical sense.

4.5.1 The weak maximum principle

Let $Q_T = \Omega \times (0, T]$ for $T > 0$ and $\partial_p Q_T := \partial\Omega \times (0, T] \bigcap \{(x, 0) : x \in \Omega\}$.

Theorem 4.5.1. *(The weak maximum principle) Suppose $u(x,t)$ satisfies the heat equation*

$$u_t = k^2 \Delta u, \qquad (x,t) \in Q_T.$$

Then,

$$\max_{(x,t)\in\bar{Q}_T} u(x,t) = \max_{(x,t)\in\partial_p Q_T} u(x,t), \qquad \min_{(x,t)\in\bar{Q}_T} u(x,t) = \min_{(x,t)\in\partial_p Q_T} u(x,t).$$

Proof. The idea of the proof is based on a simple fact from calculus: when a function attains a maximum (or minimum) value in a region, then the Hessian matrix of the function must be nonpositive-definite. However, we must make a modification in order to give a rigorous proof.

Let $\varepsilon > 0$ be a small constant. Set

$$v(x,t) = u(x,t) - \varepsilon t.$$

We see that $v(x,t)$ satisfies

$$v_t = k^2 \Delta v - \varepsilon, \qquad (x,t) \in Q_T.$$

Suppose $v(x,t)$ attains the maximum value at an interior point $(x_0, t_0) \in Q_T$. Then, we see that

$$v_t|_{(x_0,t_0)} \geq 0, \qquad \Delta v|_{(x_0,t_0)} \leq 0,$$

which is a contradiction with the equation of $v(x,t)$. It follows that the maximum of $v(x,t)$ must attain either at the initial time $t = 0$ or on the lateral boundary of the region Q_T. If we set $\varepsilon \to 0$, we see that $u(x,t)$ has the same conclusion. □

Corollary 4.5.1. *Suppose $g(x,t), h(x,t) \in C(\bar{Q}_T)$ and $f(x) \in C(\bar{\Omega})$. Let $u(x,t)$ be the solution of the following heat equation:*

$$u_t = k^2 \Delta u + g(x,t), \qquad (x,t) \in Q_T, \tag{4.5.1}$$

$$u(x,t) = h(x,t), \qquad x \in \partial\Omega \times (0,T], \tag{4.5.2}$$

$$u(x,0) = f(x), \qquad x \in \Omega. \tag{4.5.3}$$

Then,

$$||u||_0 \leq C(T)\left[||g||_0 + ||h||_0 + ||f||_0\right], \tag{4.5.4}$$

where $C(T)$ is a constant that depends only on an upper bound of T. In particular, the continuous dependence of $u(x,t)$ holds under the maximum norm.

Proof. We only prove the case for the maximum value for $u(x,t)$ in Q_T. Let $a>0$ be a constant to be determined later. We define a new function

$$v(x,t)=e^{-at}u(x,t), \qquad (x,t)\in Q_T.$$

It is clear that $v(x,t)$ satisfies

$$v_t-k^2\Delta v+av=e^{-at}g(x,t), \qquad (x,t).$$

If $v(x,t)$ attains the maximum value M at an interior point $(x_0,t_0)\in Q_T$, then at this point (x_0,t_0),

$$v_t-k^2\Delta v\big|_{(x_0,t_0)}\geq 0.$$

It follows that

$$aM\leq e^{-at_0}g(x_0,t_0).$$

Hence, we obtain the maximum estimate for $v(x,t)$ on $\bar{Q}_T$. On the other hand, if $v(x,t)$ attains its maximum on the lateral boundary of Q_T or at the initial moment, then we have the desired estimate (4.5.4). This concludes the desired estimate for v in $\bar{Q}_T$, which immediately implies the maximum estimate for $u(x,t)$ in $\bar{Q}_T$. The case for the minimum value of $u(x,t)$ is similar. We combine both cases to conclude the proof of the desired estimate (4.5.4). □

It is clear that the method in the proof of Corollary 4.5.1 fails if the boundary condition (4.5.2) is of the Neumann type. This leads to a different type of maximum principle that is often referred as the strong maximum principle. The proof of the strong maximum principle relies on the following lemma.

4.5.2 The strong maximum principle

Let Ω be a bounded domain in R^n with $\partial\Omega\in C^2$. Let $S_T=\partial\Omega\times(0,T]$.

Lemma 4.5.1. *(Hopf's lemma) Suppose $u(x,t)$ satisfies equation:*

$$u_t\leq k^2\Delta u, \qquad (x,t)\in Q_T. \tag{4.5.5}$$

If $u(x,t)$ attains its maximum M over Q_T at a boundary point $(x_0,t_0)\in S_T$ and $u(x,t)<M$ in a neighborhood of (x_0,t_0) inside Q_T. Then,

$$\frac{\partial u}{\partial \nu}\Big|_{(x_0,t_0)}>0,$$

where ν is the outward unit normal at (x_0,t_0).

Proof. Since $\partial\Omega \in C^2$, we choose R and η such that a small paraboloid frustum

$$A := \{(x,t) : |x - x^*|^2 + \eta^2(t_0 - t) < R^2, t_0 - \frac{R^2}{\eta^2} < t < t_0\} \subset Q_T,$$

with $(x_0, t_0) \in \partial A \bigcap S_T$. Moreover, we choose R sufficiently small such that

$$u(x,t) < M, \qquad (x,t) \in A.$$

Let

$$B = \{(x,t) \in A; |x - x^*| > \frac{R}{2}\}.$$

Define

$$r(x,t) = \sqrt{|x - x^*|^2 + \eta^2(t_0 - t)}, \qquad (x,t) \in A.$$

Construct an auxiliary function

$$h(x,t) = e^{-\alpha r(x,t)^2} - e^{-\alpha R^2},$$

where $\alpha > 0$ will be determined later.

It is clear that

$$h(x,t) \geq 0, \forall (x,t) \in A, \qquad h(x_0, t_0) = 0.$$

Moreover, in B,

$$\begin{aligned}
&h_t - k^2 \Delta h \\
&= e^{-\alpha r^2}\left[\eta^2 - 4\alpha^2 k^2 |x - x^*|^2 + 2\alpha k^2\right] \\
&\leq e^{-\alpha r^2}\left[\eta^2 - 2\alpha^2 k^2 R^2 + 2\alpha k^2\right] \leq 0,
\end{aligned}$$

provided that α is chosen to be sufficiently large.

Now, the parabolic boundary of B consists of S_1 and S_2, where

$$S_1 = \{(x,t) : r(x,t) = R, t_0 - \frac{R^2}{\eta^2} \leq t \leq t_0\},$$

$$S_2 = \{(x,t) : r(x,t) = \frac{R}{2}, t_0 - \frac{R^2}{\eta^2} \leq t \leq t_0\}.$$

On S_1, $h(x,t) \geq 0$ and $u(x,t) \leq u(x_0, t_0)$. On S_2, since $u(x,t) - u(x_0, t_0) < 0$, we can choose $\varepsilon > 0$ sufficiently small such that

$$u(x,t) - u(x_0, t_0) + \varepsilon h(x,t) \leq 0, \qquad (x,t) \in S_2.$$

Define

$$w(x,t) := u(x,t) - u(x_0,t_0) + \varepsilon h(x,t), (x,t) \in B.$$

Then,

$$w_t - k^2 \Delta w \le 0, \qquad (x,t) \in B.$$

It follows that $w(x,t)$ attains its maximum at (x_0,t_0). Consequently, we have

$$\nabla_\nu w|_{(x_0,t_0)} \ge 0.$$

Note that the outward unit normal of $B_R(x^*)$ at (x_0,t_0) is equal to

$$\nu = \frac{x_0 - x^*}{|x_0 - x^*|}.$$

It follows that

$$\nabla_\nu h(x,t)|_{(x_0,t_0)} < 0,$$

which implies that

$$\nabla_\nu u(x,t)|_{(x_0,t_0)} > 0. \qquad \square$$

With the help of Hopf's lemma, we can prove the following strong maximum principle.

Theorem 4.5.2. *(The strong maximum principle) Let $u(x,t)$ be a solution of the heat equation*

$$u_t - k^2 \Delta u = 0, \qquad (x,t) \in Q_T.$$

If $u(x,t)$ attains a maximum value at an interior point $(x^*,t^*) \in Q_T$, then $u(x,t)$ must be a constant.

Proof. Indeed, suppose $u(x,t)$ attains the maximum at $(x^*,t^*) \in Q_T$ and $u(x,t)$ is not a constant, then we can construct a small cone C such that

$$u(x,t) < u(x^*,t^*), \qquad (x,t) \in C.$$

Hopf's lemma in C implies

$$\frac{\partial u}{\partial \nu}|_{(x^*,t^*)} > 0,$$

a contradiction since we know $(x^*,t^*) \in Q_T$ and $\nabla u|_{(x^*,t^*)} = 0$. $\square$

4.6 Long-time behaviors of solutions

A challenging question in the study of an evolution PDE or system is to find the long-time behavior or the pattern of the solution as time evolves. Many physical phenomena such as turbulence and shock waves can be observed from the long-time behaviors of the solution for a PDE problem. In this section, we use an elementary method to prove some interesting results for the heat equation.

4.6.1 Long-time behavior of a solution for the heat equation

Let $Q = \Omega \times (0, \infty)$ and $S = \partial\Omega \times (0, \infty)$. Consider the heat equation subject to the Dirichlet boundary condition:

$$u_t = k^2 \Delta u, \qquad (x, t) \in Q, \tag{4.6.1}$$

$$u(x, t) = 0, \qquad (x, t) \in S, \tag{4.6.2}$$

$$u(x, 0) = f(x), \qquad x \in \Omega, \tag{4.6.3}$$

where $f(x) \in L^2(\Omega)$.

Since there is no heat generator or cold sink on the inside of the object, naturally, we expect that the temperature will approach the same temperature as that on the boundary after a long time. We can actually obtain a more precise result.

Theorem 4.6.1. *Let $u(x, t)$ be the solution of the problem (4.6.1)–(4.6.3). Then,*

$$u(x, t) = O(e^{-\lambda_1 k^2 t}), \qquad \text{as } t \to \infty, \tag{4.6.4}$$

where λ_1 is the first eigenvalue of the Laplace operator on Ω.

Proof. Suppose $\{\lambda_n\}_{n=1}^{\infty}$ and $\{\phi_n(x)\}_{n=1}^{\infty}$ are the eigenvalues and the corresponding orthogonal eigenfunctions for the Laplace operator. We can further choose all $\phi_n(x)$ to be uniformly bounded in Ω. Then, we know that the solution of (4.6.1)–(4.6.3) can be expressed as a series solution:

$$u(x, t) = \sum_{n=1}^{\infty} a_n e^{-\lambda_n k^2 t} \phi_n(x),$$

where

$$a_n = \frac{< f, \phi_n >}{< \phi_n, \phi_n >}, \qquad \forall n \geq 1.$$

Since

$$0 < \lambda_1 < \lambda_2 < \cdots,$$

we see that

$$u(x, t) = e^{-\lambda_1 k^2 t} \left[a_1 \phi_1(x) + \sum_{n=2}^{\infty} a_n e^{-(\lambda_n - \lambda_1) k^2 t} \phi_n \right].$$

Since $\phi_n(x)$ are uniformly bounded in Ω and $\lambda_n - \lambda_1 \geq \lambda_2 - \lambda_1$ for all $n = 2, 3, \cdots$, we see that the series is convergent to 0 when $t \geq t_0 > 0$ for any $t_0 > 0$. This concludes the desired estimate for $u(x,t)$ when $t \to \infty$. □

For example, in one-space dimension where $\Omega = [0, L]$, we see that

$$\lambda_1 = \left(\frac{\pi}{L}\right)^2.$$

It follows that $u(x,t) \to 0$ exponentially with the precise decay rate $e^{-\lambda_1 k^2 t}$ as $t \to \infty$.

In two-space dimensions where $\Omega = [0, L] \times [0, H]$,

$$\lambda_1 = \left(\frac{\pi}{L}\right)^2 + \left(\frac{\pi}{H}\right)^2.$$

It follows that $u(x,t) \to 0$ exponentially with the precise decay rate $e^{-\lambda_1 k^2 t}$ as $t \to \infty$.

We can extend the problem (4.6.1)–(4.6.3) to a more general problem with nonhomogeneous equation:

$$u_t = k^2 \Delta u + h(x,t), \qquad x \in \Omega, t > 0, \tag{4.6.5}$$

$$u(x,t) = g(x,t), \qquad x \in \partial\Omega, t > 0, \tag{4.6.6}$$

$$u(x,0) = f(x), \qquad x \in \Omega. \tag{4.6.7}$$

Let $h(x,t)$ and $g(x,t)$ be in $C(\bar{Q})$. Moreover,

$$\lim_{t\to\infty} [||h(x,t) - H(x)||_0 + ||g(x,t) - G(x)||_0] = 0.$$

Let $U(x)$ be a solution for the steady-state problem:

$$-k^2 \Delta U = H(x), \qquad x \in \Omega, \tag{4.6.8}$$

$$U(x) = G(x), \qquad x \in \partial\Omega. \tag{4.6.9}$$

Theorem 4.6.2. *Let $h(x,t)$ and $g(x,t)$ be in $L^\infty(Q)$. Moreover,*

$$\lim_{t\to\infty} \left[||h(x,t) - H(x)||_{L^\infty(\Omega)} + ||g(x,t) - G(x)||_{L^\infty(\Omega)}\right] = 0.$$

Then,

$$\lim_{t\to\infty} ||u(x,t) - U(x)||_{L^\infty(\Omega)} = 0,$$

where $U(x)$ is a solution of the corresponding steady-state problem:

$$-k^2 \Delta U = H(x), \qquad x \in \Omega, \tag{4.6.10}$$

$$U(x) = G(x), \qquad x \in \partial\Omega. \tag{4.6.11}$$

Proof. First, the maximum principle implies that there exists a constant M such that

$$\sup_Q |u(x,t)| \le M, \qquad \sup_\Omega |U(x)| \le M.$$

Define

$$w(x,t) := u(x,t) - U(x), \qquad (x,t) \in Q.$$

Then, $w(x,t)$ satisfies

$$w_t - k^2 \Delta w = h(x,t) - H(x), \qquad (x,t) \in Q, \tag{4.6.12}$$

$$w(x,t) = g(x,t) - G(x), \qquad x \in \partial\Omega, t > 0, \tag{4.6.13}$$

$$w(x,0) = f(x) - U(x), \qquad x \in \Omega. \tag{4.6.14}$$

For any $\varepsilon > 0$, there exists $T_0 > 0$ such that

$$||g(x,t) - G(x)||_0 + ||h(x,t) - H(x)||_0 < \varepsilon, \qquad \forall t \ge T_0.$$

From the series representation of $w(x,t)$ we see that

$$|w(x,t)| \le C\varepsilon, \qquad \forall t \ge T_0,$$

where C is a constant that depends only on M and other known constants.

This concludes the desired limit. □

Now, we consider Eqs. (4.6.1) and (4.6.3) subject to a Neumann boundary condition:

$$\nabla_\nu u(x,t) = 0, \qquad (x,t) \in \partial\Omega \times (0,\infty). \tag{4.6.15}$$

Clearly,

$$u_0 := \frac{1}{|\Omega|} \int_\Omega u(x,t)dx = \frac{1}{|\Omega|} \int_\Omega f(x)dx.$$

It follows that $u(x,t)$ does not decay at $t \to \infty$. However, the next theorem shows a more precise behavior as $t \to \infty$.

Theorem 4.6.3. *Let $u(x,t)$ be a solution of Eq. (4.6.1) subject to the initial condition (4.6.3) and the Neumann boundary condition (4.6.15). Then, $u(x,t) - u_0$ converges to 0 in the $L^2(\Omega)$-sense.*

Proof. The proof relies on a different version of Poincare's inequality. Let

$$v(x,t) = u(x,t) - u_0, \qquad x \in \Omega, t > 0.$$

Then, $v(x,t)$ satisfies the same heat equation and homogeneous Neumann boundary condition (4.6.15). Moreover,

$$\int_\Omega v(x,t)dx = 0, \qquad t \ge 0.$$

Let

$$E(t) = \frac{1}{2}\int_{\Omega} v(x,t)^2 dx.$$

Then,

$$E'(t) + k^2 \int_{\Omega} |\nabla v|^2 dx = 0.$$

Poincare's inequality (II) implies that there exists a constant c_0 such that

$$\int_{\Omega} |v|^2 dx \le c_0 \int_{\Omega} |\nabla v|^2 dx.$$

It follows that

$$E'(t) + aE(t) \le 0, \qquad t > 0,$$

where $a = \frac{k^2}{c_0}$.

This implies that $E(t)$ converges to 0 exponentially as $t \to \infty$. □

4.6.2 Turing instability

From Section 4.6.1 we see that the diffusion in heat conduction stabilizes the temperature distribution. However, this may not be the case for a reaction–diffusion system. In this subsection we discuss how the diffusion may cause the instability.

Consider the following 2×2 ODE system:

$$y_1'(t) = a_{11}y_1(t) + a_{12}y_2(t), \qquad t > 0, \tag{4.6.16}$$

$$y_2'(t) = a_{21}y_1(t) + a_{22}y_2(t), \tag{4.6.17}$$

subject to an initial value.

Suppose we add diffusion for each equation with a diffusion coefficient in a bounded domain $\Omega \in R^n$.

$$u_{1t} - d_1\Delta u_1 = a_{11}u_1 + a_{12}u_2, \qquad x \in \Omega, t > 0, \tag{4.6.18}$$

$$u_{2t} - d_2\Delta u_2 = a_{21}u_1 + a_{22}u_2, \qquad x \in \Omega, t > 0. \tag{4.6.19}$$

We impose a Dirichlet boundary condition for u_1 and u_2:

$$u_1(x,t) = u_2(x,t) = 0, \qquad x \in \partial\Omega, t > 0.$$

Suppose we know that the solution of the ODE system is stable as $t \to \infty$, we expect that (u_1, u_2) is also stable as t evolves. It turns out that this may not always be the case.

We know from the theory of ODEs that the solution of the ODE system (4.6.16)–(4.6.17) is stable if the real part of each eigenvalue for the matrix A is negative, where

$$A = \begin{pmatrix} a_{11} & a_{12} \\ a_{21} & a_{22} \end{pmatrix}.$$

Suppose σ_1 and σ_2 are eigenvalues of the matrix A. Then,

$$\sigma_1 + \sigma_2 = a_{11} + a_{22} < 0,$$
$$\sigma_1 \cdot \sigma_2 = a_{11}a_{22} - a_{12}a_{21} > 0.$$

Let λ_1 be the principal eigenvalue for the Laplace operator associated with a Dirichlet boundary condition and $\psi(x)$ be the corresponding positive eigenfunction. Define

$$y_1^*(t) = \int_\Omega u_1(x,t)\psi(x)dx, \qquad y_2^*(t) = \int_\Omega u_2(x,t)\psi(x)dx.$$

Then, we see that $y_1^*(t)$ and $y_2^*(t)$ satisfy the following ODE system:

$$y_1^*(t)' = (a_{11} - \lambda_1 d_1)y_1(t) + a_{12}y_2(t), \qquad t > 0, \tag{4.6.20}$$
$$y_2^*(t)' = a_{21}y_1(t) + (a_{22} - \lambda_1 a_2)y_2(t). \tag{4.6.21}$$

Let

$$A^* = \begin{pmatrix} a_{11} - \lambda_1 d_1 & a_{12} \\ a_{21} & a_{22} - \lambda_1 d_2 \end{pmatrix}.$$

Suppose σ_1^* and σ_2^* are two eigenvalues of the matrix A^*. Then,

$$\sigma_1^* + \sigma_2^* = (a_{11} - \lambda_1 d_1) + (a_{22} - \lambda_2 d_2) < 0,$$
$$\sigma_1^* \cdot \sigma_2^* = (a_{11}a_{22} - \lambda_1 d_1 a_{22} - \lambda_1 d_2 a_{11} + d_1 d_2 \lambda_1^2) - a_{12}a_{21}.$$

Clearly, if we choose d_1 sufficiently large and d_2 sufficiently small such that

$$\sigma_1^* \cdot \sigma_2^* < 0,$$

then σ_1^* and σ_2^* have opposite signs, which implies $(y_1^*(t), y_2^*(t))$ is unstable as $t \to \infty$. Namely, a diffusion added in a system can lead to an instability of the solution to a reaction–diffusion system. This is called a *Turing phenomenon*.

Remark 4.6.1. From the above examples, we see that the first eigenvalue for the Laplace operator plays an important role in the study of the long-time behavior of the solution for the heat equation and systems. The dynamics for nonlinear systems depends on the spectrum analysis for the linearized system (see [2]). In differential geometry, a class of important problems is to estimate the bounds of eigenvalues for various manifolds.

4.7 The comparison principle

In this section we prove a comparison principle by using the maximum principle. This principle is very useful in deriving an *a priori* estimate, which is an essential step in the study of nonlinear heat equations. We will present an example at the end of the section. The interested reader may find many more applications in Monographs such as [3,12,24].

Let $a(x,t) \in C(\bar{Q})$. Moreover, there exists a constant a_0 such that

$$a(x,t) \le a_0, \qquad \forall (x,t) \in Q.$$

Consider the following problem:

$$u_t = k^2 \Delta u + a(x,t)u + f(x,t), \qquad (x,t) \in Q_T, \tag{4.7.1}$$
$$B[u] = g(x,t), \qquad x \in \partial\Omega \times (0,T], \tag{4.7.2}$$
$$u(x,0) = h(x), \qquad x \in \Omega, \tag{4.7.3}$$

where

$$B[u] := u, \qquad \text{or} \qquad B[u] := \nabla_\nu u.$$

Suppose the data set $(f(x,t), g(x,t), h(x))$ are continuous and the problem (4.7.1)–(4.7.3) has a unique classical solution $u(x,t)$.

Theorem 4.7.1. *(The comparison principle) Let $u_i(x,t)$ be the solution of the problem (4.7.1)–(4.7.3) corresponding to the data set $(f_i(x,t), g_i(x,t), h_i(x))$ for $i = 1, 2$. If*

$$f_1(x,t) \ge f_2(x,t), g_1(x,t) \ge g_2(x,t), h_1(x) \ge h_2(x), \forall (x,t) \subset Q,$$

then

$$u_1(x,t) \ge u_2(x,t), \qquad \forall (x,t) \in Q.$$

Proof. We prove the theorem by dividing different cases. Set

$$u(x,t) = u_1(x,t) - u_2(x,t), \qquad (x,t) \in Q.$$

Case (a): The Dirichlet boundary condition $B[u] = u(x,t) = g(x,t)$.
Then, $u(x,t)$ satisfies

$$u_t = k^2 \Delta u + a(x,t)u + f(x,t), \qquad (x,t) \in Q_T, \tag{4.7.4}$$
$$u(x,t) = g(x,t) \ge 0, \qquad x \in \partial\Omega \times (0,T], \tag{4.7.5}$$
$$u(x,0) = h(x) \ge 0, \qquad x \in \Omega, \tag{4.7.6}$$

where

$$f(x,t) := f_1(x,t) - f_2(x,t) \ge 0,$$

$$g(x,t) := g_1(x,t) - g_2(x,t),$$
$$h(x) := h_1(x) - h_2(x).$$

The maximum principle implies that

$$u(x,t) \geq 0, \qquad (x,t) \in Q.$$

Case (b): The Neumann boundary condition $B[u] = \nabla_\nu u(x,t) = g(x,t)$.
Then, $u(x,t)$ satisfies

$$u_t = k^2 \Delta u + a(x,t)u + f(x,t), \qquad (x,t) \in Q_T, \tag{4.7.7}$$
$$\nabla_\nu u(x,t) = g(x,t) \geq 0, \qquad x \in \partial\Omega \times (0,T], \tag{4.7.8}$$
$$u(x,0) = h(x) \geq 0, \qquad x \in \Omega. \tag{4.7.9}$$

By Hopf's Lemma 4.5.1, $u(x,t)$ cannot attain a minimum on the lateral boundary $\partial\Omega \times (0,T]$. The strong maximum principle implies that $u(x,t)$ cannot attain a negative minimum at an interior point of Q unless it is a constant. It follows that

$$u(x,t) \geq 0, \qquad (x,t) \in Q. \qquad \square$$

The comparison principle is also valid if the boundary condition is of the Robin type:

$$\nabla_\nu u = b(x,t)u + g(x,t), \qquad (x,t) \in S_T. \tag{4.7.10}$$

Theorem 4.7.2. *Suppose $b(x,t) \in C(\bar{S}_T)$. Let $u(x,t) \in C(\bar{Q}) \bigcap C^{2,1}(Q_T)$ be the solution of Eq. (4.7.1), (4.7.2) associated with the Robin condition (4.7.10). If $f(x,t)$, $g(x,t)$, and $h(x)$ are nonnegative in their domain, then*

$$u(x,t) \geq 0, \qquad \forall (x,t) \in \bar{Q}.$$

Proof. Let $\varepsilon > 0$ and set $h_\varepsilon(x) := h(x) + \varepsilon$. For brevity, we still use $u(x,t)$ as the solution of (4.7.1), (4.7.3), (4.7.10) with initial value $h_\varepsilon(x)$. By the continuity principle, we know that there exists a small $T_0 > 0$ such that the solution of (4.7.1), (4.7.3), and (4.7.10) must be positive in Q_{T_0}. Suppose the nonnegativity of $u(x,t)$ in Q fails. Let T^* be the first time in which $u(x,t)$ attains 0, say at (x_0, T^*). By the strong maximum principle, (x_0, T^*) cannot be an interior point of Q. On the other hand, Hopf's lemma implies that the minimum point (x_0, T^*) cannot be located on the lateral boundary $\partial\Omega \times (0,\infty)$ since $g(x,t) \geq 0$ on $\partial\Omega \times (0,\infty)$, a contradiction. Therefore we obtain

$$u_\varepsilon(x,t) \geq 0, \qquad (x,t) \in Q.$$

After taking a limit as $\varepsilon \to 0$, we obtain the desired result. $\square$

As an application, we investigate a semilinear heat equation. This type of equations is motivated by various models arising from biological sciences and chemical engineering. For example, in population dynamics, the population concentration

$u(x,t)$ satisfies a logistic growth model with growth rate r and the maximum capacity K. Then, $u(x,t)$ satisfies

$$u_t - d\Delta u = ru(1-\frac{u}{K}).$$

Example 4.7.1. Let $f(s) \in C^1[0,\infty)$ with $f(0)=f(K)=0$. Consider the following semilinear problem:

$$\begin{aligned} &u_t - d\Delta u = f(u), && (x,t)\in Q,\\ &\nabla_\nu u(x,t) = 0, && (x,t)\in \partial\Omega\times(0,\infty),\\ &u(x,0) = u_0(x), && x\in\Omega, \end{aligned}$$

where $d>0$ is the diffusion coefficient and the initial value $u_0(x)$ satisfies

$$0 \le u_0(x) \le K.$$

The goal is to prove that the above problem has a unique solution. The uniqueness question is obvious by applying the maximum principle. Moreover, any solution to the above problem must satisfy the following *a priori* estimate:

$$0 \le u(x,t) \le K, \qquad (x,t)\in Q.$$

To prove the existence we construct two sequences called lower and upper solution sequences (see [24]). The limit of each sequence is the solution of the problem.

Set $\underline{u}_0(x,t)=0$ on Q. We define a sequence $\underline{u}_n(x,t)$ as follows: for each $n\ge 0$ we solve the following linear equation $\underline{u}_{n+1}$

$$\begin{aligned} &\underline{u}_{(n+1)t} - d\Delta \underline{u}_{n+1} = f(\underline{u}_n), && (x,t)\in Q,\\ &\nabla_\nu \underline{u}_{n+1}(x,t) = 0, && (x,t)\in \partial\Omega\times(0,\infty),\\ &\underline{u}_{n+1}(x,0) = u_0(x), && x\in\Omega. \end{aligned}$$

Let $\{\underline{u}_{n+1}(x,t)\}$ be the solution sequence. Then, the comparison principle yields that

$$0 \le \underline{u}_0 \le \underline{u}_1 \le \cdots \le \underline{u}_n(x,t) \le \underline{u}_{n+1} \le \cdots \le K, \qquad (x,t)\in Q.$$

Similarly, we set $\bar{u}_0(x,t)=K$ and define $\bar{u}_n(x,t)$ to be the solution of the following linear problem:

$$\begin{aligned} &\bar{u}_{(n+1)t} - d\Delta \bar{u}_{n+1} = f(\bar{u}_n), && (x,t)\in Q,\\ &\nabla_\nu \bar{u}_{n+1}(x,t) = 0, && (x,t)\in \partial\Omega\times(0,\infty),\\ &\bar{u}_{n+1}(x,0) = u_0(x), && x\in\Omega. \end{aligned}$$

Again, the comparison principle yields

$$0 \le \bar{u}_{n+1} \le \bar{u}_n \le \cdots \le \bar{u}_0(x,t) \le K, \qquad (x,t)\in Q.$$

Moreover, the energy estimate implies that for any $T > 0$,

$$\max_{0\le t\le T} ||\underline{u}_n||_{L^2(\Omega)} + ||\nabla \underline{u}_n||_{L^2(Q_T)} \le C;$$
$$\max_{0\le t\le T} ||\bar{u}_n||_{L^2(\Omega)} + ||\nabla \bar{u}_n||_{L^2(Q_T)} \le C;$$

where C depends only on known data, but not on n.

Since each of the above function sequences is monotone, there exists a limit, denoted by $u(x,t)$ and

$$\max_{0\le t\le T} ||u||_{L^2(\Omega)} + ||\nabla u||_{L^2 Q_T} \le C.$$

Moreover, if $u_0(x) \in C^{\alpha}(\bar{\Omega})$, then the regularity theory yields $u(x,t) \in C^{\alpha,\frac{\alpha}{2}}(\bar{Q}_T) \bigcap C^{2+\alpha,1+\frac{\alpha}{2}}(Q_T)$. Since the solution of the problem is unique, we see that

$$u(x,t) = \lim_{n\to\infty} \underline{u}_n(x,t) = \lim_{n\to\infty} \bar{u}_n(x,t),$$

which is a solution of the original semilinear problem.

4.8 Notes and remarks

In this chapter we studied the solution for the heat equation. The mathematical model of heat conduction is derived in detail as a first step. Our emphasis is on the physical meaning of each parameter in the heat-conduction model. This will give students the basic method of how to extend a simple model to more complicated problems. Section 4.2 and Section 4.3 are elementary. Students should not have any difficulty to learn this part of the materials. However, one must note that the series solution for the heat equation is formal. The smoothness of the series solution is very difficult to prove when the space dimension is higher than one. Section 4.4 is more theoretical and beginners may skip this section. The method can be used in dealing with a general parabolic equation. The asymptotic behavior of the solution is included in this chapter to show students why the sign of the eigenvalue is important. The Turing phenomenon shows that the system of reaction–diffusion equations has a much more complicated dynamics than that of the corresponding ODE system.

The maximum principle is the most important property for the heat equation. In addition to its own intrinsic interest, it is also a powerful tool in dealing with nonlinear problems. One will find many more applications in research fields (see [3,7,10,14,19,24]).

4.9 Exercises

1. Find the series solution for the following one-dimensional heat equation with mixed boundary conditions and $f(x) \in L^2(0, L)$:

$$
\begin{aligned}
&u_t = k^2 u_{xx}, && 0 < x < L, t > 0,\\
&u(0,t) = 0,\ u_x(L,t) = 0, && t > 0,\\
&u(x,0) = f(x), && 0 < x < L.
\end{aligned}
$$

2. Let $h_1(t), h_2(t) \in C^1[0, \infty)$ and $f(x) \in L^2(0, L)$. Consider the following one-dimensional heat equation:

$$
\begin{aligned}
&u_t = k^2 u_{xx}, && 0 < x < L, t > 0,\\
&u(0,t) = h_1(t), && t > 0,\\
&u_x(L,t) = h_2(t), && t > 0,\\
&u(x,0) = f(x), && 0 < x < L.
\end{aligned}
$$

(a) Find a function $w(x, t)$ such that

$$
w(0,t) = h_1(t), \qquad w_x(L,t) = h_2(t), t > 0.
$$

(b) Find the series solution for the above problem.

3. Find the solution for the following one-dimensional heat equation:

$$
\begin{aligned}
&u_t = k^2 u_{xx} + \alpha u, && 0 < x < L, t > 0,\\
&u(0,t) = 0,\ u_x(L,t) = 0, && t > 0,\\
&u(x,0) = f(x), && 0 < x < L,
\end{aligned}
$$

where $f(x) \in L^2(0, L)$ and α is a constant. Does the solution converge to 0 as $t \to \infty$? Give a physical interpretation of the result.

4. Find the long-time behavior of the solution to Exercise 1 above and also give the physical interpretation of the result.

5. Let $h_1(t)$ and $h_2(t)$ be continuous functions in $[0, \infty)$. Find a differentiable function $w(x, t)$ such that

$$
\begin{aligned}
&w_x(0,t) + a w(0,t) = h_1(t), && t > 0,\\
&w_x(L,t) + b w(L,t) = h_2(t), && t > 0,
\end{aligned}
$$

where a and b are positive constants.

6. Let $g(t) \in C[0, \infty)$ and $f(x) \in L^2(0, L)$. Let $u(x, t)$ be a solution of the following problem:

$$
\begin{aligned}
&u_t = k^2 u_{xx}, && 0 < x < L, t > 0,\\
&u(0,t) = g(t),\ u(L,t) = 0, && t > 0,
\end{aligned}
$$

$$u(x,0) = f(x), \qquad 0 < x < L.$$

Prove $u(x,t)$ decays to 0 if $g(t) \to 0$ as $t \to \infty$.

7. Let $g(x) \in L^2(0, L)$. Consider the following one-dimensional backward heat equation:

$$\begin{aligned} &u_t = k^2 u_{xx}, && 0 < x < L, 0 < t < T, \\ &u(0,t) = u(L,t) = 0, && 0 < t < T, \\ &u(x,T) = g(x), && 0 < x < L. \end{aligned}$$

Find a series solution $u(x,t)$.

8. Let $f(x) \in L^2(0, L)$. Consider the following one-dimensional heat equation:

$$\begin{aligned} &u_t = k^2 u_{xx}, && 0 < x < L, t > 0, \\ &u(0,t) = 0, && t > 0, \\ &u(x,0) = f(x), && 0 < x < L. \end{aligned}$$

Let $h(t) \in C^1[0, \infty)$. Suppose at $x = x_0 \in (0, L)$:

$$u(x_0, t) = h(t), \qquad t > 0.$$

Is it possible to determine the value of a solution $u(x,t)$ at $x = L$?

9. Let $R = [0, L] \times [0, H]$. Consider the following problem:

$$\begin{aligned} &u_t = \Delta u - t^2, && (x,y) \in R, t > 0, \\ &u_\nu(x,y,t) = 0, && (x,y) \in \partial R, t > 0, \\ &u(x,y,0) = x^2 + y^2, && (x,y) \in R, \end{aligned}$$

where u_ν represents the normal derivative. Find the series solution. Does the solution decay to 0 as $t \to \infty$?

10. Let $a(x,t) \in C(\bar{Q})$. Prove the weak maximum principle holds for the heat equation

$$u_t = \Delta u + a(x,t)u, \qquad (x,t) \in Q,$$

where $Q = \Omega \times (0, \infty)$ and $a(x,t)$ is a bounded function in $\bar{Q}$.

11. Let $u(x,t)$ be the solution for the following problem:

$$\begin{aligned} &u_t = k^2 u_{xx}, && 0 < x < L, t > 0, \\ &u(0,t) = -1, u(L,t) = 1, && t > 0, \\ &u(x) = \sin(\frac{2\pi x}{L}), && 0 < x < L. \end{aligned}$$

Let

$$\Gamma(t) := \{(x,t) \in Q : u(x,t) = 0\}.$$

Prove that $\Gamma(t)$ is the graph of a smooth curve $x = s(t)$.

12. Let $u(x,t)$ be the solution for the following problem:

$$
\begin{aligned}
&u_t = k^2 u_{xx}, && 0 < x < L, t > 0,\\
&u(0,t) = 0, u_x(L,t) + \alpha u(L,t) = 0, && t > 0,\\
&u(x,0) = f(x), && 0 < x < L,
\end{aligned}
$$

where $f(x) \in L^2(0,L)$ and α is a constant.
Find a condition on α such that the solution decays to 0 as $t \to \infty$.

13. Let $q(x,t) \geq 0$ on $\Omega \in R^n$. Consider the following problem:

$$
\begin{aligned}
&u_t = k^2 \Delta u - q(x,t)u, && x \in \Omega, t > 0,\\
&u(x,t) = 0, && x \in \partial\Omega, t > 0,\\
&u(x,0) = f(x) \geq 0, && x \in \Omega.
\end{aligned}
$$

Prove that $u(x,t)$ decays to 0 exponentially at least with the decay rate $e^{-\lambda_1 k^2 t}$ as $t \to \infty$. Explain the reason for the result from the physical model.

14. Prove the uniqueness for the backward heat problem:

$$
\begin{aligned}
&u_t = k^2 \Delta u + au, && x \in \Omega, 0 < t < T,\\
&u(x,t) = 0, && x \in \partial\Omega, 0 < t < T,\\
&u(x,T) = f(x), && x \in \Omega,
\end{aligned}
$$

where a is a constant and $f(x) \in L^2(\Omega)$.

15. Let $g(x,t), h(x,t)$ and $f(x)$ be given known continuous functions and nonnegative. Let $\partial\Omega$ be divided into two parts:

$$\partial\Omega = \bar{\Gamma}_1 \bigcup \bar{\Gamma}_2; \Gamma_1 \bigcap \Gamma_2 = \phi \text{ (empty)}.$$

Suppose $u(x,t)$ is the solution to the following problem:

$$
\begin{aligned}
&u_t = k^2 \Delta u + g(x,t), && x \in \Omega, t > 0,\\
&u(x,t) = h(x,t), && x \in \Gamma_1, t > 0,\\
&\nabla_\nu u(x,t) = 0, && x \in \Gamma_2, t > 0,\\
&u(x,0) = f(x), && x \in \Omega.
\end{aligned}
$$

Prove

$$u(x,t) \geq 0, \qquad (x,t) \in \Omega.$$

16. Suppose $f(s) \in C^1(R^+)$ and $f(0) \geq 0$. Prove the solution of the following nonlinear equation is nonnegative:

$$u_t = k^2 \Delta u + f(u), \qquad x \in \Omega, t > 0,$$

$$u(x,t)=0, \qquad x \in \partial\Omega, t>0,$$
$$u(x,0)=0, \qquad x \in \Omega.$$

17. Let $f(x) \in L^2(\Omega)$ and $u(x,t)$ be a solution of the following heat equation:

$$u_t = \Delta u + au, \qquad x \in \Omega, t>0,$$
$$u(x,t)=0, \qquad x \in \partial\Omega, t>0,$$
$$u(x,0)=f(x), \qquad x \in \Omega,$$

where a is a constant. Find the range of a such that $u(x,t)$ decays to 0 as $t \to \infty$.

CHAPTER

The wave equation

5

5.1 The mathematical model of a vibrating string

In this section we derive the mathematical model for the motion of a vibrating string. The model derivation is based on the classical Newton's second law.

5.1.1 The mathematical model of a string vibration

Consider an elastic string with length L attached on two fixed pegs. Suppose a small force is acting on the string vertically, which causes the string to vibrate in the vertical direction, see Fig. 5.1 below. The task is to find the position of the string at any time if one knows the initial position and initial velocity of the string.

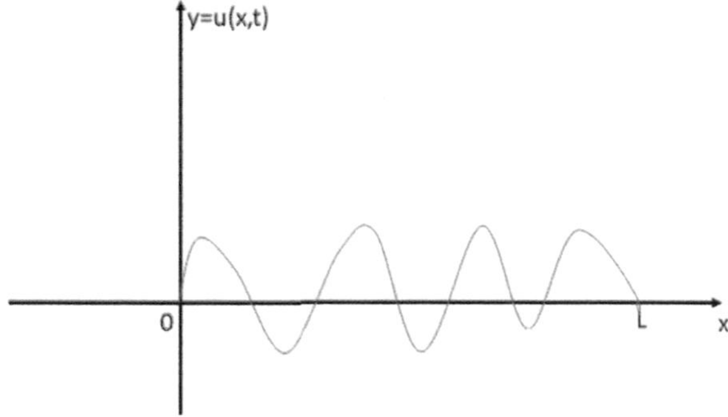

FIGURE 5.1

To derive the mathematical model, we suppose that the string lies between $[0, L]$ on the x-axis. Let ρ be the mass density of the string and f be the external force per unit length acting on the string in the vertical direction.

Let $u(x,t)$ be the displacement of the string from the x-axis at position x and time $t \geq 0$.

The physical foundation of the model is based on the classical Newton's second law:

$$\mathbf{F} = m\mathbf{a}.$$

Partial Differential Equations and Applications. https://doi.org/10.1016/B978-0-44-318705-6.00011-2

Consider a small segment of the string, say $[x, x+\Delta x]$ with $\Delta x > 0$. Then, the total mass for the string segment $[x, x+\Delta x] \approx \rho(x)\Delta x$.

First, we know the velocity and the acceleration of the string at $x \in [0, L]$ vertically are, respectively, equal to

$$\mathbf{v} = \frac{\partial u(x,t)}{\partial t}, \quad \mathbf{a} = \frac{\partial^2 u(x,t)}{\partial t^2}.$$

Now, we analyze the total force acting on the string. There are three types of force acting on the string segment $[x, x+\Delta x]$:

(1) External force density, denoted by $f_0(x,t)$, in the vertical direction.
(2) Tensile force, denoted by $\mathbf{T}$, at each end of the string segment $[x, x+\Delta x]$.
(3) Resistance force (damping force), which is assumed to be proportional to the velocity.

Assume that $\alpha(x,t)$ is the angle between the x-axis and the tangent line of the string at x, see Fig. 5.2. Since the tensile force acting on the string is in the tangential direction of the string, in the vertical direction the tensile forces at x and $x+\Delta x$, respectively, are equal to

$$\begin{aligned} \mathbf{T}(x,t)\cdot\mathbf{j} &= |\mathbf{T}(x,t)| cos(\mathbf{T}(x,t),\mathbf{j}) = |\mathbf{T}(x,t)| cos(\frac{\pi}{2}+\alpha(x,t)) \\ &= -|\mathbf{T}(x,t)| sin(\alpha(x,t)); \\ \mathbf{T}(x+\Delta x,t)\cdot\mathbf{j} &= |\mathbf{T}(x+\Delta x,t)| cos(\mathbf{T}(x+\Delta x,t),\mathbf{j}) \\ &= |\mathbf{T}(x+\Delta x,t)| cos(\frac{\pi}{2}-\alpha(x+\Delta x,t)) \\ &= |\mathbf{T}(x+\Delta x,t)| sin(\alpha(x+\Delta x,t)), \end{aligned}$$

where $\mathbf{j}$ represents the unit vector in the u-direction.

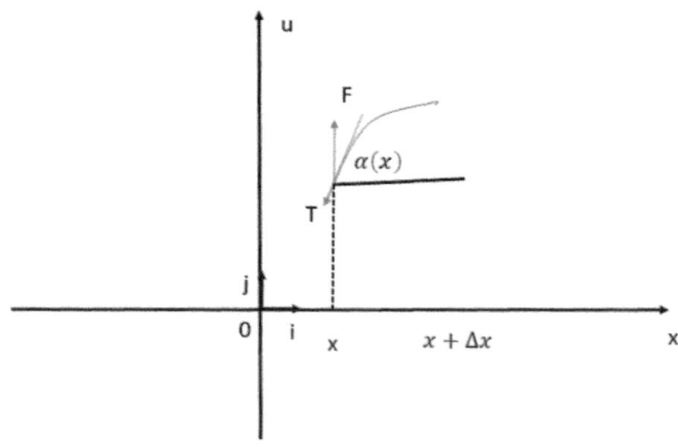

FIGURE 5.2

Since the vibration of the string is small, we may approximate $\sin(\alpha(x,t))$ and $\sin(\alpha(x+\Delta x,t))$ by the tangent:

$$\sin(\alpha(x+\Delta x,t)) \approx tan(\alpha(x+\Delta x,t)),$$
$$\sin(\alpha(x,t)) \approx tan(\alpha(x,t)).$$

On the other hand,

$$\tan(\alpha(x+\Delta x,t)) = u_x(x+\Delta x,t),\ \tan(\alpha(x,t)) = u_x(x,t).$$

Let $T(x,t) = |\mathbf{T}(x,t)|$ be the magnitude of the tensile force. Then, the total tensile force in the u-direction is approximately equal to

$$T(x+\Delta x,t)u_x(x+\Delta x,t) - T(x,t)u_x(x,t).$$

The total external and resistance forces acting on the string segment $[x, x+\Delta x]$ is approximately equal to

$$-\gamma_0 \int_x^{x+\Delta x} u_t dx + \int_x^{x+\Delta x} f_0(x,t)dx,$$

where γ_0 represents the coefficient of the resistance force.

We now apply Newton's law on the string segment $[x, x+\Delta x]$ to obtain

$$\begin{aligned}\rho_0(x)u_{tt}(x,t)\Delta x \approx\ & T(x+\Delta x,t)u_x(x+\Delta x,t)) - T(x,t)u_x(x,t)\\ & - \gamma_0 u_t(x,t)\Delta x + (\Delta x) f_0(x,t).\end{aligned}$$

After dividing by Δx in the above equation and taking the limit as $\Delta x \to 0$, we see that

$$\rho_0(x)u_{tt}(x,t) + \gamma_0 u_t = \frac{\partial}{\partial x}[T(x,t)u_x(x,t)] + f_0(x,t).$$

In particular, when the mass density and the magnitude of the tensile force are assumed to be constants, denoted by ρ_0 and T_0, respectively, we obtain the following wave equation with one space dimension:

$$u_{tt} + \gamma u_t = c^2 u_{xx} + f(x,t), \qquad 0 < x < L, t > 0, \tag{5.1.1}$$

where

$$c = \sqrt{\frac{T_0}{\rho_0}},\ \gamma = \frac{\gamma_0}{\rho_0},\ f(x,t) = \frac{f_0(x,t)}{\rho_0}.$$

If there is no resistance or external force, then $u(x,t)$ satisfies the following homogeneous wave equation in one-space dimension:

$$u_{tt} = c^2 u_{xx}, \qquad 0 < x < L, t > 0. \tag{5.1.2}$$

5.1.2 Initial and boundary conditions

To determine the position of the string in future time, one must specify an initial position and initial velocity. Moreover, one needs to know how the string is tied at each end. These are necessary initial and boundary conditions for a complete model of the vibrating string.

At the initial moment $t = 0$, the initial position and initial velocity must be specified in order to determine the motion of the string in the future. We assume:

$$u(x,0) = g_1(x), \qquad 0 \le x \le L, \tag{5.1.3}$$

$$u_t(x,0) = g_2(x), \qquad 0 \le x \le L, \tag{5.1.4}$$

where $g_1(x)$ and $g_2(x)$ are given functions.

It is also necessary to specify how the two ends of the string are set up. Namely, we need to specify the conditions at the two ends of the string. It turns out that we can classify the boundary conditions similarly to the cases for the heat equation with different physical meaning.

(a) **The first type** (Dirichlet type)

This type of boundary condition is to specify the displacement of the string at $x = 0$ and $x = L$ at any time $t \ge 0$:

$$u(0,t) = h_1(t), \qquad t \ge 0, \tag{5.1.5}$$

$$u(L,t) = h_2(t), \qquad t \ge 0, \tag{5.1.6}$$

where $h_1(t)$ and $h_2(t)$ are given functions.

In particular, for fixed pegs at the two ends of the string, $h_1(t) = h_2(t) = 0$ for all $t \ge 0$.

(b) **The second type** (Neumann type)

This type of boundary condition is to specify the tensile force at $x = 0$ and $x = L$ at any time $t \ge 0$:

$$-T_0 u_x(0,t) = h_1(t), \qquad t \ge 0, \tag{5.1.7}$$

$$T_0 u_x(L,t) = h_2(t), \qquad t \ge 0, \tag{5.1.8}$$

where $h_1(t)$ and $h_2(t)$ are given force functions.

For example, if the string is set to be free at one end $x = L$, then

$$T_0 u_x(L,t) = h_2(t) = 0, \qquad t \ge 0.$$

(c) **The third type** (Robin type)

This type of boundary condition is to specify relations between the tensile force and the displacement of the string at $x = 0$ and $x = L$ at any time $t \ge 0$.

$$-T_0 u_x(0,t) + au(0,t) = h_1(t), \qquad t \ge 0, \tag{5.1.9}$$

$$T_0 u_x(L,t) + bu(L,t) = h_2(t), \qquad t \ge 0, \tag{5.1.10}$$

where a, b are constants and $h_1(t)$ and $h_2(t)$ are given functions.

A typical example of the third type of boundary condition is when the string is tied with a spring, then Hooke's law shows the force is proportional to the displacement of the string.

(d) **Other types of boundary conditions**

There are several other types of boundary conditions from different physical models. One of these in the research literature is called a dynamical boundary condition. We give one example here.

Consider a string is attached to a dynamical system such as a spring–mass system. The position of the string at time t, $u(0,t) = y(t)$ is unknown but can be determined by Newton's law and Hooke's law:

$$m\frac{d^2y(t)}{dt^2} = -k(y(t) - y_0) + \text{other forces},$$

where k is the spring constant and y_0 is known to be the still position of the spring. The *other forces* could be a resistance force, say γu_t, or an external force, say $g(t)$.

Then, the boundary condition at $x = 0$ becomes

$$mu_{tt}(0,t) = -k(u(0,t) - y_0) + \gamma u_t(0,t) + g(t), \qquad (5.1.11)$$

where y_0 is the initial position of the string, and $g(t)$ represents the known external force acting on the object in the vertical direction.

We summarize the above discussion to obtain a completed mathematical model consisting of Eq. (5.1.1) or Eq. (5.1.2) subject to initial conditions (5.1.3)–(5.1.4) and one type of boundary conditions such as (5.1.5)–(5.1.6). It will be seen that the constant c in Eq. (5.1.1) represents the speed of the string movement.

5.2 Solutions of the wave equation in one-space dimension

In this section we find a series solution for the wave equation in one space dimension. We use the first type of boundary condition as an example.

5.2.1 Series solution of the wave equation

Consider the wave equation subject to the first type of homogeneous boundary conditions:

$$u_{tt} - c^2u_{xx} = f(x,t), \qquad 0 < x < L, t > 0, \qquad (5.2.1)$$

$$u(0,t) = u(L,t) = 0, \qquad t \geq 0, \qquad (5.2.2)$$

$$u(x,0) = g_1(x), u_t(x,0) = g_2(x), \qquad 0 \leq x \leq L. \qquad (5.2.3)$$

The basic strategy is similar to the case for the heat equation. We will solve the problem by using the method of separation of variables. As a first step, we first decompose the problem (5.2.1)–(5.2.3) into two problems:

P(a) Homogeneous equation with nonhomogeneous initial conditions.

$$u_{1tt} - c^2 u_{1xx} = 0, \qquad 0 < x < L, t > 0, \tag{5.2.4}$$

$$u_1(0,t) = u_1(L,t) = 0, \qquad t \geq 0, \tag{5.2.5}$$

$$u_1(x,0) = g_1(x), \qquad u_{1t}(x,0) = g_2(x), \qquad 0 \leq x \leq L. \tag{5.2.6}$$

P(b) Nonhomogeneous equation with homogeneous initial conditions.

$$u_{2tt} - c^2 u_{2xx} = f(x,t), \qquad 0 < x < L, t > 0, \tag{5.2.7}$$

$$u_2(0,t) = u_2(L,t) = 0, \qquad t \geq 0, \tag{5.2.8}$$

$$u_2(x,0) = u_{2t}(x,0) = 0, \qquad 0 \leq x \leq L. \tag{5.2.9}$$

It is clear that the solution of (5.2.1)–(5.2.3) is $u(x,t) = u_1(x,t) + u_2(x,t)$.

Theorem 5.2.1. *Let $g_1(x), g_2(x) \in C^2(0,L) \bigcap L^2(0,L)$. Then, the solution to the problem (5.2.4)–(5.2.6) is given by the following series:*

$$u_1(x,t) = \sum_{k=1}^{\infty} [a_k cos(\frac{k\pi ct}{L}) + b_k sin(\frac{k\pi ct}{L})] sin(\frac{k\pi x}{L}), \tag{5.2.10}$$

where a_k and b_k are derived from the Fourier coefficients of $g_1(x)$ and $g_2(x)$.

$$a_k = \frac{2}{L}\int_0^L g_1(x) sin(\frac{k\pi x}{L}) dx,$$

$$b_k = \frac{2}{k\pi c}\int_0^L g_2(x) sin(\frac{k\pi x}{L}) dx, \qquad \forall k \geq 1.$$

Proof. To find the solution $u_1(x,t)$, we use the method of separation of variables.
Let $u_1(x,t) = \phi(x)h(t)$. From Eq. (5.2.4), we have

$$\frac{h''(t)}{c^2 h(t)} = \frac{\phi''(x)}{\phi(x)},$$

which must be equal to a constant, say $-\lambda$.

Moreover, from the boundary conditions (5.2.5) we see that

$$\phi(0) = \phi(L) = 0.$$

For the eigenvalue problem:

$$-\phi'' = \lambda\phi, \qquad 0 < x < L,$$

$$\phi(0) = \phi(L) = 0.$$

All eigenvalues and the corresponding eigenfunctions are given by

$$\lambda_n = (\frac{n\pi}{L})^2, \qquad \phi_n(x) = sin(\frac{n\pi x}{L}), n = 1, 2, \cdots .$$

For $\lambda = \lambda_n$, the general solution for

$$h''(t) + c^2\lambda_n h(t) = 0, \qquad t > 0,$$

is equal to

$$h_n(t) = a_n \cos(\frac{nc\pi t}{L}) + b_n \sin(\frac{nc\pi t}{L}), \qquad t > 0,$$

where a_n and b_n are arbitrary constants.

Now, we use the superposition principle to set

$$u_1(x,t) = \sum_{k=1}^{\infty}[a_k cos(\frac{kc\pi t}{L}) + b_k sin(\frac{kc\pi t}{L})]sin(\frac{k\pi x}{L}).$$

Hence,

$$u_1(x,0) = \sum_{k=1}^{\infty} a_k sin(\frac{k\pi x}{L}),$$

$$u_{1t}(x,0) = \sum_{k=1}^{\infty} b_k \frac{kc\pi}{L} sin(\frac{k\pi x}{L}).$$

We expand the functions $g_1(x)$ and $g_2(x)$ as a Fourier sine series in $[0, L]$:

$$g_1(x) = \sum_{k=1}^{\infty} A_k sin(\frac{k\pi x}{L}),$$

$$g_2(x) = \sum_{k=1}^{\infty} B_k sin(\frac{k\pi x}{L}),$$

where

$$A_k = \frac{2}{L}\int_0^L g_1(x)\sin(\frac{k\pi x}{L})dx,$$

$$B_k = \frac{2}{L}\int_0^L g_2(x)\sin(\frac{k\pi x}{L})dx, k = 1, 2, \cdots .$$

In order to satisfy the initial (5.2.6) for the series solution $u_1(x,t)$, one must choose

$$a_k = A_k, \qquad b_k \frac{kc\pi}{L} = B_k, k = 1, 2, \cdots .$$

Consequently, we obtain

$$a_k = \frac{2}{L}\int_0^L g_1(x)\sin(\frac{k\pi x}{L})dx,$$
$$b_k = \frac{2}{kc\pi}\int_0^L g_2(x)\sin(\frac{k\pi x}{L})dx, \qquad k = 1, 2, \cdots. \qquad \square$$

Corollary 5.2.1. *(Generalized D'Alambert formula) Suppose that $g_1(x)$ and $g_2(x)$ are in $C^2(0, L)\bigcap L^2(0, L)$. Moreover, g_1 and g_2 are odd extended into $(-L, 0)$. Then, the solution of the problem P(a) can be explicitly expressed by the following formula:*

$$u(x,t) = \frac{1}{2}[\bar{g}_1(x+ct) + \bar{g}_1(x-ct)] + \frac{1}{2c}\int_{x-ct}^{x+ct} \bar{g}_2(y)dy, \tag{5.2.11}$$

where $\bar{g}_1(x)$ and $\bar{g}_2(x)$ are the odd extensions of $g_1(x)$ and $g_2(x)$ into $(-L, 0)$ with $2L$-period in R^1.

Proof. Note that

$$\sin\alpha\sin\beta = \frac{1}{2}[\cos(\alpha-\beta) - \cos(\alpha+\beta)].$$

Thus

$$\begin{aligned}
\sum_{k=1}^{\infty} b_k \sin\frac{k\pi x}{L}\sin\frac{k\pi ct}{L} &= \frac{1}{2}\sum_{k=1}^{\infty} b_k[\cos(\frac{k\pi}{L}(x-ct)) - \cos(\frac{k\pi}{L}(x+ct))] \\
&= \frac{1}{2}\sum_{k=1}^{\infty}\frac{k\pi}{L}b_k\int_{x-ct}^{x+ct} sin\frac{k\pi z}{L}dz. \\
&= \frac{1}{2c}\sum_{k=1}^{\infty} B_k\int_{x-ct}^{x+ct} sin\frac{k\pi z}{L}dz \\
&= \frac{1}{2c}\int_{x-ct}^{x+ct} g_2(z)dz.
\end{aligned}$$

Similarly, using the trigonometric identity

$$\sin\alpha\cos\beta = \frac{1}{2}[\sin(\alpha+\beta) + \sin(\alpha-\beta)],$$

one can easily derive that

$$\sum_{k=1}^{\infty} a_k \sin(\frac{k\pi x}{L})\cos(\frac{k\pi ct}{L}) = \frac{1}{2}[\hat{g}_1(x+ct) + \hat{g}_1(x-ct)]. \qquad \square$$

From the representation of the solution in Corollary 5.2.1, we see the structure of the solution in the strip $[0, L] \times [0, \infty)$, see Fig. 5.3 below.

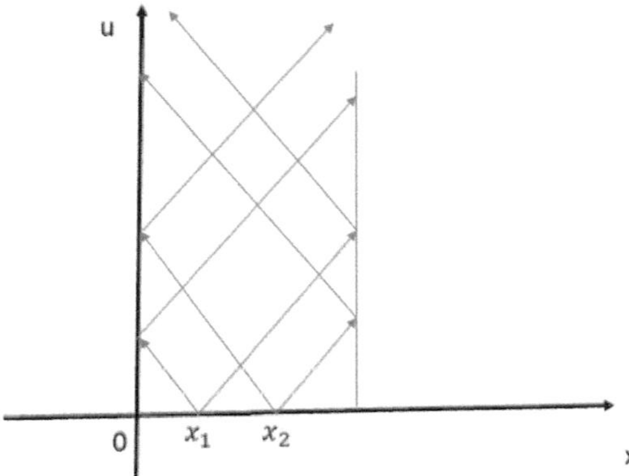

FIGURE 5.3

For Problem P(b), there are different ways to find the solution representation. One way is to find the series solution similar to the heat equation. Let

$$u(x,t) = \sum_{n=1}^{\infty} a_n(t) \sin(\frac{n\pi x}{L}),$$

$$f(x,t) = \sum_{n=1}^{\infty} \alpha_n(t) \sin(\frac{n\pi x}{L}),$$

where $\alpha_n(t)$ is the Fourier coefficient of the sine series for $f(x,t)$ over $[0, L]$ as long as $f(x,t) \in L^2(Q_T)$:

$$\alpha_n(t) = \frac{2}{L} \int_0^L f(x,t) \sin(\frac{n\pi x}{L}) dx,$$

while $a_n(t)$ is chosen to satisfy the wave equation (5.2.7) and the initial conditions:

$$\begin{aligned} &a_n''(t) + \lambda_n^2 c^2 a_n(t) = \alpha_n(t), && t > 0, \\ &a_n(0) = a_n'(0) = 0, && n = 1, 2, \cdots, n. \end{aligned}$$

Hence a series solution for Problem (b) can be obtained. Here we use a different approach. The new method will enable us to derive a compact form of the solution.

Lemma 5.2.1. *Assume $f(x,t) \in C^2(Q_T) \bigcap L^2(Q_T)$ for any $T > 0$. Let $w(x, t - \tau)$ be a solution of the following problem for any $\tau \in [0, t]$:*

$$w_{tt} - c^2 w_{xx} = 0, \qquad 0 < x < L, t > \tau,$$

$$
\begin{aligned}
& w(0, t-\tau)=w(L, t-\tau)=0, && t \geq \tau, \\
& w(x, 0)=0, \qquad t \geq \tau, \\
& w_t(x, 0)=f(x, \tau), \qquad 0 \leq x \leq L,
\end{aligned}
$$

where $0 \leq \tau \leq t$. Then,

$$
u(x, t)=\int_0^t w(x, t-\tau) d \tau \tag{5.2.12}
$$

is the solution of the problem (5.2.7)–(5.2.9).

Proof.

$$
\begin{aligned}
u_t(x, t) & =\int_0^t w_t(x, t-\tau) d \tau+\left.w(x, t-\tau)\right|_{\tau=t} \\
& =\int_0^t w_t(x, t ; \tau) d \tau,
\end{aligned}
$$

where the first initial condition $w(x, t-\tau)=0$ at $\tau=t$ is used.

$$
\begin{aligned}
u_{tt}(x, t) & =w_t(x, 0)+\int_0^t w_{tt}(x, t-\tau) d \tau \\
& =f(x, t)+\int_0^t w_{tt}(x, t-\tau) d \tau,
\end{aligned}
$$

where the second initial condition $w_t(x, t-\tau)=f(x, \tau)$ at $\tau=t$ is used.

It follows that

$$
\begin{aligned}
& u_{tt}-c^2 u_{xx}=f(x, t)+\int_0^t\left[w_{tt}-c^2 w_{xx}\right] d \tau=f(x, t), 0<x<L, t>0, \\
& u(0, t)=u(L, t)=0, \qquad t \geq 0, \\
& u(x, 0)=u_t(x, 0)=0, \qquad 0 \leq x<L,
\end{aligned}
$$

i.e., $u(x, t)$ defined as Eq. (5.2.12) is a solution of the problem (5.2.7)–(5.2.9). □

Corollary 5.2.2. *(Solution representation) Let $g_1(x), g_2(x) \in C^2(0, L) \bigcap L^2(0, L)$ and $f(x, t) \in C^2(Q) \bigcap L^2(Q_T)$ for any $T>0$. Then, the solution of (5.2.1)–(5.2.3) can be expressed by:*

$$
\begin{aligned}
u(x, t)= & \frac{1}{2}\left[\bar{g}_1(x+c t)+\bar{g}_1(x-c t)\right]+\frac{1}{2 c} \int_{x-c t}^{x+c t} \bar{g}_2(y) d y \\
& +\int_0^t \int_{x-c(t-\tau)}^{x+c(t-\tau)} f(z, \tau) d z d \tau,
\end{aligned} \tag{5.2.13}
$$

where $\bar{g}_1(x)$ and $\bar{g}_2(x)$ are defined the same as in Corollary 5.2.1. □

Remark 5.2.1. From the mathematical point of view, we can find the formal series solution for $g_1(x), g_2(x) \in L^2(0, L)$ and $f(x,t) \in L^2(Q_T)$. However, from the solution representation (5.2.13) $u(x,t)$ has the same smoothness as $g_1(x)$, $g_2(x)$ and $f(x,t)$ for $(x,t) \in (0, L) \times (0, \infty)$. This is very different from the solution of the heat equation.

Remark 5.2.2. For nonhomogeneous boundary conditions such as

$$u(0,t) = h_1(t),\ u(L,t) = h_2(t), \qquad t \geq 0,$$

set

$$w(x,t) = u(x,t) - [h_1(t) + \frac{x}{L}(h_2(t) - h_1(t))].$$

Then, $w(x,t)$ will satisfy the following nonhomogeneous wave equation:

$$w_{tt} - c^2 w_{xx} = f(x,t) - [h_1''(t) + \frac{x}{L}(h_2''(t) - h_1''(t))]$$

subject to homogeneous boundary conditions and new initial conditions

$$w(x,0) = g_1(x) - [h_1(0) + \frac{x}{L}(h_2(0) - h_1(0))],$$
$$w_t(x,0) = g_2(x) - [h_1'(0) + \frac{x}{L}(h_2'(0) - h_1'(0))].$$

Thus the solution can be found as before.

Remark 5.2.3. For the second or third type of boundary conditions

$$u_x(0,t) - \alpha u(0,t) = h_1(t), t \geq 0;$$
$$u_x(L,t) + \beta u(L,t) = h_2(t), t \geq 0.$$

One can find a series solution by using the same argument as those problems in Chapter 4 for the heat equation.

5.2.2 Natural frequency and resonance phenomenon

From the series solution of Problem (a), we define

$$N_k = \sqrt{a_k^2 + b_k^2}$$
$$\alpha_k = arctan(\frac{b_k}{a_k}).$$

Then,

$$a_k sin(\frac{ck\pi t}{L}) + b_k cos(\frac{ck\pi t}{L})$$

$$= N_k sin(\frac{ck\pi t}{L} + \alpha_k),$$

where α_k is called the *initial phase angle* and N_k is called *the magnitude*. It follows that

$$u_1(x,t) = \sum_{n=1}^{\infty} N_n sin(\frac{cn\pi t}{L} + \alpha_n) sin(\frac{n\pi x}{L}). \tag{5.2.14}$$

The above series solution can be considered as a superposition of each harmonic wave

$$W_n(x,t) = N_n sin(\frac{cn\pi t}{L} + \alpha_n) \sin(\frac{n\pi x}{L}). \tag{5.2.15}$$

It is clear that N_n is the maximum magnitude of oscillation for each wave and the period is equal to

$$p_n = \frac{2L}{cn},$$

$$\omega_n = \frac{1}{p_n} = \frac{cn}{2L} \text{ is called the } \textit{natural frequency}.$$

It is clear that $\omega_1 = \frac{c}{2L}$ is the smallest frequency, often called the *fundamental frequency*. Also, the other higher frequencies are called *overtones*, which harmonize the sound.

Recall that

$$c = \sqrt{\frac{T_0}{\rho_0}}.$$

It follows that ω_n is increasing if the tensile force T_0 is increasing or the mass density ρ_0 is decreasing.

This phenomenon is well experienced in playing musical instrument such as the violin or cello.

Suppose that at an initial moment, a string is in a still position, which implies $g_1(x) = g_2(x) = 0$ on $[0, L]$. Suppose there is an external force

$$f_0(x,t) = \sum_{n=1}^{\infty} b_n(t) \sin(\frac{n\pi x}{L})$$

acting on the string.

Let the series solution $u_2(x,t)$ be of the following form:

$$u_2(x,t) = \sum_{n=1}^{\infty} a_n(t) \sin(\frac{n\pi x}{L}). \tag{5.2.16}$$

Then, $a_n(t)$ must satisfy

$$a_n''(t) + \omega_0^2(n) a_n(t) = b_n(t), \qquad n = 1, 2, \cdots, t > 0, \tag{5.2.17}$$
$$a_n(0) = a_n'(0) = 0, \tag{5.2.18}$$

where

$$\omega_0(n) = \frac{n\pi c}{L}, \qquad n = 1, 2, \cdots.$$

Note that the general solution of the homogeneous equation for $a_n(t)$ is equal to

$$a_n(t) = c_1 \cos(\omega_0(n)t) + c_2 \sin(\omega_0(n)t),$$

where c_1 and c_2 are constants.

If there exists a number $n = n_0$ such that

$$b_{n_0}(t) = A \sin(\omega_0(n_0)t), \text{ or, } B \cos(\omega_0(n_0)t),$$

where $A > 0$, $B > 0$, then, from the ODE theory, the solution $a_n(t)$ to the problem (5.2.17)–(5.2.18) will become unbounded:

$$|a_{n_0}(t)| \to \infty, \text{ as } t \to \infty,$$

which implies

$$|u_2(x,t)| \to \infty, \qquad \text{as } t \to \infty.$$

This is called *the resonance phenomenon*. The physical interpretation is that the system will break down if there is a periodic force with a period that is close to that of $w_0(n)$ for any $n \geq 1$, even if the magnitude of the force is very small.

5.3 Wave propagation in several space dimensions

Like a vibrating string, there are many other physical phenomena that can be modeled by the wave equation. Here we briefly describe the mathematical models for a thin membrane vibration in two-space dimensions and a sound propagation in three-space dimensions.

5.3.1 The mathematical model of a vibrating membrane

Suppose a thin membrane occupies a domain $\Omega \subset R^2$ and the edge of the membrane is attached to the boundary of Ω. If there is a small external force density, denoted by $f_0 = f_0(x, y, t)$, in the vertical direction acting on the membrane, then it will vibrate. The goal is to find the position of the membrane at any time $t > 0$ if an initial position and an initial velocity are known.

Let $z = u(x, y, t)$ be the displacement of the membrane in the z-direction.

Consider a small piece S of the thin membrane and D represents the projection of S in the xy-plane. The tensile force on the edge of the small piece of the membrane in the tangential direction is

$$\mathbf{F}_T = T(\nu \times \mathbf{n}),$$

where T is the magnitude of the tensile force, ν is the tangential direction of the boundary curve of the small surface S and $\mathbf{n}$ is the unit normal of S on the boundary curve ∂S, see Fig. 5.4.

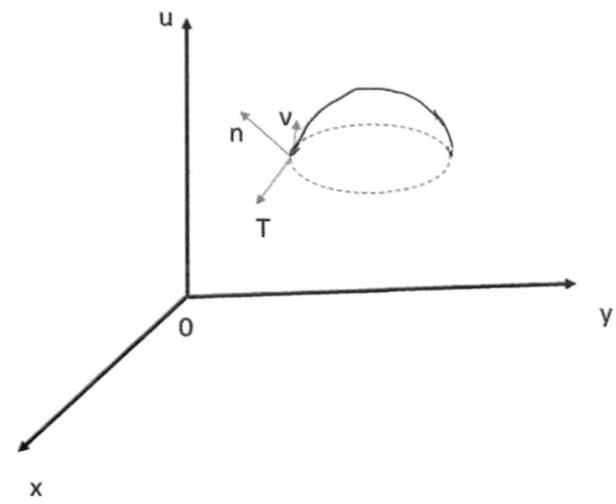

FIGURE 5.4

The component of the tensile force in the z-direction, which is the same as the motion of the membrane, equals

$$\mathbf{F}_T \cdot \mathbf{k} = T(\nu \times \mathbf{n}) \cdot \mathbf{k},$$

where $\mathbf{k}$ is the unit vector of the z-axis.

Therefore the total tensile force acting on the surface S in the z-direction is

$$\int_S T(\nu \times \mathbf{n}) \cdot \mathbf{k} ds.$$

On the other hand, the total mass of S equals

$$\int_D \rho(x, y) dA,$$

where $\rho(x, y)$ is the density of the membrane and D is the projection of S on the xy-plane.

Newton's second law implies that

$$\left(\int_D \rho dA\right) u_{tt} \approx \int_S T(\nu \times \mathbf{n}) \cdot \mathbf{k} ds + \int\int_D f_0(x, y, t) dA$$

$$= \int_S T(\mathbf{n} \times \mathbf{k}) \cdot \nu ds + \int\int_D f_0(x, y, t) dA,$$

where the vector identity

$$(\mathbf{A} \times \mathbf{B}) \cdot \mathbf{C} = (\mathbf{B} \times \mathbf{C}) \cdot \mathbf{A}$$

is used for any vectors **A**, **B**, and **C**.

Recall Gauss's theorem

$$\int\int_D \nabla \cdot \mathbf{F} dA = \int_S (\mathbf{F} \cdot \nu) ds.$$

It follows that

$$\int_S T(\mathbf{n} \times \mathbf{k}) \cdot \nu ds = \int\int_D [\nabla \cdot T(\mathbf{n} \times \mathbf{k})] dA.$$

Since S is arbitrary, it follows that

$$\rho u_{tt} = \nabla \cdot [T(\mathbf{n} \times \mathbf{k})].$$

Now, it is clear that the unit normal for $z = u(x, y, t)$ is equal to

$$\mathbf{n} = \{-\frac{u_x}{\sqrt{u_x^2 + u_y^2 + 1}}, -\frac{u_y}{\sqrt{u_x^2 + u_y^2 + 1}}, \frac{1}{\sqrt{u_x^2 + u_y^2 + 1}}\}.$$

Since the displacement of the membrane is very small, we approximate

$$\sqrt{u_x^2 + u_y^2 + 1} \approx 1.$$

It follows that

$$\mathbf{n} \times \mathbf{k} = \{-u_x, -u_y, 0\}$$

and

$$[T(\mathbf{n} \times \mathbf{k})] = \{0, (Tu_x), (Tu_y)\}.$$

Consequently,

$$\nabla \cdot [T(\mathbf{n} \times \mathbf{k})] = (Tu_x)_x + (Tu_y)_y.$$

Assume the density of the membrane and tensile force to be constants: $\rho = \rho_0$ and $T = T_0$ over Ω, respectively. Moreover, we neglect a resistance force, then the displacement function $u(x, y, t)$ satisfies

$$\rho_0 u_{tt} = T_0(u_{xx} + u_{yy}) + f_0(x, y, t),$$

which is the wave equation in two-space dimensions.

Just like a string vibration, if there exists a resistance force (damping force), which is typically proportional to the velocity, denoted by $\gamma_0 u_t$, we see that $u(x, y, t)$ satisfies

$$\rho_0 u_{tt} + \gamma_0 u_t - T_0(u_{xx} + u_{yy}) = f_0(x, y, t), \qquad (x, y) \in \Omega, t > 0.$$

Similarly, in three-space dimensions the motion of an elastic solid under a small external force f_0 can be described by the following wave equation:

$$\rho_0 u_{tt} + \gamma_0 u_t - T_0(u_{xx} + u_{yy} + u_{zz}) = f_0(x, y, z, t).$$

For brevity, we simply write

$$u_{tt} + \gamma u_t - c^2 \Delta u = f, \qquad x \in \Omega, t > 0,$$

where $x = (x_1, x_2, \cdots, x_n) \in \Omega \subset R^n$, $\gamma = \frac{\gamma_0}{\rho_0}$, $c = \sqrt{\frac{T_0}{\rho_0}}$ and $f = \frac{f_0}{\rho_0}$.

With appropriate initial and boundary conditions, we obtain a mathematical model in several space dimensions:

$$u_{tt} = c^2 \Delta u + f(x, t), \qquad (x, t) \in \Omega \times (0, \infty), \tag{5.3.1}$$

$$B[u] = h(x, t), \qquad (x, t) \in \partial\Omega \times (0, \infty), \tag{5.3.2}$$

$$u(x, 0) = g_1(x), u_t(x, 0) = g_2(x), \qquad x \in \Omega, \tag{5.3.3}$$

where B is a boundary operator.

5.3.2 The mathematical model of a sound propagation

A sound propagation is an interesting phenomenon encountered in daily life. There are many applications in health and life sciences. We will derive the mathematical model based on different physical laws.

It is a well-known experiment that sound does not propagate in vacuum. Sound can propagate in air due to the motion of the molecules in the air. We may think of the air as a compressible fluid and sound propagates in a compressible fluid.

Now, we consider a fluid that occupies a region $\Omega \subset R^3$. Let $\rho(x, t)$ be the density of the fluid at location $x = (x_1, x_2, x_3) \in R^3$ and time t. Let $\mathbf{V}(x, t)$ be the velocity field of the fluid. Suppose the trajectory of a particle in the fluid is denoted by $\mathbf{x}(t) =< x_1(t), x_2(t), x_3(t) >$. By definition, the velocity field $\mathbf{V}$ is equal to

$$\mathbf{x}'(t) = \mathbf{V}(\mathbf{x}(t), t), \qquad t \geq 0.$$

The acceleration of the fluid molecule is equal to

$$\mathbf{x}''(t) = \frac{\partial \mathbf{V}}{\partial t} + (\mathbf{V} \cdot \nabla)\mathbf{V}.$$

For any subdomain $D \subset \Omega$, the total momentum of the fluid over the domain D is equal to

$$\int_D \left[\rho(\frac{\partial \mathbf{V}}{\partial t} + (\mathbf{V} \cdot \nabla)\mathbf{V}) \right] dx.$$

Now, the change of total mass of the fluid in D must be equal to the total fluid flows out through the boundary of D in the normal direction:

$$\frac{d}{dt} \int_D \rho(x,t) dx = - \int_{\partial D} [\rho \mathbf{V} \cdot \nu(x)] ds,$$

where $\nu(x)$ is the outward unit normal at $x \in \partial D$.

Gauss's divergence theorem implies that

$$\int_D [\rho_t + div(\rho \mathbf{V})] dx = 0, t > 0.$$

Since D is arbitrary in Ω, we find that

$$\rho_t + div(\rho \mathbf{V}) = 0, \qquad x \subset \Omega, t > 0. \tag{5.3.4}$$

Eq. (5.3.4) is called the continuity equation that holds for all kinds of fluids.

Next, we analyze the force acting on the fluid molecules. There are two types of forces acting on the fluid molecules. The first is the Kelvin force or potential force due to the pressure $p(x,t)$. The other type of force is the external force, denoted by $\mathbf{F}$, such as the gravitational force and resistance force. Therefore the total momentum produced by the Kelvin force and external force in D are equal to

$$-\int_{\partial D} (p\nu) ds + \int_D \mathbf{F} dx,$$

where ν represents the outward unit normal on ∂D.

The divergence theorem implies that

$$\int_{\partial D} (p\nu) ds = \int_D (\nabla p) dx.$$

Since D is arbitrarily in Ω, we obtain

$$\rho[\mathbf{V}_t + (\mathbf{V} \cdot \nabla)\mathbf{V}] = -\nabla p + \mathbf{F}. \tag{5.3.5}$$

Eq. (5.3.5) is called Euler's equation. For a fluid such as air or gas, the pressure p is a nonlinear function of density ρ. For example, from the physics (see [25]) the relation for an ideal gas is given by

$$p = \rho^{1+\gamma} RT,$$

where $R = 8.31$ is the universal constant for an ideal gas, T is the temperature of the ideal gas and $\gamma > 0$.

Therefore we obtain the fundamental equations for the gas dynamics:

$$\rho_t + div(\rho \mathbf{V}) = 0, \qquad x \in \Omega, t > 0, \tag{5.3.6}$$

$$\rho[\mathbf{V}_t + (\mathbf{V} \cdot \nabla)\mathbf{V}] = -\nabla p(\rho) + \mathbf{F}. \tag{5.3.7}$$

Now, we make the following physical assumptions:

(a) The fluid movement is small with small velocity and the density varies little with respect to t.
(b) The density is proportional to the pressure.

With these assumptions, the nonlinear term $(\mathbf{V} \cdot \nabla)\mathbf{V}$ is negligible and $p(\rho) = c\rho$ with some constant $c > 0$. From the system (5.3.6)–(5.3.7), we find that the pressure $p(x, t)$ satisfies the following equation:

$$p_{tt} - c^2 \Delta p = f(x), \tag{5.3.8}$$

where

$$f(x, t) = c\nabla \mathbf{F}.$$

If we prescribe appropriate initial and boundary conditions, then a completed mathematical model for a sound propagation is established.

5.4 Solution of the wave equation in higher space dimension

In this section we use the method of separation of variables to find the series solution for the wave equation in several space dimensions. Since the idea is the same as for the heat equation we give one example with a different domain.

5.4.1 The series solution of the wave equation in a bounded domain in R^n

Let Ω be a bounded domain in R^n with C^1-boundary. Consider the following problem:

$$u_{tt} = c^2 \Delta u + f(x, t), \qquad (x, t) \in \Omega \times (0, \infty), \tag{5.4.1}$$

$$u(x, t) = 0, \qquad (x, t) \in \partial\Omega \times (0, \infty), \tag{5.4.2}$$

$$u(x, 0) = g_1(x), u_t(x, 0) = g_2(x), \qquad x \in \Omega, \tag{5.4.3}$$

where $f(x, t) \in C^2(Q) \bigcap L^2(Q)$ and $g_1(x), g_2(x) \in C^2(\Omega) \bigcap L^2(\Omega)$.

Similar to the one-space dimensional case, we decompose the problem (5.4.1)–(5.4.3) into two problems:

Problem (a) consists of the homogeneous equation, homogeneous boundary condition, but nonhomogeneous initial conditions. Problem (b) consists of the nonhomogeneous equation, but homogeneous initial and boundary conditions. The solution to problem (b) can be obtained from the solution for Problem (a) by using Lemma 5.2.1.

Therefore we focus on the homogeneous equation (5.4.1) where $f(x,t)=0$. Consider the eigenvalue problem:

$$-\Delta\psi=\lambda\psi, \qquad x\in\Omega, \tag{5.4.4}$$

$$\psi(x)=0, \qquad x\in\partial\Omega. \tag{5.4.5}$$

Suppose we can find all eigenvalues and all corresponding eigenfunctions (all eigenfunctions are chosen to be mutually orthogonal in $L^2(\Omega)$):

$$\{\lambda_k\}_{k=1}^{\infty}, \qquad \{\psi(x)\}_{k=1}^{\infty}.$$

Suppose $g_1(x), g_2(x)\in L^2(\Omega)$. Then, using eigenfunction expansion, we have

$$g_1(x)=\sum_{k=1}^{\infty} A_k\psi_k(x), \qquad x\in\Omega;$$

$$g_2(x)=\sum_{k=1}^{\infty} B_k\psi_k(x), \qquad x\in\Omega,$$

where

$$A_k=\frac{<g_1,\psi_k>}{<\psi_k,\psi_k>}, \qquad B_k=\frac{<g_2,\psi_k>}{<\psi_k,\psi_k>}, \forall k\geq 1.$$

Set

$$u(x,t)=\sum_{k=1}^{\infty} a_k(t)\psi_k(x),$$

where $a_k(t)$ is the solution of the following ODE:

$$a_k''(t)+c^2\lambda_k a_k=0, \qquad t>0,$$

$$a_k(0)=A_k, a_k'(0)=B_k, \qquad \forall k\geq 1.$$

We can easily solve the ODE for $a_k(t)$:

$$a_k(t)=A_k\cos(c\sqrt{\lambda_k}t)+\frac{B_k}{c\sqrt{\lambda_k}}\sin(c\sqrt{\lambda_k}t), \qquad t\geq 0.$$

Then, $u(x,t)$ is the explicit series solution of the problem (5.4.1)–(5.4.3) with $f=0$. We summarize the above derivation to obtain the following theorem.

Theorem 5.4.1. *Let $g_1(x), g_2(x) \in C^2(\Omega) \bigcap L^2(\Omega)$. Then, the series solution to the problem (5.4.1)–(5.4.3) with $f = 0$ is*

$$u(x,t) = \sum_{k=1}^{\infty} \left[A_k \cos(c\sqrt{\lambda_k} t) + \frac{B_k}{c\sqrt{\lambda_k}} \sin(c\sqrt{\lambda_k} t) \right] \psi_k(x), \tag{5.4.6}$$

where A_k and B_k are the coefficients of eigenfunction expansions of (5.4.4)–(5.4.5) for $g_1(x)$ and $g_2(x)$, respectively. □

5.4.2 The series solution of the wave equation in a disk in R^2

From Theorem 5.4.1, we see that a crucial step for finding the explicit series solution is to find all eigenvalues and corresponding eigenfunctions. We use one example here to demonstrate the procedure.

Let $a > 0$ and

$$D = \{(x, y) \in R^2 : x^2 + y^2 < a^2\}.$$

Consider the following problem:

$$u_{tt} = c^2 \Delta u + f(x, y, t), \qquad (x, y, t) \in D \times (0, \infty), \tag{5.4.7}$$

$$u(x, y, t) = 0, \qquad (x, y, t) \in \partial D \times (0, \infty), \tag{5.4.8}$$

$$u(x, y, 0) = g_1(x, y), u_t(x, y, 0) = g_2(x, y), \qquad (x, y) \in D. \tag{5.4.9}$$

We begin with the case where $f(x, y, t) = 0$.

Since D is a disk, it is convenient to introduce polar coordinates:

$$x = r\cos\theta, \; y = r\sin\theta.$$

In polar coordinates, $u = u(r, \theta, t)$ satisfies

$$u_{tt} = c^2 (u_{rr} + \frac{1}{r} u_r + \frac{1}{r^2} u_{\theta\theta}), \qquad (r, \theta) \in D, \; t > 0, \tag{5.4.10}$$

where

$$D = \{(r, \theta) : 0 < r < a, -\pi < \theta \le \pi\}.$$

Suppose

$$u(r, \theta, t) = h(t)\phi(r)\psi(\theta).$$

From Eq. (5.4.10) we see that

$$\frac{h''(t)}{c^2 h(t)} = \frac{\phi''(r) + \frac{1}{r}\phi'(r)}{\phi(r)} + \frac{\psi''(\theta)}{r^2 \psi(\theta)} := -\lambda,$$

where λ is an unknown constant.

On the other hand, we separate $\psi(\theta)$ and $\phi(r)$ as follows:

$$-\frac{\psi''(\theta)}{\psi(\theta)} = r^2\big[\lambda + \frac{\phi''(r) + \frac{1}{r}\phi'(r)}{\phi(r)}\big] := \mu,$$

where μ is an unknown constant.

Since $u(r,\theta)$ is smooth in D, we see that

$$\psi(-\pi) = \psi(\pi), \qquad \psi'(-\pi) = \psi'(\pi).$$

It follows that $\psi(\theta)$ satisfies the following eigenvalue problem with periodic boundary conditions:

$$\psi''(\theta) + \mu\psi(\theta) = 0, \qquad -\pi < \theta < \pi, \tag{5.4.11}$$

$$\psi(-\pi) = \pi(\pi), \qquad \psi'(-\pi) = \psi'(\pi). \tag{5.4.12}$$

Hence, the eigenvalues and the corresponding eigenfunctions from Chapter 3 are

$$\mu_0 = 0, \qquad \psi_0(\theta) = 1$$

and

$$\mu_m = m^2, \qquad \psi_m(\theta) = \cos(m\theta),\ \sin(m\theta), \qquad m = 1, 2, \cdots.$$

Once μ_m is found, we can find $\phi(r)$:

$$\phi''(r) + \frac{1}{r}\phi'(r) + (\lambda - \frac{m^2}{r^2})\phi(r) = 0, \qquad 0 < r < a, \tag{5.4.13}$$

$$\phi(a) = 0, |\phi(r)| < \infty. \tag{5.4.14}$$

This is the Bessel equation and its solution is

$$\phi_m(r) = J_m(r\sqrt{\lambda}),$$

where λ is chosen such that $J_m(a\sqrt{\lambda}) = 0$.

For each m, the Bessel function has an infinite number of roots (see [26]), which are all positive.

Suppose $x_n^{(m)}$ are the positive zeros of the Bessel function $J_m(z)$. Then,

$$\lambda_{nm} = (\frac{x_n^{(m)}}{a})^2, \qquad n = 1, 2, \cdots, \ m = 0, 1, 2, \cdots.$$

From the equation for $h(t)$, we see the general solution for $h(t)$ is equal to

$$h_{nm}(t) = B_{1nm}\cos\left(\frac{ctx_n^{(m)}}{a}\right) + B_{2nm}\sin\left(\frac{ctx_n^{(m)}}{a}\right),$$

where B_{1nm} and B_{2nm} are determined by the initial conditions.

Hence, the solution $u(r,\theta,t)$ can be expressed as the following series:

$$u(r,\theta,t)=\sum_{n=1}^{\infty}\sum_{m=0}^{\infty}J_m(\frac{rx_n^{(m)}}{a})\big[A_{1m}\cos(m\theta)+A_{2m}\sin(m\theta)\big]$$
$$\times\big[B_{1nm}\cos(\frac{cx_n^{(m)}}{a}t)+B_{2nm}\sin(\frac{cx_n^{(m)}}{a}t)\big],$$

where coefficients B_{1nm} and B_{2nm} can be determined uniquely from the initial values.

□

5.5 The well-posedness and energy estimates

In this section we investigate the well-posedness of the wave equation. The basic idea is similar to the heat equation.

Let $\Omega\subset R^n$ be a bounded domain with C^1-boundary and $Q_T=\Omega\times(0,T)$ for any $T>0$. Throughout this section we assume that the following basic conditions hold:

H(5.1) Let $p(x,t)\geq p_0>0$ and $p_t(x,t),q(x,t)\in L^\infty(Q_T)$ and $\gamma\geq 0$.
H(5.2) Let $f(x,t)\in L^2(Q_T)$ and $g(x)\in H^1(\Omega)$, $h(x)\in L^2(\Omega)$.

Consider the following problem:

$$u_{tt}+\gamma u_t+L[u]=f(x,t),\qquad x\in\Omega,t>0,\tag{5.5.1}$$
$$B[u]=0,\qquad (x,t)\in\partial\Omega\times(0,\infty),\tag{5.5.2}$$
$$u(x,0)=g(x),u_t(x,0)=h(x),\qquad x\in\Omega,\tag{5.5.3}$$

where the operators L and B are defined by

$$L[u]:=-\nabla[p(x,t)\nabla u]+q(x,t)u,\qquad B[u]:=\alpha p(x,t)\nabla_\nu u+\beta u,$$

where $\alpha^2+\beta^2>0$.

5.5.1 The weak solution and the energy estimate

For simplicity, we focus on the Dirichlet boundary condition:

$$B[u]=u(x,t)=0,\qquad (x,t)\in\partial\Omega\times(0,\infty).$$

Other types of boundary condition can be dealt with similarly.

Definition 5.5.1. We call $u(x,t)$ a weak solution of the problem (5.5.1)–(5.5.3) if $u(x,t)\in H^1((0,T);L^2(\Omega))\bigcap L^2((0,T);H_0^1(\Omega))$ and the following integral identity holds for all $v(x,t)\in H^1(0,T);L^2(\Omega))\bigcap L^2(0,T);H_0^1(\Omega))$ with $v(x,T)=0$

in Ω:

$$\int_0^T \int_\Omega [-u_t v_t + \gamma u_t v + p\nabla u \cdot \nabla v + quv]\, dxdt = \int_\Omega h(x)v(x,0)dx.$$

Moreover,

$$\lim_{t\to 0+} ||u(x,t) - g||_{L^2(\Omega)} = 0.$$

Clearly, it is easy to verify that a classical solution to the problem (5.5.1)–(5.5.3) must be a weak solution. Conversely, the weak solution is also classical if it is smooth. The following energy estimate is similar to the case for the heat equation.

Theorem 5.5.1. *Under the assumptions H(5.1)–H(5.2) the weak solution of the problem (5.5.1)–(5.5.3) is unique and continuously depends on the known data in the L^2-sense. Moreover, the solution satisfies the following energy estimate:*

$$\sup_{0\le t\le T} \int_\Omega [u_t^2 + |\nabla u|^2] dx \le C \int_\Omega (g^2 + |\nabla g|^2 + h^2) dx + C \int_0^T \int_\Omega f^2 dxdt,$$

where C depends only on p_0, α, β, Ω, T and the L^∞-norm of p, p_t, q in their domains.

Furthermore, if $\gamma > 0$ and $q(x,t) = f(x,t) = 0$, then

$$||u_t||_{L^2(\Omega)} + ||\nabla u||_{L^2(\Omega)} \le Ce^{-\gamma t},$$

where C depends only on known data.

Proof. Let us consider the case with $\alpha = 0, \beta = 1$. We define an energy function

$$E(t) := \frac{1}{2}\int_\Omega \left[|u_t|^2 + p(x,t)|\nabla u|^2\right] dx.$$

Then,

$$\begin{aligned}
E'(t) &= \int_\Omega [u_t u_{tt} + p(x,t)\nabla u \cdot \nabla u_t] dx + \frac{1}{2}\int_\Omega (p_t|\nabla u|^2) dx \\
&= \int_\Omega [u_t(u_{tt} - \nabla(p(x,t)\nabla u)] dx + \frac{1}{2}\int_\Omega (p_t|\nabla u|^2) dx \\
&= \int_\Omega [-\gamma u_t^2 + u_t(-qu + f)] dx + \frac{1}{2}\int_\Omega (p_t|\nabla u|^2) dx \\
&:= I + J.
\end{aligned}$$

We use Cauchy–Schwarz's inequality and Poincare's inequality to see that

$$|I| \le C_1 \int_\Omega (u_t^2 + u^2) dx + C_2 \int_\Omega f^2 dx$$

$$\leq C_3 \int_\Omega (u_t^2 + |\nabla u|^2) dx + C_2 \int_\Omega f^2 dx.$$

For J, since p_t is bounded,

$$|J| \leq C_3 \int_\Omega (|\nabla u|^2) dx.$$

It follows that

$$E'(t) \leq C_4 E(t) + C_2 \int_\Omega f^2 dx dt,$$

where C_2, C_4 depend only on known data.

Gronwall's inequality yields

$$E(t) \leq C(T)E(0) + C(T) \int_0^T \int_\Omega f^2 dx dt.$$

When $\gamma > 0$ and $q(x,t) = f(x,t) = 0$, we have

$$E'(t) + \gamma \int_\Omega u_t^2 dx \leq 0.$$

To obtain the decay estimate, we need a result from Chapter 7 that

$$\lim_{t\to\infty} \int_\Omega u_t^2 dx = \lim_{t\to\infty} \int_\Omega |\nabla u|^2 dx.$$

It follows that

$$E(t) \leq C e^{-\frac{\gamma}{2} t},$$

where C depends only on known data. □

5.5.2 The existence and regularity of a weak solution

Similar to the heat equation, we use the finite-element method to prove the existence of a weak solution to the problem (5.5.1)–(5.5.3). We use the Dirichlet boundary condition as an example and briefly sketch the basic steps.

Theorem 5.5.2. *Under the assumptions of H(5.1)–H(5.2), there exists a unique weak solution to the problem (5.5.1)–(5.5.3).*

Proof. The proof of the existence can be obtained from the following steps.

Step 1. Construct an approximate solution.

We choose an orthonormal basis $\{\phi_n(x)\}_{n=1}^\infty$ for $L^2(\Omega)$ and $\phi_k(x) \in C^\infty(\Omega)$ with $\phi(x) = 0$ on $\partial\Omega$. We employ the eigenfunction expansion to have

$$f(x,t) = \sum_{k=1}^\infty f_k(t)\phi_k(x),$$

$$g(x) = \sum_{k=1}^{\infty} g_k \phi_k(x),$$
$$h(x) = \sum_{k=1}^{\infty} h_k \phi_k(x),$$

where

$$f_k(t) = < f, \phi_k >, \ g_k = < g, \phi_k >, \ h_k = < h, \phi_k >, \qquad \forall k \geq 1.$$

Let $N \geq 1$ and set

$$u_N(x,t) = \sum_{k=1}^{N} a_k^N(t) \phi_k(x), \qquad (x,t) \in \Omega \times (0, \infty),$$

where $a_k(t)$ is the solution of the following system of ODE:

$$\frac{d^2 a_k^N(t)}{dt^2} + \gamma \frac{d a_k^N(t)}{dt} + B[u_N, \phi_k] = f_k(t), \qquad k = 1, 2 \cdots, t > 0, \tag{5.5.4}$$
$$a_k(0) = g_k, \qquad a_k'(0) = h_k, \tag{5.5.5}$$

while

$$B[u_N, \phi_k] := \int_\Omega [p(\nabla u_N) \cdot (\nabla \phi_k) + q u_N \phi_k] dx.$$

Under the assumptions H(5.1)–H(5.2), the theory of ODEs ensures that the system (5.5.4)–(5.5.5) has a global solution $\{a_k^N(t)\}$.

Step 2. Derive the uniform estimate for the approximate solution.

This estimate is very much like the energy estimate in Theorem 5.5.1. By using the positivity of p and boundedness of p_t, q we have

$$\sup_{0 \leq t \leq T} \left[||(u_N)_t||_{L^2(\Omega)} + ||\nabla u_N||_{L^2(\Omega)} \right] \leq C,$$

where C depends only on known data, but not on N.

Step 3. Prove that the limit of a convergent subsequence of $u_N(x,t)$ is a weak solution.

First, we use the weak compactness of $L^2(Q_T)$ and the compact embedding theorem in Sobolev space (see [5]) to extract a convergent subsequence (still denoted by $u_N(x,t)$). Then, from the system of ODEs, we obtain an integral identity (5.5.4) for $u_N(x,t)$. Finally, we take the limit to conclude that $u(x,t)$ is a weak solution to the problem (5.5.1)–(5.5.3).

To prove the uniqueness, we only need to show that $u(x,t)=0$ when $f=g=h=0$. Choose the test function $v(x,t)$ as follows: for any $s\in[0,T]$,

$$v(x,t)=\begin{cases}\int_t^s u(x,\tau)d\tau & \text{if } 0\le t\le s,\\ 0 & \text{if } s\le t\le T.\end{cases}$$

Then, we can easily see that $u(x,t)=0$ is the only weak solution (see [8]). □

The regularity of the weak solution is a difficult topic for the wave equation. For this book we just state that the weak solution is indeed classical if the coefficients and known data are smooth and the compatibility conditions hold on $\partial\Omega\times\{t=0\}$. A detailed proof for a general hyperbolic equation can be found in the book [8]. We omit it here.

5.6 A qualitative property: The finite propagation speed

A distinct feature for the solution of a wave equation is that the propagation speed of the wave is finite. Let (x_0,t_0) be a fixed point with $t_0>0$. Define a cone with vertex (x_0,t_0):

$$C(x_0,t_0)=\{(x,t)\in Q:|x-x_0|<t_0-ct, 0<t<\frac{t_0}{c}\}.$$

Let $B(x_0,t)$ be the ball centered at x_0 with radius t. It is clear that $B(x_0,t_0)$ is the bottom of the cone $C(x_0,t_0)$, see Fig. 5.5.

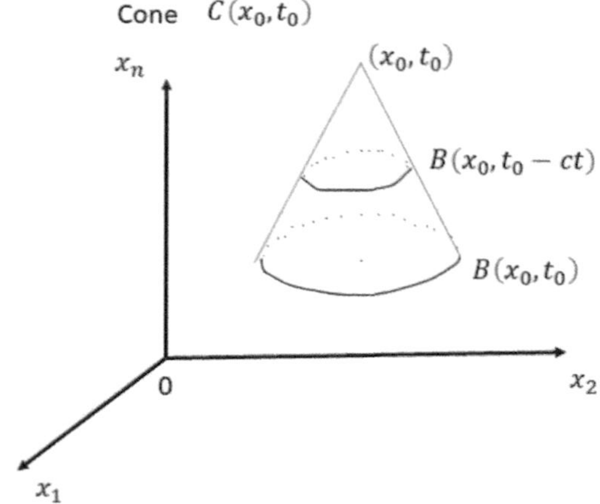

FIGURE 5.5

Theorem 5.6.1. *Let $u(x,t)$ be a solution of the wave equation*

$$u_{tt}-c^2\Delta u=0, \qquad x\in\Omega, t>0.$$

Let $(x_0, t_0) \in \Omega \times (0, \infty)$ *such that* $C(x_0, t_0) \subset \Omega \times (0, \infty)$.

If

$$u(x, 0) = u_t(x, 0) = 0, \qquad x \in B(x_0, t_0),$$

then

$$u(x, t) = 0, \qquad (x, t) \in C(x_0, t_0).$$

Proof. Let $B(x_0, t_0 - ct)$ be the intersection ball of the cone $C(x_0, t_0)$ and the hyperplane $T = t_0 - ct$. Define an energy function

$$E(t) = \frac{1}{2}\int_{B(x_0,t_0-ct)} [u_t^2 + c^2|\nabla u|^2]dx, \qquad 0 < t < \frac{t_0}{c}.$$

Then,

$$\begin{aligned} E'(t) &= \int_{B(x_0,t_0-ct)} [u_t u_{tt} + c^2 \nabla u \cdot \nabla u_t]dx - \frac{c}{2}\int_{\partial B(x_0,t_0-ct)} [u_t^2 + c^2|\nabla u|^2]ds \\ &= \int_{B(x_0,t_0-ct)} [u_t(u_{tt} - c^2\Delta u)]dx + c^2\int_{\partial B(x_0,t_0-ct)} u_t \nabla_\nu u ds \\ &\quad - \frac{c}{2}\int_{\partial B(x_0,t_0-ct)} [u_t^2 + c^2|\nabla u|^2]ds \\ &= c^2\int_{\partial B(x_0,t_0-ct)} u_t \nabla_\nu u ds - \frac{c}{2}\int_{\partial B(x_0,t_0-ct)} [u_t^2 + c^2|\nabla u|^2]ds. \end{aligned}$$

Now, Cauchy–Schwarz's inequality yields

$$c^2\int_{\partial B(x_0,t_0-ct)} u_t \nabla_\nu u ds \le \frac{c}{2}\int_{\partial B(x_0,t_0-ct)} [u_t^2 + c^2|\nabla u|^2]ds.$$

It follows that

$$E'(t) \le 0, \qquad 0 < t < \frac{t_0}{c}.$$

Consequently,

$$0 \le E(t) \le E(0) = 0, \qquad 0 < t < \frac{t_0}{c}. \qquad \square$$

The physical meaning of Theorem 5.6.1 is interesting. The value of $u(x, t)$ at (x_0, t_0) depends only on the value of the initial position and initial velocity in the disk $B(x_0, t_0)$. This is due to the fact that the speed of the wave propagation is finite.

5.7 The Cauchy problem for the wave equation

In this section we derive the solution representation of a Cauchy problem for the wave equation.

5.7.1 Duhamel's principle

Let $f(x,t) \in C^1(Q), g(x), h(x) \in C^2(\bar{\Omega})$. The following consistency condition holds:

$$-c^2 \Delta g(x) = f(x,0), \qquad x \in R^n.$$

Consider the following Cauchy problem for the wave equation:

$$u_{tt} - c^2 \Delta u = f(x,t), \qquad (x,t) \in R^n \times R^1_+, \tag{5.7.1}$$

$$u(x,0) = g(x),\ u_t(x,0) = h(x), \qquad x \in R^n. \tag{5.7.2}$$

We first decompose the problem (5.7.1)–(5.7.2) into the following three problems in which each problem has only one nonhomogeneous term:

Problem (I):

$$u_{tt} - c^2 \Delta u = 0, \qquad (x,t) \in R^n \times R^1_+, \tag{5.7.3}$$

$$u(x,0) = g(x),\ u_t(x,0) = 0, \qquad x \in R^n. \tag{5.7.4}$$

Problem (II):

$$u_{tt} - c^2 \Delta u = 0, \qquad (x,t) \in R^n \times R^1_+, \tag{5.7.5}$$

$$u(x,0) = 0,\ u_t(x,0) = h(x), \qquad x \in R^n. \tag{5.7.6}$$

Problem (III):

$$u_{tt} - c^2 \Delta u = f(x,t), \qquad (x,t) \in R^n \times R^1_+, \tag{5.7.7}$$

$$u(x,0) = 0,\ u_t(x,0) = 0, \qquad x \in R^n. \tag{5.7.8}$$

It is clear that the solution of problem (5.7.1)–(5.7.2) is the sum of the solutions for P(I), P(II), and P(III).

Theorem 5.7.1. *(Duhamel's principle) Let u_1, u_2, and u_3 be the solutions of P(I), P(II), and P(III). Suppose $u_2(x,t) = M_g(x,t)$ is the solution of P(II) associated with an initial velocity $g(x)$. Then,*

$$(a)\quad u_1(x,t) = \frac{\partial}{\partial t} M_g(x,t), \qquad (x,t) \in R^n \times R^1_+,$$

$$(b)\quad u_3(x,t) = \int_0^t M_{f(\tau)}(x, t-\tau)\,d\tau, \qquad (x,t) \in R^n \times R^1_+.$$

Proof. We first prove (a). We assume that $u_1(x,t)$ is smooth and satisfies the wave equation up to $t = 0$.

Clearly, $u_1(x,t) = M_g(x,t)$ satisfies the wave equation (5.7.1). By definition,

$$u_1(x,0) = \frac{\partial}{\partial t} M_g(x,t)\Big|_{t=0} = g(x), \qquad x \in R^n.$$

We use the wave equation to see that

$$u_{1t}(x,0)=\frac{\partial^2}{\partial t^2}M_g(x,t)=c^2\Delta M_g(x,0)\big|_{t=0}=0, \qquad x\in R^n.$$

It follows that $u_1(x,t)$ is a solution of the problem (I).

To probe (b), we note that, for any fixed $\tau\geq 0$,

$$w(x,t):=M_{f(\tau)}(x,t-\tau)$$

solves the following problem:

$$\begin{aligned}
&w_{tt}-c^2\Delta w=0, \qquad (x,t)\in R^n\times(\tau,\infty),\\
&w(x,0)=0, \qquad x\in R^n,\\
&w_t(x,0)=f(x,\tau), \qquad x\in R^n.
\end{aligned}$$

Now,

$$u_3(x,t):=\int_0^t M_{f(\tau)}(x,t-\tau)d\tau.$$

It follows that

$$u_3(x,0)=0, \qquad x\in R^n,$$

$$\begin{aligned}
u_{3t}&=M_{f(\tau)}(x,t-\tau)\big|_{\tau=t}+\int_0^t\frac{\partial}{\partial t}M_{f(\tau)}(x,t-\tau)d\tau\\
&=M_{f(\tau)}(x,0)+\int_0^t\frac{\partial}{\partial t}M_{f(\tau)}(x,t-\tau)d\tau\\
&=\int_0^t\frac{\partial}{\partial t}M_{f(\tau)}(x,t-\tau)d\tau.
\end{aligned}$$

Hence,

$$u_{3t}(x,0)=0, \qquad x\in R^n.$$

Moreover,

$$\begin{aligned}
u_{3tt}&=\frac{\partial}{\partial t}M_{f(\tau)}(x,t-\tau)\big|_{\tau=t}+\int_0^t\frac{\partial^2}{\partial t^2}M_{f(\tau)}(x,t-\tau)d\tau\\
&=f(x,t)+\int_0^t c^2\Delta M_{f(\tau)}(x,t-\tau)d\tau\\
&=f(x,t)+c^2\Delta\int_0^t M_{f(\tau)}(x,t-\tau)d\tau\\
&=f(x,t)+c^2\Delta u_3.
\end{aligned}$$

The proof of (b) is completed. □

5.7.2 Representation for $n = 1$: D'Alembert's formula

Consider a special case where the space dimension is equal to 1:

$$u_{tt} - c^2 u_{xx} = f(x,t), \qquad (x,t) \in R^1 \times (0,\infty), \tag{5.7.9}$$

$$u(x,0) = g(x), \qquad x \in R^1, \tag{5.7.10}$$

$$u_t(x,0) = h(x), \qquad x \in R^1. \tag{5.7.11}$$

We begin with the homogeneous equation by setting $f = 0$.

Introduce new variables:

$$\xi = x + ct, \eta = x - ct.$$

Set

$$v(\xi,\eta) := u(x,t).$$

It is easy to see that Eq. (5.7.9) is equivalent to

$$v_{\xi\eta}(\xi,\eta) = 0.$$

It follows by a direct integration that

$$v(\xi,\eta) = p(\xi) + q(\eta),$$

where $p(\xi)$ and $q(\eta)$ are two arbitrarily differentiable functions.

Namely,

$$u(x,t) = p(x+ct) + q(x-ct).$$

From the initial conditions (5.7.10)–(5.7.11), we have

$$p(x) + q(x) = g(x), \qquad x \in R^1, \tag{5.7.12}$$

$$cp'(x) - cq'(x) = h(x), \qquad x \in R^1. \tag{5.7.13}$$

From Eq. (5.7.12), we see that

$$p'(x) + q'(x) = g'(x), \qquad x \in R^1.$$

It follows that

$$p'(x) = \frac{1}{2} g'(x) + \frac{1}{2c} h(x).$$

Hence,

$$\begin{aligned} p(x) &= p(0) + \frac{1}{2}\int_0^x g'(y)dy + \frac{1}{2c}\int_0^x h(y)dy \\ &= p(0) + \frac{1}{2}[g(x) - g(0)] + \frac{1}{2c}\int_0^x h(y)dy. \end{aligned}$$

On the other hand,

$$q(x) = g(x) - p(x) = \frac{1}{2}g(x) + \frac{1}{2}g(0) - \frac{1}{2c}\int_0^x h(y)dy - p(0).$$

It follows that

$$\begin{aligned} u(x,t) &= p(x+ct) + g(x-ct) \\ &= \frac{1}{2}[g(x-ct) + g(x+ct)] + \frac{1}{2c}\int_{x-ct}^{x+ct} h(y)dy. \end{aligned}$$

Finally, we use Duhamel's principle to obtain the solution representation for the problem (5.7.9)–(5.7.11):

$$\begin{aligned} u(x,t) = &\frac{1}{2}[g(x-ct) + g(x+ct)] + \frac{1}{2c}\int_{x-ct}^{x+ct} h(y)dy \\ &+ \frac{1}{2c}\int_0^t \int_{x-c(t-\tau)}^{x+c(t-\tau)} f(y,\tau)dyd\tau. \end{aligned}$$

We sum up the above derivation to obtain the following theorem.

Theorem 5.7.2. ***(D'Alembert's formula)*** *Let $g(x), h(x) \in C^2(R^1)$ and $f(x,t) \in C(R^1 \times R^1_+)$. The consistency condition holds:*

$$-c^2 g''(x) = f(x,0), \qquad x \in R^1.$$

Then, the solution of the problem (5.7.9)–(5.7.11) is equal to

$$\begin{aligned} u(x,t) = &\frac{1}{2}[g(x-ct) + g(x+ct)] + \frac{1}{2c}\int_{x-ct}^{x+ct} h(y)dy \\ &+ \frac{1}{2c}\int_0^t \int_{x-c(t-\tau)}^{x+c(t-\tau)} f(y,\tau)dyd\tau. \quad \square \end{aligned}$$

From D'Alembert's formula, we see that if the initial values $g(x)$, $h(x)$ and $f(x,t)$ have compact support at each level t, then $u(x,t)$ must have a compact support for each $t > 0$, see Fig. 5.6. This implies that the propagation speed of sound is finite. We will see this property again in higher-space dimensions.

5.7.3 Solution representation for space dimension $n = 3$

To derive the solution representation, we introduce a spherical mean function below that will reduce the Laplace operator to a one-dimensional differential operator. For convenience, we take $c = 1$.

Suppose $u(x,t)$ satisfies

$$u_{tt} - \Delta u = 0, \qquad x \in R^n, t > 0. \tag{5.7.14}$$

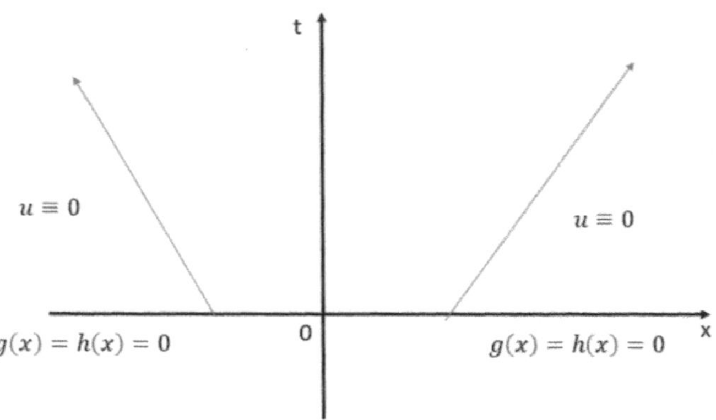

FIGURE 5.6

$$u(x,0) = g(x),\ u_t(x,0) = h(x), \qquad x \in R^n. \tag{5.7.15}$$

Let $x \in R^n$ be fixed. Define

$$U(x,r,t) = \fint_{\partial B_r(x)} u(y,t)ds(y) = \fint_{\partial B_1(0)} u(x+rz,t)ds(z), \qquad r>0, t>0,$$

where $y = x + rz \in B_r(x)$ if and only if $z = \frac{y-x}{r} \in B_1(0)$. Then,

$$\begin{aligned} U_r(x,r,t) &= \fint_{\partial B_1(0)} \nabla u(x+rz,t)\cdot z ds(z) \\ &= \fint_{\partial B_r(x)} \nabla u(y,t)\cdot \frac{y-x}{r} ds(y) \\ &= \fint_{\partial B_r(x)} \nabla_\nu u(y,t) ds(y) \\ &= \frac{r}{n}\fint_{B_r(x)} \Delta u dy = \frac{1}{n\omega_n r^{n-1}} \int_{B_r(x)} \Delta u(y,t) dy. \end{aligned}$$

Here at the final step, we used Green's theorem. Consequently, we find that

$$\lim_{r\to 0+} U_r(x,r,t) = 0.$$

Recall the coarea formula: For any integrable function $f(y)$,

$$\frac{d}{dr}\int_{B_r(x)} f(y)dy = \int_{\partial B_r(x)} f(y)ds(y).$$

Thus

$$U_{rr} = \frac{1}{n\omega_n r^{n-1}} \int_{\partial B_r(x)} \Delta u(y)ds(y) - \frac{n-1}{n\omega_n r^n}\int_{B_r(x)} \Delta u(y)dy$$

$$
\begin{aligned}
&= \fint_{\partial B_r(x)} \Delta u(y) ds(y) + (\frac{1-n}{n}) \fint_{B_r(x)} \Delta u dy \\
&\to \Delta u(x,t) + (\frac{1-n}{n}) \Delta u(x,t), \qquad \text{as } r \to 0.
\end{aligned}
$$

It follows that

$$
\lim_{r \to 0+} U_{rr}(x,r,t) = \frac{1}{n} \Delta u(x,t).
$$

On the other hand, from the equation of $u(x,t)$ we see that

$$
\begin{aligned}
U_r &= \frac{r}{n} \fint_{B_r(x)} \Delta u dy = \frac{r}{n} \fint_{B_r(x)} u_{tt} dy \\
&= \frac{1}{n\omega_n r^{n-1}} \int_{B_r(x)} u_{tt}(y,t) dy.
\end{aligned}
$$

It follows that

$$
(r^{n-1}U_r)_r = \frac{1}{n\omega_n} \int_{B_r(x)} u_{tt} dy = r^{n-1} \fint_{\partial B_r(x)} u_{tt} ds(y) = r^{n-1} U_{tt}.
$$

Moreover, from the initial conditions we see that

$$
U(x,r,0) = G(x,r), U_t(x,r,0) = H(x,r),
$$

where

$$
G(x,r) = \fint_{\partial B_r(x)} g(y) ds(y), \qquad H(x,r) = \fint_{\partial B_r(x)} h(y) ds(y).
$$

Consider the case where $n = 3$. Define

$$
\hat{U}(x,r,t) = rU(x,r,t), \qquad \hat{G}(x,r) = rG(x,r), \qquad \hat{H}(x,r) = rH(x,r).
$$

It is easy to see $\hat{U}(x,r,t)$ is the solution of the following problem:

$$
\begin{aligned}
&\hat{U}_{tt} - \hat{U}_{rr} = 0, \qquad (r,t) \in R^1_+ \times (0,\infty), && (5.7.16) \\
&\hat{U}(x,r,0) = \hat{G}(x,r), \qquad r \geq 0, && (5.7.17) \\
&\hat{U}_t(x,r,0) = \hat{H}(x,r), \qquad r \geq 0, && (5.7.18) \\
&\hat{U}(x,0,t) = 0, \qquad t > 0. && (5.7.19)
\end{aligned}
$$

If we use an odd extension for $\hat{G}$ and $\hat{H}$ to $(-\infty, 0]$ (extended functions are denoted by the same notation), then we have from Theorem 5.7.2 the solution

$$
\hat{U}(x,r,t) = \frac{1}{2}[\hat{G}(x,r+t) - \hat{G}(x,t-r)] + \frac{1}{2} \int_{t-r}^{r+t} \hat{H}(x,y) dy, \qquad 0 \leq r < t,
$$

where we used the odd extension $\hat{G}(x, r-t) = \hat{G}(x, t-r)$, since $r < t$.

Now, we are ready to state the following theorem.

Theorem 5.7.3. *(Kirchhoff's formula) Let $g(x) \in C^1(R^1)$ and $h(x) \in C(R^1)$. For $n = 3$, the solution of the problem (5.7.14)–(5.7.15) can be expressed as*

$$u(x,t) = \fint_{\partial B_t(x)} [th(y) + g(y) + \nabla g(y) \cdot (y-x)]\, ds(y), \qquad x \in R^3, t \geq 0.$$

Proof. By the definition of $\hat{U}(x,r,t)$, we see that

$$\begin{aligned} u(x,t) &= \lim_{r \to 0+} \frac{\hat{U}(x,r,t)}{r} \\ &= \lim_{r \to 0+} \left[\frac{\hat{G}(t+r) - \hat{G}(t-r)}{2r} + \frac{1}{2r} \int_{t-r}^{t+r} \hat{H}(x,y)dy \right] \\ &= \hat{G}'(t) + \hat{H}'(t). \end{aligned}$$

It follows that

$$\begin{aligned} u(x,t) &= \frac{\partial}{\partial t}\left(t \fint_{\partial B_t(x)} g(y)ds(y) \right) + t \fint_{B_t(x)} h(y)ds(y) \\ &= \fint_{\partial B_t(x)} g(y)ds(y) + t\frac{\partial}{\partial t}\left(\fint_{\partial B_t(x)} g(y)ds(y) \right) + t \fint_{B_t(x)} h(y)ds(y). \end{aligned}$$

Now,

$$\begin{aligned} &\frac{\partial}{\partial t}\left(\fint_{\partial B_t(x)} g(y)ds(y) \right) \\ &= \frac{\partial}{\partial t}\left(\fint_{\partial B_1(0)} g(x+tz)ds(z) \right) \\ &= \fint_{\partial B_1(0)} [\nabla g(x+tz) \cdot z] ds(z) \\ &= \fint_{\partial B_t(x)} [\nabla g(y) \cdot \frac{y-x}{t}] ds(z). \end{aligned}$$

We combine the above calculations to obtain the expression of $u(x,t)$. □

5.7.4 Solution representation for space dimension $n = 2$: Method of descent

The method in derivation of Kirchhoff's formula for $n = 3$ fails for $n = 2$. Here, we simply consider the problem (5.7.14)–(5.7.15) as a special case where the solution of the wave equation does not depend on the third variable.

Theorem 5.7.4. *(Poisson's formula) Let $g(x) \in C^1(R^2)$ and $h(x) \in C(R^2)$. For $n = 2$, the solution of the problem (5.7.11)–(5.7.12) can be expressed by*

$$u(x,t) = \frac{1}{2}\fint_{B_t(x)} \frac{tg(y) + t^2h(y) + t\nabla g(y)\cdot(y-x)}{\sqrt{t^2 - |y-x|^2}}dy.$$

Proof. The basic idea is that we consider the problem (5.7.14)–(5.7.15) in R^3 with special initial data that depend only on two variables. Let $u(x_1, x_2, t)$ be a solution of the wave problem (5.7.14)–(5.7.15) in two-space dimensions. Set

$$x = (x_1, x_2),\ \bar{x} = (x_1, x_2, 0),\ \bar{u}(x_1, x_2, x_3, t) := u(x,t).$$

Then, $\bar{u}$ satisfies the following problem in $R^3 \times R^1_+$:

$$\begin{aligned} &\bar{u}_{tt} - \Delta\bar{u} = 0, && \in R^3 \times R^1_+, \\ &\bar{u}(x,0) = \bar{g}(x),\ \bar{u}_t(x,0) = \bar{h}(x), && x \subset R^3, \end{aligned}$$

where

$$\bar{g}(x) = g(x_1, x_2), \qquad \bar{h}(x) = h(x_1, x_2), \qquad x \in R^3.$$

By Kirchhoff's formula, we have

$$\begin{aligned} u(x,t) &= \bar{u}(\bar{x}, t) \\ &= \frac{\partial}{\partial t}\left(t\fint_{\partial\bar{B}_t(\bar{x})} \bar{g}(\bar{y})ds(\bar{y})\right) + t\fint_{\partial\bar{B}_t(\bar{x})} \bar{h}(\bar{y})ds(\bar{y}), \end{aligned}$$

where

$$\bar{B}_t(\bar{x}) = \{(y_1, y_2, y_3) : (y_1 - x_1)^2 + (y_2 - x_2)^2 + y_3^2 < t^2\}.$$

To calculate the surface integral, we express the surface $\partial\bar{B}_t(\bar{x})$ in explicit form:

$$y_3 = \pm\sqrt{t^2 - |y-x|^2}, \qquad y \in B_t(x).$$

Then, on $\partial\bar{B}_t(\bar{x})$, $|y-x|^2 = t^2$, we find that

$$ds(\bar{y}) = \sqrt{1 + |\nabla y_3|^2}dy = \frac{t}{\sqrt{t^2 - |y-x|^2}}dy.$$

It follows that

$$\begin{aligned} \fint_{\partial\bar{B}_t(\bar{x})} \bar{g}(\bar{y})ds(\bar{y}) &= \frac{1}{2\pi t}\int_{B_t(x)} \frac{g(y)}{\sqrt{t^2 - |y-x|^2}}dy \\ &= \frac{t}{2}\fint_{B_t(x)} \frac{g(y)}{\sqrt{t^2 - |y-x|^2}}dy. \end{aligned}$$

It also follows that

$$u(x,t)=\frac{1}{2}\frac{\partial}{\partial t}\left(t^2\fint_{B_t(x)}\frac{g(y)}{\sqrt{t^2-|y-x|^2}}dy\right)+\frac{t^2}{2}\fint_{B_t(x)}\frac{h(y)}{\sqrt{t^2-|y-x|^2}}dy.$$

Note that

$$t^2\fint_{B_t(x)}\frac{g(y)}{\sqrt{t^2-|y-x|^2}}dy=t\fint_{B_1(0)}\frac{g(x+tz)}{\sqrt{1-|z|^2}}dz.$$

We find that

$$\begin{aligned}&\frac{\partial}{\partial t}\left(t^2\fint_{B_t(x)}\frac{g(y)}{\sqrt{t^2-|y-x|^2}}dy\right)\\&=\fint_{B_1(0)}\frac{g(x+tz)}{\sqrt{1-|z|^2}}dz+t\fint_{B_1(0)}\frac{\nabla g(x+tz)\cdot z}{\sqrt{1-|z|^2}}dz\\&=t\fint_{B_t(x)}\frac{g(y)}{\sqrt{t^2-|y-x|^2}}dy+t\fint_{B_t(x)}\frac{\nabla g(y)\cdot(y-x)}{\sqrt{t^2-|y-x|^2}}dy.\end{aligned}$$

Consequently, we combine the above calculation to obtain

$$u(x,t)=\frac{1}{2}\fint_{B_t(x)}\frac{tg(y)+t^2h(y)+t\nabla g(y)\cdot(y-x)}{\sqrt{t^2-|y-x|^2}}dy,\qquad x\in R^2,t\geq 0.\quad\square$$

Corollary 5.7.1. *Suppose $g(x)\in C^1(R^n)$ and $h(x),f(x,t)\in C(R^2\times R^1_+)$. Let $u(x,t)$ be a solution of the following nonhomogeneous problem:*

$$u_{tt}-\Delta u=f(x,t),\qquad x\in R^n,t>0,\tag{5.7.20}$$

$$u(x,0)=g(x),\qquad u_t(x,0)=h(x),\qquad x\in R^n.\tag{5.7.21}$$

Then,

(a) *if $n=2$,*

$$\begin{aligned}u(x,t)=&\frac{1}{2}\fint_{B_t(x)}\frac{tg(y)+t^2h(y)+t\nabla g(y)\cdot(y-x)}{\sqrt{t^2-|y-x|^2}}dy\\&+\int_0^t\fint_{B_{t-\tau}(x)}\frac{(t-\tau)^2f(y,t-\tau)}{\sqrt{(t-\tau)^2-|y-x|^2}}dyd\tau.\end{aligned}$$

(b) *if $n=3$,*

$$\begin{aligned}u(x,t)=&\fint_{\partial B_t(x)}[th(y)+g(y)+\nabla g(y)\cdot(y-x)]ds(y)\\&+\int_0^t\fint_{\partial B_{t-\tau}(x)}[(t-\tau)f(y,t-\tau)]ds(y).\end{aligned}$$

5.7.5 Domain dependence, data disturbances, and Huygens principle

For the homogeneous problem where $f(x,t)=0$, from the representation of $u(x,t)$ in R^2 and R^3, we find a very interesting phenomenon. When the space dimension $n=3$, the value of u at (x_0,t_0) depends only on the value of known data along the surface of $B_{t_0-ct}(x_0)$. On the other hand, when the space dimension $n=2$, the value of u at (x_0,t_0) depends on the known data in $B_{t_0}(x_0)$. The ball $B_{t_0-ct}(x_0)$ is called the *domain dependence*. $\partial B_{t_0-ct}(x_0)$ is called the *data disturbance*. This phenomenon for $n=3$ is often referred to as the Huygens principle.

For a more physical interpretation, the front water wave propagation in a lake depends on all the waves inside of the front wave disk. On the other hand, the sound propagation in R^3 depends only on the surface of the ball $B_t(x)$. The sound waves do not interfere with each other when t increases.

5.8 Notes and remarks

The materials in Section 5.1 to Section 5.3 are elementary. Great effort was made in these sections to establish several mathematical models for the wave-propagation phenomenon. These mathematical models will help students and researchers in applied fields to better understand wave propagation. The nonlinear Euler's equation is still a very active research topic in compressible fluid mechanics. Sections 5.4 and 5.5 are basic materials for theoretical study for the wave equation. The regularity of the weak solution for the wave equation is extremely challenging, since the regularity of the solution depends on the regularity of the initial and boundary values; even the coefficients in the equations are smooth. This is very different from the heat equation and the Laplace equation. It is known in advanced PDEs that the solution to the heat equation is smooth as long as the coefficients of the equation is smooth. However, the smoothness of the weak solution for the wave equation depends on the smoothness of the initial and boundary values as well as the coefficients of the equation.

The representation of the solution for the Cauchy problem is very elegant. The presentation in Section 5.7 closely follows Evans' book ([8]). It reveals many interesting phenomena for the solution of the wave equation.

5.9 Exercises

1. Assume a rope with length L and density $\rho_0=1$ is hanging on the roof top. At the initial moment it is still. Assume a wind starts at one hour later with the force $f_0=10\,N/s^2$ acting on the string uniformly on the rope in an easterly direction. Let the coefficient of the resistance force be equal to 20 and the tensile force be equal to 4 N/S^2. Formulate the complete mathematical model.
2. Find the series solution for Exercise 1 above.

3. The position $u(x,t)$ of a vibrating string satisfies

$$\begin{aligned} &u_{tt} - 25u_{xx} = 5\sin(\omega t), && 0 < x < L, t \geq 0, \\ &u(0,t) = u(L,t) = 0, && t > 0, \\ &u(x,0) = u_t(x,0) = 0, && 0 < x < L. \end{aligned}$$

(a) Find the natural frequency.
(b) What ω should be avoided for the external force to prevent the break down of the string.

4. Find the series solution for the following wave equation:

$$\begin{aligned} &u_{tt} - c^2 u_{xx} + \gamma u_t = 0, && 0 < x < L, t \geq 0, \\ &u(0,t) = u(L,t) = 0, && t > 0, \\ &u(x,0) = f(x), u_t(x,0) = g(x), && 0 < x < L, \end{aligned}$$

where $\gamma > 0$, $f(x), g(x) \in L^2(0,L) \bigcap C^2(0,L)$.

5. Find the series solution for the following wave equation:

$$\begin{aligned} &u_{tt} - u_{xx} = 0, && 0 < x < L, t \geq 0, \\ &u(0,t) = u(L,t) = 0, && t > 0, \\ &u(x,0) = \delta(x - \frac{L}{2}), u_t(x,0) = 0, && 0 < x < L, \end{aligned}$$

where $\delta(x)$ is the Dirac-delta function with the following property: for any continuous function $f(x)$ in R^1,

$$\int_{-\infty}^{\infty} f(x)\delta(x)dx = f(0).$$

6. Let $\delta(t)$ be the Dirac-delta function. The position $u(x,t)$ of a vibrating string satisfies

$$\begin{aligned} &u_{tt} - u_{xx} = \delta(t-1) + \delta(t-2), && 0 < x < L, t \geq t, \\ &u(0,t) = u(L,t) = 0, && t > 0, \\ &u(x,0) = u_t(x,0) = 0, && 0 < x < L. \end{aligned}$$

(a) Find the series solution.
(b) Does the solution decay in time? Explain the physical interpretation of your result.

7. Let $R = \{(x,y) : 0 < x < L, 0 < y < H\}$. Find the solution to the following wave equation:

$$\begin{aligned} &u_{tt} - c^2 \Delta u = au, && x \in R, t > 0, \\ &u(x,t) = 0, && x \in \partial R, t > 0, \end{aligned}$$

$$u(x,0)=f(x), u_t(x,0)=g(x), x\in R,$$

where $f(x), g(x)\in L^2(\Omega)$ and a is a constant.

8. Let $g(t)\in C^2[0,\infty)$. Find the solution representation for the following wave equation in a half-space:

$$\begin{aligned}
&u_{tt}-u_{xx}=0, \quad && x\in[0,\infty, t>0,\\
&u(0,t)=g(t), && x\in\partial R, t>0,\\
&u(x,0)=u_t(x,0)=0, x\in R,
\end{aligned}$$

9. Let Ω be a bounded domain in R^n and $u(x,t)$ be the solution to the following wave equation:

$$\begin{aligned}
&u_{tt}-c^2\Delta u+\gamma u_t=0, \quad && x\in\Omega, t>0,\\
&u(x,t)=0, && x\in\partial\Omega, t>0,\\
&u(x,0)=f(x), u_t(x,0)=g(x), x\in\Omega,
\end{aligned}$$

where $\gamma>0$ is a constant, $f(x), g(x)\in H^1(\Omega)$.
Prove the energy functional

$$E(t):=\frac{1}{2}\int_\Omega[u_t^2+c^2|\nabla u|^2]dx$$

decays to 0 exponentially. Explain the reason for your mathematical result.

10. Let $p(x), q(x)\in L^\infty(\Omega)$ with $p(x)\ge p_0>0$ and $q(x)\ge 0$ on Ω. Prove the following wave problem is well-posed:

$$\begin{aligned}
&u_{tt}-\nabla[p(x)\nabla u]+q(x)u=0, \quad && x\in\Omega, t>0,\\
&u_\nu(x,t)=\alpha u, && x\in\partial\Omega, t>0,\\
&u(x,0)=f(x), u_t(x,0)=g(x), x\in\Omega,
\end{aligned}$$

where $u_\nu=\nabla_\nu u$ represents the normal derivative on $\partial\Omega$.

11. Let $p(x), q(x)\in L^\infty(\Omega)$ with $p(x)\ge p_0>0$ and $q(x)\ge 0$ on Ω. Prove that the solution of the following wave problem is unique:

$$\begin{aligned}
&u_{tt}-\nabla[p(x)\nabla u]+q(x)u=0, \quad && x\in\Omega, t>0,\\
&u_\nu(x,t)=a(x)u, && x\in\partial\Omega, t>0,\\
&u(x,0)=f(x), u_t(x,0)=g(x), x\in\Omega.
\end{aligned}$$

12. Let $u(x,t)$ be the solution of the Cauchy problem

$$\begin{aligned}
&u_{tt}-\Delta u=0, \quad && x\in R^3, t>0\\
&u(x,0)=f(x), u_t(x,0)=g(x), && x\in R^3.
\end{aligned}$$

Suppose that $f(x)$ and $g(x)$ have compact support in R^3. Prove that

$$\int_{R^3} u_t^2 dx = \int_{R^3} |\nabla u|^2 dx.$$

13. Let $u(x,t)$ be a solution of the wave equation

$$u_{tt} + \gamma(x)u_t - \Delta u = 0, \qquad x \in R^3, t > 0$$
$$u(x,0) = f(x), u_t(x,0) = g(x), \qquad x \in R^3,$$

where $f(x), g(x) \in L^2(R^3)$ and $\gamma(x) \geq 0$ in R^3. Suppose there exists a ball $B_a(0)$ such that

$$\gamma(x) \geq \gamma_0 > 0.$$

Prove that the energy function $E(t)$ defined in Section 5.6 decays to 0 as $t \to \infty$.

14. Let $a(t) \in C^1[0,\infty)$. Let $u(x,t)$ be a solution of the wave equation

$$u_{tt} - \Delta u + a(t)u = 0, \qquad x \in R^3, t > 0$$
$$u(x,0) = f(x), u_t(x,0) = g(x), \qquad x \in R^3,$$

where $f(x), g(x) \in L^2(R^3)$.
Find the condition for $a(t)$ such that the energy function $E(t)$ defined in Section 5.6 decays to 0 as $t \to \infty$.

CHAPTER

The Laplace equation 6

6.1 Some mathematical models of the Laplace equation

In this section we derive some mathematical models governed by the Laplace equation and its generalization. All of these models arise from mathematical and physical sciences.

6.1.1 The steady-state equations of evolution equations in the physical sciences

The first example is that the Laplace equation may be considered as the steady-state equation of an evolution equation in the physical sciences. The long-time behavior of the solution to an evolution equation relies on the deep understanding of the solution of the steady-state equation. Here, we give a few examples as motivation.

Example 6.1.1. The steady-state equation for some evolution equations:

(a) The steady-state equation for the heat conduction with a heat source $f(x)$:

$$-k^2\Delta u = f(x), \qquad x \in \Omega.$$

(b) The steady-state equation for the wave propagation with an internal force $f(x)$:

$$-c^2\Delta u = f(x), \qquad x \in \Omega.$$

(c) The steady-state equation of the Schrödinger equation with potential $v(x)$:

$$-\Delta u + v(x)u = 0, \qquad x \in \Omega.$$

(d) The steady-state equation of an electric field.

When an electric field $\mathbf{E}$ is static, physical experiments show that it can be expressed as the gradient of a potential function:

$$\mathbf{E} := -\nabla u.$$

If $\rho(x)$ is the charge density in Ω, then the conservation of charge implies

$$\nabla \cdot \mathbf{E} = \rho(x).$$

Partial Differential Equations and Applications. https://doi.org/10.1016/B978-0-44-318705-6.00012-4

It follows that

$$-\Delta u = \rho(x).$$

The second motivation of the Laplace equation comes from the minimization problem, which is referred to as the *calculus of variation*.

6.1.2 The minimization of kinetic energy: Calculus of variation

Consider a bounded domain Ω in R^3. For a physical quantity v such as velocity in a channel, the total kinetic energy in Ω is given by

$$\int_\Omega |\nabla v|^2 dx.$$

If there is an external force, denoted by $f(x) \in L^2(\Omega)$, it is natural to study an energy functional:

$$E(v) := \frac{1}{2}\int_\Omega |\nabla v|^2 dx - \int_\Omega f(x) v dx.$$

A natural question in physics is to find the minimum of such a kinetic-energy functional. This leads to a mathematical question: Find $u(x)$ in a function space A such that

$$E(u) = \min_{v \in A} E(v).$$

The set A is typically a subset of $C^1(\bar{\Omega})$. We have to impose a condition on $\partial\Omega$ in order to determine $u(x)$ uniquely. For example, we may assume that

$$A = \{v(x) \in C^1(\bar{\Omega}) : v(x) = 0, \qquad x \in \partial\Omega\}.$$

One can impose different types of condition on the boundary. The set A is called an admissible set in optimization theory. With the mathematical formulation, how do we find such a minimum?

We first derive a necessary condition. Suppose $E(v)$ attains its minimum at $u \in A$. Let t be a parameter. For any $v \in A$, consider

$$I(t) := E(u + tv) \geq E(u), \qquad t \in R^1.$$

It follows that $I(t)$ attains its minimum at $t = 0$. Hence, $I'(0) = 0$.

Now,

$$I'(0) = \lim_{t \to 0} \frac{I(t) - I(0)}{t} = \int_\Omega (\nabla u \cdot \nabla v) dx - \int_\Omega f(x) v(x) dx.$$

If $u \in C^1(\Omega)$, we use Gauss's divergence theorem to obtain

$$\int_\Omega [(\Delta u + f(x)) v(x)] dx = 0, \qquad v \in A.$$

Since $v(x) \in A$ is arbitrary and A is dense in $C^1(\bar{\Omega})$, it follows that $u(x)$ is a solution of the following equation:

$$
\begin{aligned}
-\Delta u &= f(x), && x \in \Omega, \\
u(x) &= 0, && x \in \partial\Omega.
\end{aligned}
$$

By a similar calculation, it is easy to see that

$$I''(0) \geq 0,$$

which ensures that the functional $E(v)$ attains a minimum at $u \in A$. This procedure is called the *Calculus of Variation*.

6.1.3 The minimum-surface equation

We give another interesting example that is important in physics and mechanics as well as in differential geometry. Suppose a simple closed curve C in R^3 is given by

$$C : z = z(x, y), (x, y) \in \partial\Omega,$$

where Ω is a region in R^2.

The surface with the boundary C is denoted by $v(x, y)$. Then, the surface area is given by

$$J(v) := \int\int_{\Omega} \sqrt{1 + v_x^2 + v_y^2} dxdy.$$

We are interested in finding a surface with the minimum surface area. Mathematically, this is equivalent to finding the minimum of $J(v)$ among all possible differentiable functions with $v = z(x, y)$ on the boundary of Ω.

If we define the admissible set

$$A = \{v(x, y) \in C^1(\bar{\Omega}) : v(x, y) = z(x, y), \quad (x, y) \in \partial\Omega\}.$$

We can perform the same calculation as for the kinetic-energy functional $J(v)$ in Section 6.1.2 to obtain that the minimum surface $u(x, y)$ satisfies

$$
\begin{aligned}
\nabla\left[\frac{\nabla u}{\sqrt{1 + |\nabla u|^2}}\right] &= 0, && (x, y) \in \Omega, \\
u(x, y) &= z(x, y), && (x, y) \in \partial\Omega.
\end{aligned}
$$

When $|\nabla u|$ is small, we can approximate the minimum surface by the Laplace equation:

$$
\begin{aligned}
\Delta u &= 0, && (x, y) \in \Omega, \\
u(x, y) &= z(x, y), && (x, y) \in \partial\Omega.
\end{aligned}
$$

6.1.4 The Cauchy–Riemann equation in complex analysis

As a fourth example, we consider the fundamental equation in complex analysis. Recall that a complex function $f(z)$ is called analytic in a region Ω, if the Cauchy–Riemann equation holds in Ω

$$\frac{\partial f(z)}{\partial \bar{z}} = 0,$$

where $z = x + iy$ and $\bar{z} = x - iy$.

A fundamental problem in complex analysis is to determine whether or not a complex $f(z)$ is analytic in a domain $\Omega \subset R^2$.

Now, we rewrite the complex function $f(z)$ as real and imaginary parts:

$$f(z) = u(x, y) + iv(x, y), \qquad (x, y) \in \Omega.$$

Note that

$$\frac{\partial f(z)}{\partial \bar{z}} = (\frac{\partial}{\partial x} - i\frac{\partial}{\partial y})[u(x, y) + iv(x, y)].$$

It follows that

$$(u_x + v_y) + i(-u_y + v_x) = 0, \qquad (x, y) \in \Omega.$$

Consequently,

$$u_x + v_y = 0, \qquad -u_y + v_x = 0, \qquad (x, y) \in \Omega,$$

which is equivalent to

$$u_{xx} + u_{yy} = 0, \qquad v_{xx} + v_{yy} = 0, \qquad (x, y) \in \Omega.$$

This shows that $f(z)$ is analytic in Ω if and only if $Re f(z)$ and $Im f(z)$ are harmonic functions. Many properties for analytic functions can be derived from the harmonic functions.

6.2 Series solution for the Laplace equation

In this section we first present a general result for the Laplace equation associated with appropriate boundary conditions. We then give several examples on how to find a series solution in terms of the eigenfunctions.

6.2.1 A general boundary value problem for the Laplace equation

When the dimension is equal to 1, the Laplace equation becomes an ODE, which is the Sturm–Liouville theory discussed in Chapter 3. We now focus on the problem in a domain in R^n with $n \geq 2$.

Let Ω be a bounded domain in R^n with a smooth boundary $\partial\Omega$. Suppose $f(x) \in L^2(\Omega)$ and $g(x) \in L^2(\partial\Omega)$. Consider the following boundary value problem:

$$-\Delta u = f(x), \qquad x \in \Omega, \tag{6.2.1}$$

$$B[u] = g(x), \qquad x \in \partial\Omega, \tag{6.2.2}$$

where B is a general boundary operator:

$$B[u] = \alpha \nabla_\nu u + \beta u, \qquad \alpha^2 + \beta^2 > 0,$$

the vector ν is the outward unit normal on $\partial\Omega$.

Obviously, by proper choice of the parameters α and β the boundary operator B contains all three types of boundary conditions that are discussed in previous chapters.

We are interested in the well-posedness for the problem and the solution representation. The examples in Chapter 1 indicate that the problem (6.2.1)–(6.2.2) may not have a solution even with a classical boundary condition.

To find a solution to the problem (6.2.1)–(6.2.2), we may split the problem (6.2.1)–(6.2.2) into two problems:

Problem (I) (Nonhomogeneous equation with a homogeneous boundary condition):

$$-\Delta u = f(x), \qquad x \in \Omega, \tag{6.2.3}$$

$$B[u] = 0, \qquad x \in \partial\Omega. \tag{6.2.4}$$

Problem (II) (Homogeneous equation with nonhomogeneous boundary condition):

$$-\Delta u = 0, \qquad x \in \Omega, \tag{6.2.5}$$

$$B[u] = g(x), \qquad x \in \partial\Omega. \tag{6.2.6}$$

It will be seen that the solvability for Problem (I) depends on the eigenvalue problem (6.2.7)–(6.2.8), while the solvability for Problem (II) depends on the boundary eigenvalue problem (6.2.9)–(6.2.10) below.

Now, we consider the following eigenvalue problem:

$$-\Delta \psi = \lambda \psi, \qquad x \in \Omega, \tag{6.2.7}$$

$$B[\psi] = 0, \qquad x \in \partial\Omega. \tag{6.2.8}$$

If we know there exist an infinite number of eigenvalues $\{\lambda_n\}_{n=1}^{\infty}$ and all corresponding eigenfunctions $M = \{\phi_n(x)\}_{n=1}^{\infty}$ form a mutually orthogonal basis for $L^2(\Omega)$, then we can find a series solution for Problem (I).

We need to divide the eigenvalues for the operator in Problem (I) into two cases.

Case 1: $\lambda = 0$ is not an eigenvalue for the eigenvalue problem (6.2.7)–(6.2.8).

Let

$$f(x) = \sum_{n=1}^{\infty} a_n \phi_n(x),$$

where

$$a_n = \frac{< f, \phi_n >}{< \phi_n, \phi_n >}, \qquad n = 1, 2, \cdots .$$

Suppose a solution to the problem (6.2.3)–(6.2.4) is expressed by the eigenfunction series:

$$u(x) = \sum_{n=1}^{\infty} c_n \phi_n(x), \qquad x \in \Omega,$$

where c_n will be specified later.

From Eq. (6.2.3), we see that

$$-\Delta u = -\sum_{n=1}^{\infty} c_n \Delta\phi_n = \sum_{n=1}^{\infty} c_n \lambda_n \phi_n .$$

If we choose

$$c_n = \frac{a_n}{\lambda_n}, \qquad n = 1, 2, \cdots ,$$

then, $u(x)$ is a series solution of the problem (6.2.1)–(6.2.2).

Case 2: $\lambda = 0$ is an eigenvalue of (6.2.7)–(6.2.8).

Without loss of generality, we assume that $\lambda_0 = 0$ is an eigenvalue. Let

$$N_0(B) = span\{\text{all mutually orthogonal eigenfunctions corresponding to } \lambda_0 = 0\}.$$

We would like to see the structure of the subspace $N_0(B)$. From the Rayleigh quotient that $\alpha \neq 0$. Hence, the boundary condition (6.2.4) becomes

$$\nabla_\nu u = -\frac{\beta}{\alpha} u := bu, \qquad x \in \partial\Omega,$$

where $b = -\frac{\beta}{\alpha}$.

(a) If $b = 0$, the boundary condition (6.2.4) becomes

$$\nabla_\nu u = 0, \qquad x \in \partial\Omega.$$

From the Rayleigh quotient, the corresponding eigenfunction ϕ_0 must satisfy

$$\int_\Omega |\nabla\phi_0|^2 dx = 0.$$

Therefore the corresponding eigenfunction must be a constant. We choose the corresponding eigenfunction $\phi_0(x) = 1$. It follows that $dim\, N_0(B) = 1$ and

$$N_0(B) = span\{\phi_0(x) = 1\}.$$

By applying the Fredholm Alternative, we see that the problem (I) has a solution for $f(x) \in L^2(\Omega)$ if and only if

$$< f, \phi_0 >= \int_\Omega f(x)dx = 0.$$

(b) If $b \neq 0$, then from the Rayleigh quotient, the corresponding eigenfunction $\phi_h(x)$ must satisfy

$$\int_\Omega |\nabla \phi_h|^2 dx - b \int_{\partial\Omega} \phi_h^2 ds = 0.$$

It follows that only when

$$b > 0,$$

an eigenfunction possibly exist. In this case, if an eigenfunction $\phi_h(x)$ does exist, it must satisfy the following necessary condition:

$$\int_{\partial\Omega} \phi_h(x)ds = 0.$$

Next, we want to find value of b such that the space $N_0(B)$ is not empty. This leads to a different eigenvalue problem on the boundary.

Consider the following boundary eigenvalue problem: Find $b > 0$ and $\psi(x)$ such that

$$\Delta\psi = 0, \qquad x \in \Omega, \tag{6.2.9}$$

$$\nabla_\nu \psi = b\psi, \qquad x \in \partial\Omega. \tag{6.2.10}$$

From Chapter 3 we know that there exist only one eigenvalue and one eigenfunction when the space dimension is equal to 1. If the space dimension is greater than 1, then there exists a set of eigenvalues on the boundary, denoted by,

$$\Sigma_b := \{0 = b_0 < b_1 < b_2 < \cdots < b_n, \cdots\},$$

where

$$\lim_{n\to\infty} b_n = \infty.$$

Similar to the eigenvalue problem (6.2.7)–(6.2.8) in a bounded domain, the first positive eigenvalue for the boundary eigenvalue problem (6.2.9)–(6.2.10) can be found as follows:

$$b_1 = \min_{u(x)\in H^1(\Omega)} \frac{\int_\Omega |\nabla u|^2 dx}{\int_{\partial\Omega} u^2 ds}.$$

One can follow the same procedure to find $b_2, b_3, \cdots$. Let $N_m(B)$ be the subspace of the span of all eigenfunctions corresponding to the eigenvalue b_m.

By applying the Fredholm Alternative, we see that the problem I has a solution for every $f(x) \in L^2(\Omega)$ if $b \notin \Sigma_b$. On the other hand, if $b \in \Sigma_b$, say, $b = b_m$ for some $m \geq 0$, then the problem I has a solution for $f(x) \in L^2(\Omega)$ if and only if

$$< f, \phi_m >= 0, \qquad \forall \phi_m \in N_m(B).$$

A special case is that for $b_0 = 0$ and the boundary condition (6.2.4) is given by the homogeneous Neumann type:

$$\nabla_\nu u(x) = 0, \qquad x \in \partial\Omega.$$

In this case, every eigenfunction in $N_0(B)$ must be a constant and $dim\, N_0(B) = 1$.

Let

$$c_0 = \frac{1}{|\Omega|} \int_\Omega u(x)dx.$$

Then, Poincare's inequality yields:

$$\int_\Omega (\phi_h - c_0)^2 \leq s_0 \int_\Omega |\nabla \phi_h|^2 dx,$$

where s_0 is a positive constant (called the Sobolev constant). It follows that a necessary condition for b_n for $n \geq 1$ is

$$b_n \geq \frac{1}{s_0}, \qquad n = 1, 2, \cdots.$$

Since the set of all eigenfunctions is mutually orthogonal in $L^2(\Omega)$ and forms a basis for $L^2(\Omega)$, except for c_0, we follow the same procedure to obtain

$$u(x) = c_0 + \sum_{n=1}^{\infty} c_n \phi_n(x),$$

where

$$c_n = \frac{a_n}{\lambda_n}, \qquad n = 1, 2, \cdots.$$

To find c_0, we take an inner product $u(x)$ with $\phi_0 \in N_0$. Note that

$$< \phi_h, \phi_0 >= 0, \qquad n = 1, 2, \cdots.$$

We see that

$$< u, \phi_0 >= c_0|\Omega|.$$

It follows that

$$c_0 = \frac{< u, \pi_0 >}{|\Omega|} = \frac{1}{|\Omega|} \int_\Omega u(x)dx.$$

As long as we can find c_0, we can uniquely determine the solution.

Moreover, from the Fredholm Alternative, the problem (6.2.3)–(6.2.4) has a solution if and only if $f(x) \in N_0^\perp$

$$< f(x), \phi_h >= 0, \qquad \forall \phi_n \in N_n(B), n = 0, 1, 2, \cdots.$$

In particular, for the Neumann boundary condition, in order for the problem (I) to have a solution, a necessary and sufficient condition is

$$\int_\Omega f(x)dx = 0.$$

We summarize the above analysis to obtain the following theorem.

Theorem 6.2.1. *Suppose all eigenvalues and corresponding eigenfunctions of (6.2.7)–(6.2.8) are given by*

$$\{\lambda_n\}_{n=1}^\infty, \qquad \{\phi_n(x)\}_{n=1}^\infty.$$

Moreover, we choose $\{\phi_n(x)\}$ to be mutually orthogonal in $L^2(\Omega)$ and form a basis of $L^2(\Omega)$. Let $f(x) \in L^2(\Omega)$ and

$$f(x) = \sum_{n=1}^\infty a_n \phi_n(x).$$

(a) If $\lambda = 0$ is not an eigenvalue to (6.2.7)–(6.2.8), then the problem (I) has a unique series solution:

$$u(x) = \sum_{n=1}^\infty \frac{a_n}{\lambda_n} \phi_n(x).$$

(b) If $\lambda = 0$ is an eigenvalue to (6.2.7)–(6.2.8), there are two cases:

Case 1: $b \notin \Sigma_b$ with $b := -\frac{\beta}{\alpha}$.
Then, the problem (I) has a solution for all $f(x) \in L^2(\Omega)$.

Case 2: $b \in \Sigma_b$. Suppose $b = b_m \in \Sigma_b$ for some $m \geq 0$.
Then, the problem (I) has a solution for every $f(x) \in L^2(\Omega)$ if and only if

$$< f, \phi_m >= 0, \qquad \forall \phi_m(x) \in N_m(B),$$

where $N_m(B)$ is the eigenspace corresponding the boundary eigenvalue b_m.

In particular, when $\beta = 0$, the necessary and sufficient condition of the solvability for the problem (I) is

$$\int_\Omega f(x)dx = 0.$$

The series solution is given by

$$u(x) = c_0 + \sum_{n=1}^{\infty} \frac{a_n}{\lambda_n}\phi_n,$$

where

$$c_0 = \frac{1}{|\Omega|}\int_\Omega u(x)dx.$$

The solution is unique in the sense that the difference between any two solutions to (6.2.3)–(6.2.4) must be a constant.

Proof. The regularity theory for elliptic equations ([12]) shows that all eigenfunctions are smooth in Ω. With the smooth result, one can easily verify that

$$u(x) = \sum_{n=1}^{\infty} \frac{a_n}{\lambda_n}\phi_n(x)$$

solves the problem (6.2.3)–(6.2.4).

The proof for other cases is similar. □

For Problem (II) with a nonhomogeneous Robin boundary condition, we have a similar result. Let $g(x) \in L^2(\partial\Omega)$ with $n \geq 2$. Consider the following problem:

$$\Delta u = 0, \qquad x \in \Omega, \tag{6.2.11}$$

$$\nabla_\nu u = au + g(x), \qquad (x, y) \in \partial\Omega. \tag{6.2.12}$$

Theorem 6.2.2. *Suppose all eigenvalues and corresponding eigenfunctions of (6.2.9)–(6.2.10) are given by*

$$\Sigma_b := \{b_n\}_{n=1}^{\infty}, \qquad \{\psi_n(x)\}_{n=1}^{\infty}.$$

Moreover, we choose $\{\psi_n(x)\}$ to be mutually orthogonal in $L^2(\partial\Omega)$ and form a basis of $L^2(\partial\Omega)$. Let $g(x) \in L^2(\partial\Omega)$ and

$$g(x) = \sum_{n=1}^{\infty} a_n\psi_n(x), \qquad x \in \partial\Omega.$$

Then,

(a) if $a \notin \Sigma_b$, then the problem (6.2.11)–(6.2.12) has a unique solution for any $g(x) \in L^2(\partial\Omega)$.

(b) if $a = b_m$ for some $m \geq 0$, then the problem (6.2.11)–(6.2.12) has a solution if and only if $g(x)$ is orthogonal to $\phi_h(x)$ for all $\psi_h \in N_m(B)$ under the inner product

$$< v_1, v_2 >:= \int_{\partial\Omega} v_1(x) v_2(x) ds(x),$$

where $N_m(B)$ is the subspace of $L^2(\partial\Omega)$ that is the span of all eigenfunctions corresponding to the eigenvalue b_m. In particular, when $a = 0$, the problem (II) has a solution for every $g(x) \in L^2(\partial\Omega)$ if and only if

$$\int_{\partial\Omega} g(x) ds = 0. \quad \square$$

6.2.2 Series solution for the Laplace equation in some special domains

When a domain has a special structure, we can find the series solution for the problem (6.2.1)–(6.2.2) explicitly. The basic steps are as follows:

Step 1: Find all eigenvalues and corresponding eigenfunctions for the eigenvalue problem (6.2.7)–(6.2.8) for Problem (I) (or the eigenvalue problem (6.2.9)–(6.2.10) for Problem (II)). Use the Gram–Schmidt process to choose all eigenfunctions corresponding to each eigenvalue to be mutually orthogonal.

Step 2: Determine whether or not $\lambda = 0$ is an eigenvalue for the corresponding eigenvalue problem. We use Theorem 6.2.1 or Theorem 6.2.2 to express u as a series of eigenfunctions and find these unknown coefficients.

Example 6.2.1. Let $R = [0, L] \times [0, H]$ and $g(x) \in L^2(0, L)$. Find the series solution for the following problem:

$$-\Delta u = 0, \qquad (x, y) \in R, \tag{6.2.13}$$

$$u(0, y) = u(L, y) = 0, \qquad 0 < y < H, \tag{6.2.14}$$

$$u_y(x, 0) = 0, u_y(x, H) = g(x), \qquad 0 < x < L. \tag{6.2.15}$$

Solution. First, we need to determine whether or not $\lambda = 0$ is an eigenvalue. Consider the eigenvalue problem:

$$-\Delta u = \lambda u, \qquad x \in R, \tag{6.2.16}$$

$$u(0, y) = u(L, y) = 0, \qquad 0 < y < H, \tag{6.2.17}$$

$$u_x(x, 0) = 0, u_x(x, H) = 0, \qquad 0 < x < L. \tag{6.2.18}$$

By using the separation of variables, we know from Example 3.4.1 in Chapter 3 that the eigenvalues are

$$\lambda_{nm} = (\frac{n\pi}{L})^2 + (\frac{m\pi}{H})^2, \qquad n = 1, 2, \cdots, m = 0, 1, 2, \cdots.$$

The corresponding eigenfunctions are

$$\phi_{nm}(x, y) = \sin(\frac{n\pi x}{L})\cos(\frac{m\pi y}{H}), \qquad n = 1, 2, \cdots, m = 0, 1, 2, \cdots.$$

It follows that $\lambda = 0$ is not an eigenvalue for the eigenvalue problem (6.2.4)–(6.2.5) and the problem (6.2.13)–(6.2.15) has a unique solution for any $g(x) \in L^2(0, L)$.

We may use the general method introduced above to transform the problem into the problem I with a homogeneous boundary condition. Here, we use a direct way to find the series solution.

Let

$$u(x, y) = \phi(x)h(y), \qquad (x, y) \in R.$$

The Laplace equation (6.2.1) becomes

$$\phi''(x)h(y) + \phi(x)h''(y) = 0, \qquad (x, y) \in R.$$

Equivalently, we have

$$\frac{\phi''(x)}{\phi(x)} = -\frac{h''(y)}{h(y)} = -\lambda,$$

where λ is an unknown constant.

It follows that

$$\phi''(x) + \lambda\phi(x) = 0, \qquad 0 < x < L,$$
$$\phi(0) = \phi(L) = 0.$$

Moreover,

$$h''(y) - \lambda h(y) = 0, \qquad 0 < y < H,$$
$$h'(0) = 0.$$

We already know that all eigenvalues and corresponding eigenfunctions for $\phi(x)$ are

$$\lambda_n = (\frac{n\pi}{L})^2, \qquad \phi_n(x) = \sin(\frac{n\pi x}{L}), n = 1, 2, \cdots.$$

On the other hand, once λ_n is found, the general solution for $h(y)$ is equal to

$$h_n(y) = Ae^{\frac{n\pi y}{L}} + Be^{-\frac{n\pi y}{L}},$$

where A and B are arbitrary constants.

The boundary condition $h'(0) = 0$ implies $A = B$. It follows that

$$h_n(y) = A\big[e^{\frac{n\pi}{L}} + e^{-\frac{n\pi}{L}}\big] = 2A\cosh(\frac{n\pi y}{L}).$$

Consequently, for all $n \geq 1$,

$$u_n(x, y) = \sin(\frac{n\pi x}{L}) \cosh(\frac{n\pi y}{L})$$

satisfies the Laplace equation and all boundary conditions except on $y = H$.

We set a series solution in the following form:

$$u(x, y) = \sum_{n=1}^{\infty} A_n \sin(\frac{n\pi x}{L}) \cosh(\frac{n\pi y}{L}),$$

where A_n will be chosen later.

Then,

$$u_y(x, H) = \sum_{n=1}^{\infty} \frac{n\pi A_n}{L} \sin(\frac{n\pi x}{L}) \sinh(\frac{n\pi H}{L}).$$

On the other hand, we can express $g(x)$ over $[0, L]$ as a sine series:

$$g(x) = \sum_{n=1}^{\infty} g_n \sin(\frac{n\pi x}{L}),$$

where

$$g_n = \frac{2}{L} \int_0^L g(x) \sin(\frac{n\pi x}{L}) dx, \qquad n = 1, 2, \cdots .$$

If we choose A_n such that

$$g_n = \frac{n\pi A_n}{L} \sinh(\frac{n\pi H}{L}), \qquad n = 1, 2, \cdots ,$$

then,

$$u_y(x, H) = g(x), \qquad 0 < x < L.$$

Hence, the series solution to the problem (6.2.13)–(6.2.15) is equal to

$$u(x, y) = \sum_{n=1}^{\infty} A_n \sin(\frac{n\pi x}{L}) \cosh(\frac{n\pi y}{L}),$$

where

$$A_n = \frac{g_n L}{n\pi \sinh(\frac{n\pi H}{L})}, \qquad n = 1, 2, \cdots .$$

One can easily extend the method to a rectangular domain in R^3 (see Exercise 9 in Exercises 6.8).

The next example deals with the Laplace equation in a disk in R^2. Let $a > 0$ and

$$B_a(0) = \{(x, y) : x^2 + y^2 < a\} = \{(r, \theta) : 0 < r < a, -\pi < \theta \leq \pi\}.$$

For this case it is convenient to express the Laplace equation in polar coordinates. Note that

$$x = r\cos\theta, \qquad y = r\sin\theta,$$

$$\begin{aligned}\nabla &=< \frac{\partial}{\partial x}, \frac{\partial}{\partial y} > \\ &=< \cos\theta\frac{\partial}{\partial r} - r\sin\theta\frac{\partial}{\partial\theta}, \sin\theta\frac{\partial}{\partial r} + r\cos\theta\frac{\partial}{\partial\theta} > .\end{aligned}$$

It follows that $v(r, \theta) := u(x, y)$ satisfies

$$\Delta u = \frac{1}{r}\frac{\partial}{\partial r}[r\frac{\partial v}{\partial r}] + \frac{1}{r^2}\frac{\partial^2 v}{\partial\theta^2} = 0.$$

Since $u(x, y) = v(r, \theta)$ is smooth in $B_a(0)$, we see that

$$v(r, -\pi) = v(r, \pi), \qquad v_\theta(r, -\pi) = v_\theta(r, \pi).$$

Moreover, $v(r, \theta)$ must be bounded in $B_a(0)$. This leads to the following problem.

Example 6.2.2. Let $f(\theta) \in L^2(-\pi, \pi)$. Find the series solution for the following problem:

$$\frac{1}{r}\frac{\partial}{\partial r}[r\frac{\partial u}{\partial r}] + \frac{1}{r^2}\frac{\partial^2 u}{\partial\theta^2} = 0, \qquad (r, \theta) \in B_a(0), \tag{6.2.19}$$

$$u(r, -\pi) = u(r, \pi),\ u_\theta(r, -\pi) = u_\theta(r, \pi), \qquad 0 < r < a, \tag{6.2.20}$$

$$u(a, \theta) = f(\theta), \qquad -\pi \leq \theta \leq \pi, \tag{6.2.21}$$

$$|u(0, \theta)| < \infty.$$

Solution. We follow the same idea by using the method of separating variables. Let

$$u(r, \theta) = \phi(\theta)G(r), \qquad (r, \theta) \in B_a(0).$$

The Laplace equation for u is equivalent to

$$\frac{r}{G(r)}\frac{d}{dr}\left(r\frac{dG}{dr}\right) = -\frac{\phi''(\theta)}{\phi(\theta)} = \lambda, \qquad (r, \theta) \in B_a(0),$$

where λ is an unknown constant.

Equivalently, $\phi(\theta)$ satisfies

$$-\phi''(\theta) = \lambda\phi(\theta), \qquad -\pi < \theta < \pi, \tag{6.2.22}$$

$$\phi(-\pi) = \phi(\pi), \phi'(-\pi) = \phi'(\pi). \tag{6.2.23}$$

Moreover, $G(r)$ satisfies

$$\frac{d}{dr}\left(r\frac{dG}{dr}\right) - \frac{\lambda}{r}G(r) = 0, \qquad 0 < r < a,$$
$$|G(0)| < \infty.$$

We already know the eigenvalues and corresponding eigenfunctions to (6.2.22)–(6.2.23):

$$\lambda_0 = 0, \phi_0(\theta) = 1$$

and

$$\lambda_n = n^2, \ \phi_n(\theta) = \cos(n\theta), \sin(n\theta), \ n = 1, 2, \cdots .$$

Now, $G(r)$ satisfies

$$r^2 G''(r) + rG'(r) - n^2 G = 0, \qquad 0 < r < a.$$

To find the general solution for $G(r)$, we set

$$G(r) = r^p, \qquad 0 < r < a.$$

Then, we have

$$[p(p-1) + (p - n^2)]r^p = 0,$$

which implies that $p = \pm n$, $n = 1, 2, \cdots,$.

Hence, the general solution is

$$G(r) = c_1 r^n + c_2 r^{-n}, \qquad n = 1, 2, \cdots .$$

For $n = 0$, we obtain $G(r) = 1$ is one solution. To find another solution that is linearly independent of 1, we note that

$$\frac{d}{dr}\left(r\frac{dG(r)}{dr}\right) = 0, 0 < r < a.$$

It follows that

$$r\frac{dG(r)}{dr} = c_1, 0 < r < a.$$

Thus

$$G(r) = c_1 lnr + c_2, \qquad 0 < r < a,$$

where c_1 and c_2 are arbitrary constants.

Note that

$$|u(0,\theta)| < \infty.$$

It follows that all linearly independent solutions for $G(r)$ with $|G(0)| < \infty$ must have the following form:

$$G_n(r) = r^n, \qquad n = 0, 1, 2, \cdots.$$

Namely,

$$u_n(r,\theta) = A_n r^n \cos(n\theta) + B_n r^n \sin(n\theta), \qquad n = 0, 1, 2, \cdots.$$

We set the series solution in the following form:

$$u(r,\theta) = A_0 + \sum_{n=1}^{\infty} r^n \left[A_n \cos(n\theta) + B_n \sin(n\theta)\right].$$

To satisfy the boundary condition $u(a,\theta) = f(\theta)$, we only need to choose

$$\begin{aligned}
A_0 &= \frac{1}{2\pi}\int_{-\pi}^{\pi} f(x)dx, \\
A_n &= \frac{a^n}{\pi}\int_{-\pi}^{\pi} f(\theta)\cos(n\theta)d\theta, \qquad n = 1, 2, \cdots; \\
B_n &= \frac{a^n}{\pi}\int_{-\pi}^{\pi} f(\theta)\sin(n\theta)d\theta, \qquad n = 1, 2, \cdots.
\end{aligned}$$

6.3 The well-posedness of the Laplace equation

In this section we study the well-posedness of a self-adjoint elliptic operator L defined in Chapter 3. The basic tool is based on the Lax–Milgram theorem in Chapter 2. The result provides the theoretical foundation that justifies the results obtained in Section 6.2.

6.3.1 The well-posedness

Let Ω be a bounded domain in R^n with Lipschitz boundary $\partial\Omega$. We first focus on the homogeneous Dirichlet boundary condition. The well-posedness for a nonhomogeneous case and other boundary conditions are parallel to the heat equation in Section 4.3 in Chapter 4. The existence theory is based on the Lax–Milgram theorem.

H(3.1) Let $p(x), q(x) \in L^\infty(\Omega)$ with $p(x) \geq p_0 > 0$ and $q(x) \geq 0$ on Ω.

For $f(x) \in L^2(\Omega)$, we consider the following problem:

$$L[u] := -\nabla[p(x)\nabla u] + q(x)u = f(x), \qquad x \in \Omega, \tag{6.3.1}$$

$$u(x) = 0, \qquad x \in \partial\Omega. \tag{6.3.2}$$

Definition 6.3.1. We call $u(x)$ a weak solution of the problem (6.3.1)–(6.3.2) if $u(x) \in H_0^1(\Omega)$ and

$$\int_\Omega [p(x)\nabla u \cdot \nabla v + q(x)uv]\,dx = \int_\Omega f(x)v(x)dx, \tag{6.3.3}$$

for all $v(x) \in H_0^1(\Omega)$.

Clearly, we have the following proposition.

Proposition 6.3.1. *A classical solution of (6.3.1)–(6.3.2) must be a weak solution. Conversely, if a weak solution is smooth, then it is a classical solution of (6.3.1)–(6.3.2).*

Theorem 6.3.1. *Under the condition H(3.1) for every $f(x) \in L^2(\Omega)$ the problem (6.3.1)–(6.3.2) has a unique solution $u(x) \in H_0^1(\Omega)$. Moreover, the solution is smooth in Ω if $p(x)$ and $q(x)$ are smooth in Ω.*

Proof. Define a bilinear form on $H_0^1(\Omega) \times H_0^1(\Omega)$:

$$B[u, v] := \int_\Omega [p(x)\nabla u \cdot \nabla v + q(x)uv]\,dx.$$

By H(3.1) and Cauchy–Schwarz's inequality, we see that

$$|B[u, v]| \le C||u||_{H_0^1(\Omega)}||v||_{H_0^1(\Omega)}, \qquad \forall (u, v) \in H_0^1(\Omega) \times H_0^1(\Omega),$$

where C depends only on the upper bounds of $p(x)$ and $q(x)$.

Moreover, since $q(x) \ge 0$ and the norm of $H_0^1(\Omega)$ is equivalent to

$$||\nabla u||_{L^2(\Omega)},$$

we find that

$$B[u, u] \ge p_0||u||_{H_0^1(\Omega)}, \qquad \forall u \in H_0^1(\Omega).$$

The Lax–Milgram theorem yields that for every $f(x) \in L^2(\Omega)$ there exists a unique $u(x) \in H_0^1(\Omega)$ such that

$$B[u, v] = < f, v >, \qquad \forall v(x) \in H_0^1(\Omega).$$

The uniqueness is obvious since for any two weak solutions u_1 and u_2, $u = u_1 - u_2$ satisfies

$$B[u, v] = 0, \forall v \in H_0^1(\Omega).$$

We choose $v = u$ to see that $u(x) = 0$ on Ω.

The regularity of the weak solution can be proved in an advanced PDE book, such as [12]. □

Note that the nonnegativtity of $q(x)$ is important when employing the Lax–Milgram theorem. Next, we investigate the solvability for the operator L subject to the general Robin condition.

Let $g(x) \in L^2(\partial\Omega)$. Consider the following problem:

$$L[u] := -\nabla[p(x)\nabla u] + q(x)u = 0, \qquad x \in \Omega, \tag{6.3.4}$$

$$B[u] := p(x)\nabla_\nu u + c(x)u(x) = g(x), \qquad x \in \partial\Omega. \tag{6.3.5}$$

A weak solution to the problem (6.3.4)–(6.3.5) can be defined similarly.

Definition 6.3.2. We call $u(x)$ to be a weak solution of the problem (6.3.4)–(6.3.5) if $u(x) \in H^1(\Omega)$ and

$$\int_\Omega [p(x)\nabla u \cdot \nabla v + q(x)uv]\,dx + \int_{\partial\Omega} c(x)u(x)v(x)ds = \int_{\partial\Omega} g(x)v(x)ds, \tag{6.3.6}$$

for all $v(x) \in H^1(\Omega)$.

Theorem 6.3.2. *Let the assumption H(3.1) be in force and $c(x) \geq c_0 > 0$ over $\partial\Omega$ for some $c_0 > 0$. Then, for every $g(x) \in L^2(\partial\Omega)$ the problem (6.3.4)–(6.3.5) has a unique solution $u(x) \in H^1(\Omega)$.*

Proof. For the problem (6.3.4)–(6.3.5) we define the bilinear form:

$$B_1[u, v] = \int_\Omega [(\nabla u) \cdot (\nabla v)]\,dx + \int_{\partial\Omega} u(x)v(x)ds, \forall u, v \in H^1(\Omega).$$

Note that the norm of $H^1(\Omega)$ is equivalent to (see [23])

$$||\nabla u||_{L^2(\Omega)} + ||u||_{L^2(\partial\Omega)}.$$

Since $c(x) \geq c_0 > 0$, one can easily verify that the bilinear form $B_1[u, v]$ satisfies the Lax–Milgram conditions. It follows that the problem (6.3.4)–(6.3.5) has a unique solution $u(x)$ for every $g(x) \in L^2(\partial\Omega)$. □

Remark 6.3.1. For the existence result in Theorem 6.3.2, the positivity condition for $c(x)$ can be dropped. However, one needs a condition for the sign of $c(x)$ for the uniqueness.

By using the energy method, we can obtain the continuous dependence on $f(x)$ in $L^2(\Omega)$-space.

Corollary 6.3.1. *Let H(3.1) hold.*

(a) If $u_i(x)$ is the solution of (6.3.1)–(6.3.2) corresponding to $f_i(x)$ for $i = 1, 2$, then,

$$||\nabla(u_1 - u_2)||_{L^2(\Omega)} \leq C||f_1 - f_2||_{L^2(\Omega)};$$

(b) Let $c(x) \geq c_0 > 0$. If $u_i(x)$ is a solution of (6.3.4)–(6.3.5) corresponding to $g_i(x)$ for $i = 1, 2$, then

$$||\nabla(u_1 - u_2)||_{L^2(\Omega)} \leq C||g_1 - g_2||_{L^2(\partial\Omega)},$$

where C depends only on the known data.

Proof. We give a proof for (a) with the Dirichlet boundary condition. Let $u(x) := u_1(x) - u_2(x)$. Then,

$$\int_\Omega [p(x)|\nabla u|^2 + q(x)u^2]dx = \int_\Omega (f_1 - f_2)udx.$$

Since $p(x) \geq p_0 > 0$ and $q(x) \geq 0$, we have

$$\begin{aligned} p_0 \int_\Omega |\nabla u|^2 dx &\leq \varepsilon \int_\Omega u^2 dx + C(\varepsilon) \int_\Omega (f_1 - f_2)^2 dx \\ &\leq \varepsilon s_0 \int_\Omega |\nabla u|^2 dx + C(\varepsilon) \int_\Omega (f_1 - f_2)^2 dx. \end{aligned}$$

If we choose ε sufficiently small, then the desired estimate holds in $L^2(\Omega)$. □

6.4 A qualitative property: The mean-value formula

There are many interesting properties for a harmonic function. We introduce one of them here.

Let Ω be a domain in R^n and let $u(x)$ satisfy

$$\Delta u = 0, \qquad x \in \Omega.$$

6.4.1 The mean-value formula and subharmonic functions

Theorem 6.4.1. *Let $u(x)$ be a harmonic function in Ω. Then, for any ball $B_r(x) \subset \Omega$,*

$$u(x) = \fint_{\partial B_r(x)} u(y)ds(y) = \fint_{B_r(x)} u(y)dy.$$

Proof. Let $x \in \Omega$ be fixed. Define

$$\phi(r) = \fint_{\partial B_r(x)} u(y)ds(y) = \fint_{B_1(0)} u(x + rz)ds(z).$$

It follows that

$$\phi'(r) = \fint_{\partial B_1(0)} \nabla u(x + rz) \cdot zds(z).$$

Note that the outward unit normal on $\partial B_r(0)$ is equal to $\frac{x}{r}$.

Hence,

$$\begin{aligned}\phi'(r) &= \fint_{\partial B_r(x)} \nabla u(y) \cdot \frac{y-x}{r} ds(y) \\ &= \fint_{\partial B_r(x)} \nabla_\nu u(y) ds(y) \\ &= \frac{r}{n} \fint_{B_r(x)} \Delta u dy = 0.\end{aligned}$$

Here at the final step, the divergence theorem is used.

Therefore $\phi(r)$ must be a constant in Ω.

On the other hand,

$$\lim_{r \to 0} \phi(r) = \fint_{\partial B_r(x)} u(y) ds(y) = u(x).$$

It follows that

$$u(x) = \fint_{\partial B_r(x)} u(y) ds(y).$$

Note that

$$\int\int_{B_r(x)} u(y) dy = \int_0^r \int_{\partial B_t(x)} u(y) ds(y) dt = u(x) \int_0^r n\alpha(n) t^{n-1} dt = \alpha(n) r^n u(x).$$

This concludes the proof of Theorem 6.4.1. □

The significance of Theorem 6.4.1 shows that the value of a harmonic function at the center of a ball in a region depends only on the value on the surface of the ball. Conversely, one can show that the function must be a harmonic function if this property holds in any ball in the region.

Definition 6.4.1. A function $u(x)$ is called a subharmonic function on Ω if

$$-\Delta u \leq 0, \qquad x \in \Omega.$$

From the proof of the mean-value formula, we obtain a direct consequence.

Corollary 6.4.1. *If $u(x)$ satisfies*

$$\Delta u \geq 0 \qquad (or \leq 0), \qquad x \in \Omega,$$

then, for any $B_r(x) \subset \Omega$,

$$u(x) \leq \fint_{\partial B_r(x)} u(y) ds(y) = \fint_{B_r(x)} u(y) dy, \qquad x \in \Omega. \quad \square$$

6.5 The maximum principle and applications

Similar to the heat equation, the maximum principle holds for the Laplace equation. This property is fundamentally different from the hyperbolic equation. It is a very powerful tool in the study of general elliptic equations.

6.5.1 The maximum principle

The idea of the maximum principle comes from the calculus. When a smooth function has a local maximum at an interior point $x = x_0$, then the Hessian matrix of the function at $x = x_0$ is seminegative-definite. This gives a way to estimate the maximum and minimum values of the function over the domain.

Theorem 6.5.1. *(The weak maximum principle) Let $c(x) \geq 0$ on Ω. Suppose $u(x) \in C^2(\Omega) \bigcap C(\bar{\Omega})$ satisfies*

$$L[u] := -\Delta u + c(x)u \leq 0, \qquad x \in \Omega.$$

Then,

$$u(x) \leq max\{0, \max_{\partial\Omega} u(x)\}, \qquad x \in \Omega.$$

Proof. We first assume that $c(x) \geq c_0 > 0$ on Ω. If $u(x)$ attains a positive maximum at an interior point $x_0 \in \Omega$, then at this point,

$$-\Delta u \geq 0, c(x_0)u(x_0) > 0.$$

Hence,

$$L[u]|_{x=x_0} > 0,$$

a contradiction with the assumption. It follows that

$$u(x) \leq max\{0, \max_{\partial\Omega} u(x)\}.$$

For a general case, we just use $c_\varepsilon(x) := c(x) + \varepsilon$ to replace $c(x)$. Then, we obtain

$$u_\varepsilon(x) \leq max\{0, \max_{\partial\Omega} u(x)\}.$$

By taking the limit as $\varepsilon \to 0$, we conclude the desired estimate. □

A direct consequence of Theorem 6.5.1 is the following a priori estimate.

Corollary 6.5.1. *(The weak maximum principle) Let $c(x) \geq 0$ on Ω. Suppose $u(x) \in C^2(\Omega) \bigcap C(\bar{\Omega})$ satisfies*

$$L[u] := -\Delta u + c(x)u = 0, \qquad x \in \Omega.$$

Then,

$$min\{0, \min_{\partial\Omega} u(x)\} \leq u(x) \leq max\{0, \max_{\partial\Omega} u(x)\}, \qquad x \in \Omega. \quad \square$$

A strong version of the maximum principle is the following theorem.

Theorem 6.5.2. *(The strong maximum principle) Let Ω be a bounded and connected domain in R^n with C^2-boundary. Suppose $u(x) \in C^2(\Omega) \bigcap C(\bar{\Omega})$ satisfies*

$$-\Delta u \leq 0 \qquad (or \geq 0), \qquad x \in \Omega.$$

Then, $u(x)$ cannot attain its maximum (or minimum) in the interior of Ω unless it is a constant in Ω. Thus

$$\max_{\bar{\Omega}} u(x) = \max_{\partial\Omega} u(x), \qquad x \in \bar{\Omega}.$$

In particular, if u is a harmonic function in Ω, then

$$\min_{\bar{\Omega}} u(x) = \min_{\partial\Omega} u(x), \qquad \max_{\bar{\Omega}} u(x) = \max_{\partial\Omega} u(x).$$

Proof. Let $M = \max_{\bar{\Omega}} u(x)$. Suppose $u(x)$ attains its maximum at a point $x_0 \in \Omega$. Define

$$\Omega_M = \{x \in \Omega : u(x) = M\}.$$

Clearly, Ω_M is closed and nonempty. On the other hand, for any $x_0 \in \Omega_M$ such that $B_r(x_0) \subset \Omega$, then, we have

$$0 = u(x_0) - M \leq \frac{1}{\alpha(n)r^n} \int_{B_r(x_0)} (u - M)dx \leq 0,$$

which yields $u(x) = M$ in $B_r(x_0)$. Consequently, Ω_M is open and also relatively closed in Ω. Hence, $\Omega_M = \Omega$. $\square$

In dealing with problems with a Neumann boundary condition, we may need a different type of maximum principle.

Theorem 6.5.3. *(Hopf's lemma) Let Ω be a bounded domain with C^2-boundary and let $u(x)$ satisfy*

$$-\Delta u \geq 0, \qquad x \in \Omega.$$

If $u(x)$ attains a maximum M at a boundary point $x_0 \in \partial\Omega$ and $u(x) < M$ in a neighborhood $N(x_0) \bigcap \Omega$. Then,

$$\frac{\partial u}{\partial \nu}\Big|_{x=x_0} > 0,$$

where ν is the outward unit normal at $x = x_0$.

Proof. First, we note that for any C^2-function $w(x)$, if $w(x)$ attains a maximum at $x_0 \in \partial\Omega$ in a neighborhood of x_0, then

$$\frac{\partial w(x)}{\partial \nu}|_{x=x_0} \geq 0,$$

where ν is the outward unit normal at x_0.

Since $\partial\Omega \in C^2$, there exists a small ball $B_r(y_0) \subset \Omega$ and $x_0 \in \partial B_r(y_0)$ such that

$$u(x) < M, \qquad x \in B_r(y_0).$$

For simplicity, we assume $y_0 = 0$. Define an auxiliary function

$$v(x) = e^{-\lambda|x|^2} - e^{-\lambda r^2},$$

where λ will be chosen later. Then,

$$-\Delta v = e^{\ \lambda|x|^2}[-4\lambda^2|x|^2 + 2n\lambda].$$

In $R := B_r(0) - B_{r/2}(0)$, if we choose λ sufficiently large, then

$$-\Delta v \leq 0, \qquad x \in R.$$

Note that

$$u(x_0) > u(x), \qquad x \in N(x_0) \bigcap \Omega.$$

We can choose ε sufficiently small to obtain

$$u(x_0) \geq u(x) + \varepsilon v(x), \qquad x \in \partial B_{r/2}(0).$$

Since $v(x) = 0$ on $\partial B_r(x_0)$, we see that

$$u(x_0) \geq u(x) + \varepsilon v(x), \qquad x \in \partial R.$$

Let $w(x) := u(x) + \varepsilon v(x) - u(x_0)$. Then,

$$-\Delta w \geq 0, \qquad x \in R$$

and $w \leq 0$ on ∂R. The maximum principle implies

$$u(x) + \varepsilon v(x) - u(x_0) \leq 0, \qquad x \in R.$$

Note that

$$[u(x) + \varepsilon v(x) - u(x_0)]\big|_{x=x_0} = 0.$$

It follows that

$$\frac{\partial w(x)}{\partial \nu}\Big|_{x=x_0} \geq 0.$$

Consequently,

$$\frac{\partial u(x)}{\partial \nu}|_{x=x_0} \geq -\varepsilon \frac{\partial v(x)}{\partial \nu}|_{x=x_0} = 2\lambda \varepsilon r e^{-\lambda r^2} > 0. \qquad \square$$

6.5.2 Applications

The maximum principle can be used in many applications. We give an application in theoretical analysis as an example.

Theorem 6.5.4. *Let Ω be a bounded domain in R^n with C^2-boundary and $c(x) \geq 0$ and $f(x), g(x) \in C(\bar{\Omega})$. Suppose that $u(x)$ is a solution of the following equation:*

$$L[u] := -\Delta u + c(x)u = f(x), \qquad x \in \Omega, \tag{6.5.1}$$

$$u(x) = g(x), \qquad x \in \partial\Omega. \tag{6.5.2}$$

Then,

$$||u||_0 \leq C[||f||_0 + ||g|_0],$$

where C is a constant that depends only on Ω.

Proof. Suppose Ω lies in the slab $0 < |\Omega| < d_0$. Define an auxiliary function

$$v(x) := \max_{\partial\Omega} |g(x)| + (e^{d_0} - e^{x_1})||f||_0, \qquad x \in \Omega.$$

Then,

$$-\Delta(v-u) \geq 0, \qquad x \in \Omega.$$

On $\partial\Omega$,

$$v(x) - u(x) \geq 0.$$

The maximum principle yields

$$\max_{\Omega} u(x) \leq \max_{\Omega} v(x). \qquad \square$$

The method used in the proof of Theorem 6.5.4 is quite powerful. As a direct consequence, we immediately obtain the uniqueness and continuous dependence for the solution of the problem (6.5.1)–(6.5.2). The crucial step is to construct a suitable comparison function. Another direct consequence is the following comparison principle. We will show more applications in dealing with nonlinear equations in Chapter 8.

Corollary 6.5.2. *(Comparison principle) Let $c(x) \geq 0$ on Ω. Let u_i be the solution of (6.5.1)–(6.5.2) corresponding to $f_i(x)$ and $g_i(x), i = 1, 2$. If*

$$f_i(x) \leq f_2(x), \qquad g_i(x) \leq g_2(x), \qquad \forall x \in \Omega,$$

then,

$$u_1(x) \le u_2(x), \qquad \forall x \in \Omega.$$

Example 6.5.1. Let Ω be a bounded domain in R^n with C^2-boundary. Find all solutions for the following nonlinear problem:

$$-\Delta u = u(1-u), \qquad x \in \Omega,$$
$$\nabla_\nu u = 0, \qquad x \in \partial\Omega.$$

Proof. Obviously, $u_1(x) = 0$ and $u_2(x) = 1$ are two solutions. We claim that there is no other solution. Indeed, the maximum principle shows that any solution must satisfy

$$0 \le u(x) \le 1, \qquad x \in \Omega.$$

On the other hand, from the equation we find that

$$\int_\Omega u(1-u)dx = 0.$$

It follows that there is no solution other than $u(x) = 0$ or $u(x) = 1$. □

6.6 An $L^\infty(\Omega)$-estimate: Moser's iteration method

In this section we present an iteration method to derive an $L^\infty(\Omega)$-estimate without using the maximum principle under a weaker assumption on $f(x)$. The method can be used to derive the DiGorgi–Nash's estimate for a general elliptic equation. We begin with an elementary iteration lemma.

Lemma 6.6.1. *Let $g(t)$ be a nonnegative and monotone decreasing function on $[k_0, \infty)$. Moreover, for some positive constants $M > 0$ and $\alpha > 0$, $\beta > 1$, $g(t)$ satisfies*

$$g(h) \le \left(\frac{M}{(h-k)^\alpha}\right) g(k)^\beta, \ \forall h > k \ge k_0. \tag{6.6.1}$$

Then,

$$g(d + d_0) = 0,$$

where

$$d = M^{\frac{1}{\alpha}} g(k_0)^{\frac{\beta-1}{\alpha}} 2^{\frac{\beta}{\beta-1}}.$$

Proof. Define a sequence

$$k_n = k_0 + d - \frac{d}{2^n}, \qquad n = 0, 1, 2, \cdots.$$

Then, we have

$$g(k_{n+1}) \le \frac{M2^{(n+1)\alpha}}{d^\alpha} g(k_n)^\beta, \qquad n = 0, 1, 2, \cdots. \tag{6.6.2}$$

We claim that there exists a number $r > 1$ such that

$$g(k_n) \le \frac{g(k_0)}{r^n}, \qquad n = 0, 1, \cdots.$$

Indeed, the iteration inequality (6.6.2) yields

$$\begin{aligned} g(k_{n+1}) &\le \frac{M2^{(n+1)\alpha}}{d^\alpha}\left[\frac{g(k_0)}{r^n}\right]^\beta \\ &= \frac{g(k_0)}{r^{n+1}} \frac{M2^{(n+1)\alpha}}{d^\alpha r^{s(\beta-1)-1}} g(k_0)^{\beta-1}. \end{aligned}$$

If we choose

$$r = 2^{\frac{\alpha}{\beta-1}} > 1,$$

and note the choice of d, we see that

$$g(k_n) \le \frac{g(d_0)}{r^n}, \qquad n = 0, 1, 2, \cdots.$$

From the definition of k_n, we find $k_n \to k_0 + d$ and $g(k_0 + d) \to 0$ as $n \to \infty$. ☐

Define

$$u^+(x) = \max\{u(x), 0\}, \qquad u^- = \min\{u(x), 0\}.$$

Theorem 6.6.1. *Let $q(x) \ge 0$ in Ω and $f(x) \in L^p(\Omega)$ with $p > n$. Then, a solution $u(x)$ of (6.3.1) satisfies the following estimate:*

$$||u||_{L^\infty(\Omega)} \le \sup_{\partial\Omega} u^+ + C||f||_{L^{\frac{np}{n+p}}(\Omega)},$$

where C depends only on p_0 and $|\Omega|$.

Proof. Let $n > 2$ (the case for $n = 1, 2$ is trivial). Let $k_0 = \sup_{\partial\Omega} u(x)$. For $k \ge k_0$, define

$$A(k) = \{x \in \Omega : u(x) \ge k\}, \qquad g(k) = |A(k)|.$$

Clearly, $g(k) \ge 0$ and $g(k)$ is monotone decreasing on $[k_0, \infty)$. Set

$$\phi(x) := (u - k)^+, \qquad x \in \Omega.$$

We multiply Eq. (6.3.1) by $\phi(x)$ and integrate over Ω to obtain:

$$p_0 \int_\Omega |\nabla\phi|^2 dx \le \int_\Omega |f\phi| dx,$$

where we have used $p(x) \geq p_0$ and $q(x) \geq 0$ on Ω.

The Hölder inequality with $p_1 = p^* := \frac{np}{n+p}$, $p_2 = 2^* := \frac{2n}{n-2}$, $p_3 = \frac{2p}{p-2}$ implies

$$\int_\Omega |f\phi(x)|dx \leq ||f||_{p^*}||\phi||_{2^*}g(k)^{\frac{1}{2}-\frac{1}{p}}.$$

We then obtain

$$||\nabla\phi||^2_{L^2(\Omega)} \leq CF_0^2 g(k)^{1-\frac{2}{p}},$$

where

$$F_0 = ||f||_{p^*}.$$

Recall Sobolev's embedding:

$$||\phi||_{2^*} \leq c_0||\nabla\phi||_2.$$

We find that

$$||\phi||_{2^*} \leq CF_0 g(k)^{\frac{1}{2}-\frac{1}{p}}.$$

On the other hand, if $h > k$ we have

$$||\phi||_{2^*} \geq (h-k)g(h)^{\frac{1}{2^*}}.$$

It follows that

$$g(h) \leq \frac{(CF_0)^{2^*}}{(h-k)^{2^*}} g(k)^{\frac{n(p-2)}{p(n-2)}}.$$

By using Lemma 6.6.1, we obtain

$$g(k_0 + d) = 0,$$

where

$$d := CF_0 g(k_0)^{\frac{1}{n}-\frac{1}{p}} 2^{\frac{n(p-2)}{2(p-n)}}.$$

Consequently,

$$\sup_\Omega u \leq k_0 + \hat{C}F_0. \qquad \square$$

Remark 6.6.1. The crucial step in the proof of Theorem 6.6.1 is the Sobolev embedding. This is also an essential step for the DiGorgi–Nash–Moser's estimate (Hölder continuity for weak solution) for a general elliptic equation in divergence form.

6.7 Notes and remarks

In applied fields, the Laplace equation and its generalization are used to understand the long-time behavior of the solution to an evolution equation. This is particularly used for reaction–diffusion systems. There are several ways to establish the well-posedness associated with the Laplace equation. We used a method that can be used to deal with a general elliptic equation in divergence form. In Chapter 8 we will use a different way based on the fundamental solution and Green's function to obtain the solution representation. The maximum principle is a powerful tool in the study of nonlinear elliptic equations (see [12,20]).

The study for the solution of the Laplace equation (harmonic function) is a very active research topic in harmonic and complex analyses. Many problems in different geometries are equivalent to the solvability for nonlinear elliptic equations. The materials in this chapter and Chapter 8 will provide a good starting point to conduct further research in this direction.

For the beginners, Sections 6.1 to 6.4 are sufficient. The rest of the sections are for graduate students interested in research. The method used in Section 6.5 and Section 6.6 is elegant.

6.8 Exercises

1. Let a and b be known constants. Consider the following boundary value problem:

$$\begin{aligned} &u''(x)=0, \qquad 0<x<L,\\ &-u'(0)=au(0)+b, \quad u'(L)=au(L)+b. \end{aligned}$$

(a) Does the problem have a solution?
(b) Let $L=2$ and $a=1$, prove the problem has a solution for any b.

2. Let $R=\{(x,y):0<x<L, 0<y<H\}$ and $f(x,y)\in L^2(R)$. Find the series solution for the following problem:

$$\begin{aligned} &-\Delta u=f(x,y), && (x,y)\in R,\\ &u(0,y)=u(L,y)=0, && 0<y<H,\\ &u_y(x,0)=u_y(x,H)=0, && 0<x<L. \end{aligned}$$

3. Let $R=\{(x,y):0<x<L, 0<y<H\}$ and $g(x),h(x)\in L^2(0,L)$. Find the series solution for the following problem:

$$\begin{aligned} &-\Delta u=0, && (x,y)\in R,\\ &u(0,y)=u(L,y)=0, && 0<y<H,\\ &u_y(x,0)=g(x), \quad u_y(x,H)=h(x), && 0<x<L. \end{aligned}$$

4. Let $R = \{(x, y) : 0 < x < L, 0 < y < H\}$ and $g(x) \in L^2(0, L)$. Let c be a constant. Find the series solution for the following problem:

$$\begin{aligned} &-\Delta u + cu = 0, && (x, y) \in R, \\ &u(0, y) = u(L, y) = 0, && 0 < y < H, \\ &u(x, 0) = g(x), \quad u(x, H) = 0, && 0 < x < L. \end{aligned}$$

5. Let

$$D = \{(x, y) : 0 < x < L, y > 0\}.$$

Let $f(x), g(x) \in L^2(0, L)$. Find the series solution for the following problem:

$$\begin{aligned} &-\Delta u = 0, && (x, y) \in D, \\ &u(0, y) = u(L, y) = 0, && y \geq 0, \\ &u(x, 0) = f(x), u_y(x, 0) = g(x), && 0 < x < L. \end{aligned}$$

6. Let

$$D = \{(x, y) : 0 < x < L, y > 0\}.$$

Let $f(x) \in L^\infty(0, L)$. Find a bounded solution for the following problem:

$$\begin{aligned} &-\Delta u = 0, && (x, y) \in D, \\ &u(0, y) = u(L, y) = 0, && y \geq 0, \\ &u(x, 0) = f(x), && 0 < x < L, \\ &|u(x, y)| < \infty, && (x, y) \in D. \end{aligned}$$

7. Let $D = \{(x, y) : 0 < x < L, 0 < y < H\}$. Find all eigenvalues and corresponding eigenfunctions for the following eigenvalue problem:

$$\begin{aligned} &-\Delta u = \lambda u, && (x, y) \in D, \\ &u(0, y) = u(L, y) = 0, && 0 \leq y \leq H, \\ &u(x, 0) = u(x, H), u_y(x, 0) = u_y(x, H), && 0 < x < L. \end{aligned}$$

8. Let $a > 0$ and $D = \{(r, \theta) : 0 < r < a, -\pi < \theta < \pi\}$ be a disk. Let $f(r, \theta) \in L^2(D)$. Find the series solution for the following problem:

$$\begin{aligned} &\frac{1}{r}\frac{\partial}{\partial r}[r\frac{\partial u}{\partial r}] + \frac{1}{r^2}\frac{\partial^2 u}{\partial \theta^2} = f(r, \theta), && (r, \theta) \in D, \\ &u(r, -\pi) = u(r, \pi), u_\theta(r, -\pi) = u_\theta(r, \pi), 0 < r < a, \\ &u(a, \theta) = 0, |u(0, \theta)| < \infty, && -\pi < \theta < \pi. \end{aligned}$$

9. Let $D = [0, L] \times [0, H] \times [0, K]$ be a rectangle in R^3 and $g(x, y) \in L^2([0, L] \times [0, H])$. Find the series solution for the following problem:

$$\begin{aligned}
&-\Delta u = 0, && (x, y, z) \in D,\\
&u(0, y, z) = u(L, y, z) = 0, && 0 < y < H, 0 < z < K,\\
&u(x, 0, z) = u(x, H, z) = 0, && 0 < x < L, 0 < z < K,\\
&u(x, y, 0) = 0,\ u(x, y, K) = g(x, y), && 0 < x < L, 0 < y < H.
\end{aligned}$$

10. Let $f(x) \in L^2(\Omega)$ and $A = \{v(x) \in C^1(\bar{\Omega}) : v(x) = 0 \text{ on } \partial\Omega\}$. Let

$$E(u) = \min_{v \in A}\Big[\frac{1}{p}\int_\Omega |\nabla v|^p dx - \int_\Omega f(x)v(x)dx\Big],\ p > 1.$$

Prove $u(x)$ is a solution of the following equation:

$$\begin{aligned}
&-\nabla[|\nabla u|^{p-2}\nabla u] = f(x), && x \in \Omega,\\
&u(x) = 0, && x \in \partial\Omega.
\end{aligned}$$

(Hint: Let $I(t) = E(u + tv)$ and prove $I'(0) = 0$ and $I''(0) \geq 0$.)

11. Let $\Omega \subset R^n$ be a bounded region and $g(x) \in C(\bar{\Omega})$. Let $u(x)$ be a solution of the following problem:

$$\begin{aligned}
&-\Delta u + u^3 = 0, && x \in \Omega,\\
&u(x) = g(x), && x \in \partial\Omega.
\end{aligned}$$

Derive the maximum bound for $u(x)$.

12. Let Ω be a bounded domain with C^2-boundary. Moreover, $\partial\Omega := \bar{\Gamma}_1 \bigcup \bar{\Gamma}_2$, $\Gamma_1 \bigcap \Gamma_2 = \phi$(empty set). Suppose $f(x) \in C(\bar{\Gamma}_1)$, $g(x) \in C(\bar{\Gamma}_2)$ and $c(x) \geq 0$. Let $u(x)$ be a solution of the following problem:

$$\begin{aligned}
&-\Delta u + c(x)u = 0,\ x \in \Omega,\\
&u(x) = f(x), && x \in \Gamma_1,\\
&u_\nu(x) = g(x), && x \in \Gamma_2.
\end{aligned}$$

Prove the solution must be bounded and unique.

13. Let $f(s) \in C^1(R^1)$, $f'(s) \leq 0$ for $s \in R^1$ and $f(0) \geq 0$. Prove that the solution of the following problem:

$$\begin{aligned}
&-\Delta u = f(u), && x \in \Omega,\\
&u(x) = 0, && x \in \partial\Omega
\end{aligned}$$

is nonnegative. Is the conclusion still true if the boundary condition is replaced by

$$\nabla_\nu u(x) = 0, \qquad x \in \partial\Omega?$$

14. Let $c(x), f(x), g(x) \in C(\bar{\Omega})$. Consider the following problem

$$
\begin{aligned}
-\Delta u + c(x)u = f(x), \qquad & x \in \Omega, \\
u(x) = g(x), \qquad & x \in \partial\Omega.
\end{aligned}
$$

Suppose $||c||_0 < \lambda_1$. Prove the solution of the problem is bounded by

$$||g||_0 + \frac{||f||_0}{\lambda_1 - ||c||_0},$$

where λ_1 is the principal eigenvalue of the Laplace equation associated with a Dirichlet boundary condition on Ω.

15. Let Ω be a bounded domain in R^n and $g(x) \in C(\bar{\Omega})$ with $0 \le g(x) \le 1$ over Ω. Let $u(x)$ be a solution of the following semilinear equation:

$$
\begin{aligned}
-\Delta u = u(1-u), \qquad & x \in \Omega, \\
u(x) = g(x), \qquad & x \in \partial\Omega.
\end{aligned}
$$

Prove

$$0 \le u(x) \le 1, \qquad x \in \Omega.$$

CHAPTER

The Fourier transform and applications

7

7.1 Definition of the Fourier transform

We have seen in Chapter 2 that a function in a bounded interval $[-L, L]$ can be expressed by a Fourier series. Now, we want to extend the idea for a function defined in $(-\infty, \infty)$. In this case we have to introduce an integral transform instead of a series.

As the Fourier transform involves complex functions, in this section a function may be real or complex. For a complex number $z = x + iy \in C^n$, the complex conjugate is denoted by $\bar{z} = x - iy$ for all $x, y \in R^n$. The complex conjugate of a complex function $f(z)$ is denoted by $\bar{f}(z)$.

7.1.1 A hierarchy definition of the Fourier transform

The following Euler's identity is used frequently in this chapter:

$$e^{ia} = \cos a + i \sin a = \sum_{n=0}^{\infty} \frac{(ia)^n}{n!}.$$

Recall the Fourier series for a real function $f(x) \in L^2(-L, L)$,

$$f(x) = a_0 + \sum_{n=1}^{\infty} \left[a_n \cos(\frac{n\pi x}{L}) + b_n \sin(\frac{n\pi x}{L}) \right],$$

where

$$a_0 = \frac{1}{2L} \int_{-L}^{L} f(x) dx, \qquad a_n = \frac{1}{L} \int_{-L}^{L} f(x) \cos(\frac{n\pi x}{L}) dx,$$
$$b_n = \frac{1}{L} \int_{-L}^{L} f(x) \sin(\frac{n\pi x}{L}) dx, \qquad \forall n \geq 1.$$

Now,

$$\cos(a) = \frac{e^{ia} + e^{-ia}}{2}, \qquad \sin(a) = \frac{e^{ia} - e^{-ia}}{2i}.$$

Partial Differential Equations and Applications. https://doi.org/10.1016/B978-0-44-318705-6.00013-6

We can rewrite the Fourier series of $f(x)$ in the complex form

$$f(x)=\sum_{n=-\infty}^{\infty} c_n e^{\frac{in\pi x}{L}},$$

where

$$c_0=a_0,\ c_n=\frac{1}{2}(a_n-ib_n),\ c_{-n}=\frac{1}{2}(a_{-n}+ib_{-n}), \qquad n=1,2,\cdots.$$

Equivalently,

$$c_n=\frac{1}{2L}\int_{-L}^{L} f(x)e^{\frac{-in\pi x}{L}}dx,\ \forall n\in Z.$$

Define

$$\omega_n=\frac{n\pi}{L},\ \Delta\omega_n=\omega_n-\omega_{n-1}=\frac{\pi}{L}.$$

Then,

$$f(x)=\sum_{n=-\infty}^{\infty} c(\omega_n)\Delta\omega_n e^{i\omega_n x},$$

where

$$c(\omega_n)=\frac{1}{2\pi}\int_{-L}^{L} f(x)e^{-i\omega_n x}dx.$$

Let $L\to\infty$, formally we consider the series as the Riemann summation to obtain

$$f(x)=\int_{-\infty}^{\infty} c(\omega)e^{i\omega x}d\omega$$
$$c(\omega)=\frac{1}{2\pi}\int_{-\infty}^{\infty} f(x)e^{-i\omega x}dx.$$

It is worth noting that the improper integral is different from the calculus. It is defined as

$$\int_{-\infty}^{\infty} f(x)dx=\lim_{L\to\infty}\int_{-L}^{L} f(x)dx.$$

This improper integral is called the *Cauchy principal value* of the integral. In this section, all improper integrals are understood in this definition.

This leads to the following definition of the Fourier transform.

Definition 7.1.1. Let $f(x)\in L^1(R^n)$. Define an operator $\mathcal{F}$ from $L^1(R^n)$ to $L^\infty(R^n)$ by

$$\hat{f}(\omega) := \mathcal{F}[f] = (2\pi)^{-\frac{n}{2}} \int_{R^n} e^{-ix\cdot\omega} f(x)dx.$$

$\hat{f}(\omega)$ is called the Fourier transform of $f(x)$.

It is easy to see that the operator $\mathcal{F}$ is well defined, linear, and bounded. Moreover, we will see that $\mathcal{F}$ is one-to-one and the inverse Fourier transform $\mathcal{F}^{-1}$ has an explicit form (see Theorem 7.2.1 in the next section).

Definition 7.1.2. Let $f(\omega) \in L^1(R^n)$. The inverse Fourier transform $\mathcal{F}^{-1}$ is defined as

$$\check{f}(x) := \mathcal{F}^{-1}[f] = (2\pi)^{-\frac{n}{2}} \int_{R^n} e^{ix\cdot\omega} f(\omega)d\omega.$$

Thus

$$(\hat{f})^{\vee}(\omega) = f(x), \qquad \text{if } f(x), \hat{f}(\omega) \in L^1(R^n).$$

Suppose $\hat{f}(\omega) \in L^1(R^n)$, we see that

$$\mathcal{F}[f](-x) = \mathcal{F}^{-1}[f](x).$$

Let $\mathcal{R}$ be the reflection operator from $L^1(R^n)$ to $L^1(R^n)$:

$$\mathcal{R}[f](x) = f(-x).$$

Then, from the definition, we have the following proposition:

Proposition 7.1.1. *The operators $\mathcal{F}$ and $\mathcal{R}$ satisfy the following identities:*

$$\begin{aligned} \mathcal{F}^2 &= \mathcal{R}; \\ \mathcal{F}^3 &= \mathcal{R} \circ \mathcal{F} = \mathcal{F} \circ \mathcal{R}; \\ \mathcal{F}^4 &= \mathcal{I}, \end{aligned}$$

where $\mathcal{I}$ represents the identity operator and $\circ$ is the composition of the operators.

From Proposition 7.1.1, we see that $\mathcal{F}$ is one-to-one. Therefore the inverse Fourier transform exists.

7.1.2 Some examples of the Fourier transform

Example 7.1.1. Let $a > 0$ and

$$f(x) = e^{-a|x|}, \qquad x \in R^1.$$

Find $\hat{f}(\omega) = \mathcal{F}[f]$.

Solution. By the definition, we have

$$\hat{f}(\omega) = \frac{1}{\sqrt{2\pi}} \int_{-\infty}^{\infty} e^{-ix\omega} e^{-a|x|} dx$$

$$= \frac{1}{\sqrt{2\pi}}[\frac{1}{a-i\omega} + \frac{1}{a+i\omega}]$$
$$= \frac{2a}{\sqrt{2\pi}(a^2+\omega^2)}.$$

Example 7.1.2. Let $a > 0$ and

$$f(x) = \frac{a}{a^2+x^2}, \qquad x \in R^1.$$

Find $\hat{f}(\omega) = \mathcal{F}[f]$.

Solution. From the Fourier inversion formula, we see that

$$e^{-a|x|} = \frac{\sqrt{2\pi}}{2}\mathcal{F}^{-1}[f] = \frac{\sqrt{2\pi}}{2}\frac{1}{\sqrt{2\pi}}\int_{-\infty}^{\infty} e^{ix\omega}\frac{a}{(a^2+\omega^2)}d\omega.$$

It follows that

$$\mathcal{F}[f](\omega) = \frac{\sqrt{2\pi}}{2}e^{-a|\omega|}.$$

Example 7.1.3. Find the Fourier transform for the following functions.

(a) $f_0(x) = \frac{1}{\sqrt{2\pi}}e^{-\frac{x^2}{2}}$.

(b) Let $\sigma > 0$ and $m \in R^1$ be constants. Let

$$f_\sigma(x) = \frac{1}{\sqrt{2\pi\sigma}}e^{-\frac{(x-m)^2}{2\sigma}}, \qquad x \in R^1.$$

Solution. First, for any constants a and $b > 0$ we note that

$$-bx^2 - iax = -b[(x+\frac{ia}{2b})^2] - \frac{a^2}{4b}.$$

Introduce a new variable

$$s = x + \frac{ia}{2b},$$

$$\int_{-\infty}^{\infty} e^{-bx^2-iax}dx = e^{-\frac{a^2}{4b}}\int_{-\infty+\frac{ia}{2b}}^{\infty+\frac{ia}{2b}} e^{-bs^2}ds.$$

Since the function e^{-bs^2} is analytic and decays to 0 as $s \to \infty$, we can take for the contour integral in the complex plane along the rectangle

$$R = [-A, A] \times [0, \frac{ia}{2b}]$$

and let $A \to \infty$ to obtain

$$\int_{-\infty}^{\infty} e^{-bx^2-iax} dx = e^{-\frac{a^2}{4b}} \int_{-\infty}^{\infty} e^{-bs^2} ds.$$

It follows that

$$\int_{-\infty}^{\infty} e^{-bx^2-iax} dx = \left(\frac{\pi}{b}\right)^{1/2} e^{-\frac{a^2}{4b}}.$$

(a) For $a = \omega$ and $b = \frac{1}{2}$, we find that

$$\begin{aligned} \hat{f}_0(\omega) &= \frac{1}{2\pi} \int_{-\infty}^{\infty} e^{-ix\omega} e^{-\frac{x^2}{2}} dx \\ &= \frac{1}{\sqrt{2\pi}} e^{-\frac{\omega^2}{2}}. \end{aligned}$$

(b) Introduce a new variable

$$\xi = \frac{x-m}{\sqrt{\sigma}}.$$

Then, by a similar calculation we have

$$\begin{aligned} \hat{f}_\sigma(\omega) &= \frac{1}{\sqrt{2\pi}} \int_{-\infty}^{\infty} e^{-ix\omega} f_\sigma(x) dx \\ &= \frac{1}{2\pi} \int_{-\infty}^{\infty} e^{-\frac{\xi^2}{2}} e^{-i\omega(m+\sqrt{\sigma}\xi)} d\xi \\ &= \frac{1}{2\pi} e^{-i\omega m} \int_{-\infty}^{\infty} e^{-\frac{\xi^2}{2} - i\sqrt{\sigma}\xi\omega} d\xi \\ &= \frac{1}{\sqrt{2\pi}} e^{-i\omega m - \frac{\sigma\omega^2}{2}}, \end{aligned}$$

where at the final step we have used the identity $a = \sqrt{\sigma}\omega$ and $b = \frac{1}{2}$.

We can extend the above calculation to n-space dimension.

Example 7.1.4. Let $\sigma > 0$ and $m = (m_1, \cdots, m_n) \in R^n$ be a constant vector. Let

$$f(x) = \frac{1}{(2\pi\sigma)^{n/2}} e^{-\frac{|x-m|^2}{2\sigma}}, \qquad x \in R^n.$$

Find the Fourier transform $\hat{f}(y) = \mathcal{F}[f]$ of $f(x)$.

Solution. We use the same calculation as for the one-dimensional case. For brevity, we use

$$x - m = (x_1 - m_1, \cdots, x_n - m_n).$$

Introduce a new variable

$$\xi = \frac{x-m}{\sqrt{\sigma}}.$$

Note that

$$\int_{R^n} e^{-|x|^2}dx = \prod_{i=1}^{n}\int_{-\infty}^{\infty} e^{-x_i^2}dx = (\sqrt{\pi})^n.$$

We repeat the calculation for one dimension to obtain:

$$\begin{aligned}\hat{f}(\omega) &= \frac{1}{(2\pi)^{n/2}}\int_{R^n} e^{-ix\cdot\omega} f(x)dx \\ &= \frac{1}{(2\pi)^{n/2}}\int_{R^n} e^{-\frac{|\xi|^2}{2}} e^{-i\omega\cdot(m+\sqrt{\sigma}\xi)}d\xi \\ &= \frac{1}{(2\pi)^{n/2}} e^{-im\cdot\omega-\frac{\sigma|\omega|^2}{2}}.\end{aligned}$$

7.2 Properties of the Fourier transform

In this section we first present some elementary properties of the Fourier transform. Then, we prove the Fourier inversion formula that justifies Definition 7.1.2.

For convenience, for any constant vector $a = (a_1, a_2, \cdots, a_n)$ and $x \in R^n$ we use

$$x - a = (x_1 - a, \cdots, x_n - a),\ a \cdot x = \sum_{i=1}^{n} a_i x_i.$$

7.2.1 Elementary properties of the Fourier transform

Proposition 7.2.1. *For any $a \in R^n$ and any real number,*

$$\begin{aligned}&(a)\quad \mathcal{F}[f(x-a)] = e^{-ia\cdot\omega}\hat{f}(\omega);\\ &(b)\quad \mathcal{F}[f(x)e^{ia\cdot x}] = \hat{f}(\omega - a);\\ &(c)\quad \mathcal{F}[f(ax)] = \frac{1}{|a|^n}\mathcal{F}[f](\frac{\omega}{a}),\ a \neq 0.\end{aligned}$$

Proof. The proof is elementary from the definition. □

Proposition 7.2.2. *Let $f(x), g(x) \in L^1(R^n)\bigcap L^2(R^n)$. Then,*

$$\begin{aligned}&(a)\quad \int_{R^n} \hat{f}(\omega)\bar{\hat{g}}(\omega)d\omega = \int_{R^n} f(x)\bar{g}(x)dx.\\ &(b)\quad ||f||_{L^2(R^n)} = ||\hat{f}||_{L^2(R^n)}.\end{aligned}$$

Proof. (a)

$$\begin{aligned}\int_{R^n} \hat{f}(\omega)\bar{\hat{g}}(\omega)d\omega &= (2\pi)^{-\frac{n}{2}} \int_{R^n}\int_{R^n} \hat{f}(\omega)(e^{ix\cdot\omega}\bar{g}(x)dxd\omega \\ &= (2\pi)^{-\frac{n}{2}} \int_{R^n}\int_{R^n} \hat{f}(\omega)e^{ix\cdot\omega}\bar{g}(x)dxd\omega \\ &= \int_{R^n} \mathcal{F}^{-1}[\hat{f}]\bar{g}(x)dx \\ &= \int_{R^n} f(x)\bar{g}(x)dx.\end{aligned}$$

(b) By taking $g = f$ in (a), we obtain

$$\begin{aligned}||\hat{f}||^2_{L^2(R^n)} &= \int_{R^n} \hat{f}\bar{\hat{f}}dx \\ &= \int_{R^n} f(\hat{\bar{f}})^{\vee}dx \\ &= \int_{R^n} f\bar{f}dx = ||f||^2_{L^2(R^n)}.\end{aligned}$$

□

The next proposition involves partial derivatives with a multiindex. Recall that for a multiindex $\alpha = (\alpha_1, \cdots, \alpha_n)$ and $x = (x_1, \cdots, x_n)$, we use

$$x^\alpha := x_1^{\alpha_1} \cdots x_n^{\alpha_n}.$$

Proposition 7.2.3. *Let α be a multiindex. If $D^\alpha f \in L^1(R^n)$, then*

$$\mathcal{F}[D^\alpha f] = i^{|\alpha|}\omega^\alpha \hat{f}(\omega).$$

Proof. We first assume that $f(x) \in C_0^\infty(R^n)$. Then, for any $D_k := \frac{\partial}{\partial x_k}$ with $1 \le k \le n$ we perform the integration by parts to see that

$$\begin{aligned}\mathcal{F}[D_k f] &= (2\pi)^{-n/2} \int_{R^n} e^{-ix\cdot\omega} D_k f(x)dx \\ &= (i\omega_k)(2\pi)^{-n/2} \int_{R^n} e^{-ix\cdot\omega} f(x)dx \\ &= (i\omega_k)\hat{f}(\omega).\end{aligned}$$

For any multiindex $\alpha = (\alpha_1, \cdots, \alpha_n)$, we use the same calculation repeatedly to find that

$$\mathcal{F}[D^\alpha f] = i^{|\alpha|}\omega^\alpha \hat{f}(\omega).$$

For a general case where $D^\alpha f(x) \in L^1(R^n)$, we use a smooth approximation $f_n(x) \in C_0^\infty(R^n)$ with

$$||D^\alpha f_n - D^\alpha f||_{L^1(R^n)} \to 0 \qquad \text{as } n \to \infty.$$

$$\mathcal{F}[D^\alpha f_n] = i^{|\alpha|}\omega^\alpha \hat{f}_n.$$

As the Fourier transform is one-to-one, we obtain the desired result. □

We recall the convolution for two functions $f(x)$ and $g(x)$:

$$(f * g)(x) := \int_{R^n} f(x-y)g(y)dy.$$

Proposition 7.2.4. *Let $f(x), g(x) \in L^1(R^n)\bigcap L^2(R^n)$. Then,*

$$(a) \quad \mathcal{F}[f * g] = (2\pi)^{n/2}\hat{f}(\omega)\hat{g}(\omega),$$
$$(b) \quad \mathcal{F}[fg] = (2\pi)^{-n/2}\hat{f}(\omega) * \hat{g}(\omega).$$

Proof. (a)

$$\begin{aligned}
\mathcal{F}[f * g] &= \frac{1}{(2\pi)^{n/2}}\int_{R^n}[e^{-ix\cdot\omega}\int_{R^n} f(x-s)g(s)ds]dx \\
&= \frac{1}{(2\pi)^{n/2}}\int_{R^n}\int_{R^n}[e^{-i(x-s)\cdot\omega} f(x-s)e^{-is\cdot\omega}g(s)]dxds \\
&= (2\pi)^{n/2}\hat{f}(\omega)\hat{g}(\omega).
\end{aligned}$$

(b) Similar to the proof of (a) we have

$$\mathcal{F}^{-1}[\hat{f}(\omega) * \hat{g}(\omega)] = (2\pi)^{n/2}\mathcal{F}^{-1}[\hat{f}]\mathcal{F}^{-1}[\hat{g}] = (2\pi)^{n/2} f(x)g(x).$$

It follows that

$$\mathcal{F}[fg] = (2\pi)^{-n/2}\hat{f}(\omega) * \hat{g}(\omega).$$ □

7.2.2 The Fourier inversion theorem

With the above properties for the Fourier transform, we are ready to prove the following Fourier inversion theorem.

Theorem 7.2.1. *(The Fourier inversion theorem) The inverse Fourier transform $\mathcal{F}^{-1}$ exists for every $f \in L^1(R^n)$ and it can be expressed by*

$$\check{f}(x) := \mathcal{F}^{-1}[f] = (2\pi)^{-\frac{n}{2}}\int_{R^n} e^{ix\cdot\omega} f(\omega)d\omega.$$

Thus

$$(\hat{f})^\vee = f(x), \qquad \text{if } f(x), \hat{f}(\omega) \in L^1(R^n).$$

Proof. By the assumption, we know that $\hat{f} \in L^1(R^n)$. By the dominated convergence theorem, we find that

$$\int_{R^n} e^{ix\cdot\xi}\hat{f}(\xi)d\xi = \lim_{\varepsilon\to 0}\int_{R^n} e^{-\frac{\varepsilon|\xi|^2}{2}+ix\cdot\xi}\hat{f}(\xi)d\xi.$$

Define

$$g_x(\xi) = e^{-\frac{\varepsilon|\xi|^2}{2} + ix\cdot\xi}.$$

Then,

$$\mathcal{F}[g_x](y) = \frac{1}{(\sqrt{2\pi\varepsilon})^n} e^{-\frac{\varepsilon|x-y|^2}{\sqrt{2\pi}}} = \phi_\varepsilon(y-x).$$

On the other hand, we use Property 7.2.2 to find that

$$\int_{R^n} e^{-\frac{\varepsilon|\xi|^2}{2}+ix\cdot\xi} \hat{f}(\xi)d\xi = \int_{R^n} \mathcal{F}\left[e^{-\frac{\varepsilon|\xi|^2}{2}+ix\cdot\xi}\right] f(y)dy = \phi_\varepsilon \star f(x).$$

It follows that

$$\lim_{\varepsilon\to 0} \int_{R^n} e^{-\frac{\varepsilon|\xi|^2}{2}+ix\cdot\xi} \hat{f}(\xi)d\xi = \lim_{\varepsilon\to 0} \phi_\varepsilon \star f(x) = f(x). \qquad \square$$

7.3 Applications to the Laplace equation

In this section we will illustrate how to use the Fourier transform to find the solution representation for the Laplace equation.

7.3.1 Application to the Laplace equation with a lower-order term

Let $f(x) \in L^1(R^n)\bigcap L^2(R^n)$ and $a > 0$. Consider the Laplace equation with a lower-order term:

$$-\Delta u + a^2 u = f(x), \qquad x \in R^n. \tag{7.3.1}$$

We want to find a smooth solution that is bounded as $|x| \to \infty$.

Theorem 7.3.1. *The solution of the Laplace equation (7.3.1) can be expressed by*

$$u(x) = \frac{1}{(4\pi)^{n/2}} \int_0^\infty \int_{R^n} \frac{e^{-a^2t-\frac{|x-y|^2}{4t}}}{t^{n/2}} f(y)dydt. \tag{7.3.2}$$

Proof. Set

$$\hat{u} = \mathcal{F}[u].$$

Then,

$$(a^2 + |\omega|^2)\hat{u} = \hat{f}.$$

It follows that

$$u(x) := \mathcal{F}^{-1}\left[\frac{\hat{f}}{a^2+|\omega|^2}\right] = \frac{1}{(2\pi)^{n/2}} f(x) * B(x),$$

where

$$B(x) := \mathcal{F}^{-1}[\frac{1}{a^2+|\omega|^2}].$$

To find an explicit $B(x)$, we use the fact that

$$\frac{1}{a^2+|\omega|^2} = \int_0^\infty e^{-(a^2+|\omega|^2)t}dt.$$

It follows that

$$\begin{aligned}
B(x) &= \mathcal{F}^{-1}[\frac{1}{a^2+|\omega|^2}] \\
&= \frac{1}{(2\pi)^{n/2}}\int_0^\infty e^{-a^2t}\int_{R^n} e^{ix\cdot\omega - t|\omega|^2}d\omega dt \\
&= \frac{1}{2^{n/2}}\int_0^\infty \frac{e^{-a^2t-\frac{|x|^2}{4t}}}{t^{n/2}}dt.
\end{aligned}$$

Consequently,

$$u(x) = \frac{1}{(4\pi)^{n/2}}\int_0^\infty\int_{R^n}\frac{e^{-a^2t-\frac{|x-y|^2}{4t}}}{t^{n/2}}f(y)dydt. \qquad \square$$

Remark 7.3.1. When $a=0$, the Fourier transform for $\Gamma(\omega)=\frac{1}{|\omega|^2}$ is singular at $\omega=0$, which is not in $L^1(R^n)$. The calculation of its inverse Fourier transform is very complicated since one must introduce a generalized Fourier transform.

We give some other examples to illustrate the Fourier transform method to find the explicit solution.

Example 7.3.1. Let $g(x)\in L^1(R^1)\bigcap L^\infty(R^1)$. Find a *bounded solution* in half-space R^2_+:

$$\begin{aligned}
&u_{xx}+u_{yy}=0, && x\in R^1,\ y>0,\\
&u(x,0)=g(x), && x\in R^1.
\end{aligned}$$

Solution. Normally, one needs two initial conditions $u(x,0)$ and $u_y(x,0)$ at $y=0$ in order to have a unique solution. We give one initial condition and also impose that the solution must be bounded. The boundedness of the solution in $R^1\times R^1_+$ is equivalent to a condition at $r=\sqrt{x^2+y^2}=\infty$.

Set

$$\hat{u}(\omega,y) = \frac{1}{\sqrt{2\pi}}\int_{-\infty}^\infty e^{-ix\omega}u(x,y)dx.$$

Note that we only take the Fourier transform with respect to x.

Then, we take the Fourier transform of the equation to obtain:

$$-\omega^2\hat{u} + \hat{u}_{yy} = 0.$$

The general solution for the above ODE of $\hat{u}$ is equal to

$$\hat{u}(\omega, y) = C_1(\omega)e^{-\omega y} + C_2(\omega)e^{\omega y},$$

where $C_1(\omega)$ and $C_2(\omega)$ are arbitrarily functions of ω.

Since we require that the solution is bounded, it follows that $\hat{u}$ must be bounded for $y \in R^1_+$. Therefore we rewrite the solution in the following form:

$$\hat{u}(\omega, y) = C(\omega)e^{-y|\omega|},$$

where $C(\omega)$ is any bounded function of ω.

Now, from the initial condition we obtain

$$\hat{u}(\omega, 0) = \hat{g}(\omega).$$

It follows that

$$\hat{u}(\omega, y) = \hat{g}(\omega)e^{-y|\omega|}.$$

We already found from Example 7.1.2 that the Fourier transform for the function

$$\mathcal{F}[e^{-\omega|y|}] = \frac{2}{\sqrt{2\pi}}\frac{\omega}{\omega^2 + y^2}.$$

It follows that

$$u(x, y) = \mathcal{F}^{-1}[\hat{g}(\omega)e^{-y|\omega|}] = \frac{1}{\pi}\int_{-\infty}^{\infty}\frac{\omega g(x-\omega)}{(x-\omega)^2 + y^2}d\omega.$$

Clearly, $u(x, y)$ is bounded on $(x, y) \in R^1 \times R^1_+$.

Example 7.3.2. Find a solution in half-space $R^1 \times R^1_+$:

$$\begin{aligned} &u_{xx} + u_{yy} = 0, && x \in R^1, y > 0, \\ &u(x, 0) = g(x), u_y(x, 0) = h(x), && x \in R^1. \end{aligned}$$

Solution. We take the Fourier transform with respect to the x-variable. Define

$$\hat{u}(\omega, y) = \frac{1}{\sqrt{2\pi}}\int_{-\infty}^{\infty} e^{-ix\omega}u(x, y)dx.$$

Then,

$$-\omega^2\hat{u} + \hat{u}_{yy} = 0.$$

The general solution of the ODE for $\hat{u}$ is equal to

$$\hat{u}(\omega, y) = c_1(\omega)e^{-\omega y} + c_1(\omega)e^{\omega y}.$$

From the initial condition, we have

$$\hat{u}(\omega, 0) = \hat{g}(\omega), \hat{u}_y(\omega, 0) = \hat{h}(\omega).$$

It follows that

$$\begin{aligned} &c_1(\omega) + c_2(\omega) = \hat{g}(\omega), \\ &-\omega c_1(\omega) + \omega c_2(\omega) = \hat{h}(\omega). \end{aligned}$$

By solving for c_1 and c_2, we have

$$c_1(\omega) = \frac{\omega\hat{g} - \hat{h}}{2\omega}, \qquad c_2(\omega) = \frac{\omega\hat{g} + \hat{h}}{2\omega}.$$

Let

$$G(\omega, y) = c_1(\omega)e^{-\omega y} + c_2(\omega)e^{\omega y}.$$

Thus

$$u(x, y) = \mathcal{F}^{-1}[G(\omega, y)].$$

In general, it is difficult to find the explicit form for the inverse Fourier transform for $G(\omega, y)$.

7.4 Applications to the heat equation

In this section we use the Fourier transform method to derive the solution representation for the Cauchy problem of the heat equation.

7.4.1 The solution representation for the Cauchy problem of the heat equation

The following assumption is assumed throughout this section:

H(7.4.1) Let $f(x,t) \in L^1(R^n \times R^1_+) \bigcap L^2(R^n \times R^1_+)$ and $g(x) \in L^1(R^n) \bigcap L^2(R^n)$.

Consider the heat equation

$$u_t - \Delta u = f(x,t), \qquad (x,t) \in R^n \times (0, \infty), \tag{7.4.1}$$

$$u(x, 0) = g(x), \qquad x \in R^n. \tag{7.4.2}$$

Theorem 7.4.1. *Under the assumption H(7.4.1) the solution of the problem (7.4.1)–(7.4.2) can be represented by*

$$
\begin{aligned}
u(x,t) &= \frac{1}{(4\pi t)^{n/2}} \int_{R^n} e^{-\frac{|x-y|^2}{4t}} g(y)dy \\
&+ \frac{1}{(4\pi)^{n/2}} \int_0^t \int_{R^n} \frac{e^{-\frac{|x-y|^2}{4(t-\tau)}}}{(t-\tau)^{n/2}} f(y,\tau)dyd\tau. \qquad (7.4.3)
\end{aligned}
$$

Proof. We take the Fourier transform with respect to space variables. Let

$$\hat{u} := \mathcal{F}[u].$$

If we take the Fourier transform for Eq. (7.4.1), we find that

$$
\begin{aligned}
&\hat{u}_t + |\omega|^2 \hat{u} = \hat{f}(\omega,t), \qquad t \in (0,\infty), \\
&\hat{u}(\omega,0) = \hat{g}(\omega).
\end{aligned}
$$

It follows that

$$\hat{u}(\omega,t) = e^{-t|\omega|^2}\hat{g}(\omega) + \int_0^t e^{-(t-\tau)|\omega|^2} \hat{f}(\omega,\tau)d\tau.$$

By the inverse Fourier transform, we have that

$$u(x,t) = \frac{1}{(2\pi)^{n/2}}[g * F + \int_0^t f * F d\tau],$$

where

$$F(x,t) = \mathcal{F}^{-1}[e^{-t|\omega|^2}].$$

From Section 7.1 we see that

$$F(x,t) = \frac{e^{-\frac{|x|^2}{4t}}}{(2t)^{n/2}}, \qquad (x,t) \in R^n \times (0,\infty).$$

This concludes the proof of Theorem 7.4.1. □

From Theorem 7.4.1, we have the following decay estimate.

Corollary 7.4.1. *Let $f(x,t) = 0$ and $g(x)$ satisfy H(7.4.1). Then, $u(x,t)$ decays at the rate $\frac{1}{t^{n/2}}$.*

Another direct consequence is that the speed of heat-wave propagation is infinite. Let $\varepsilon > 0$ and

$$
g_\varepsilon(x) = \begin{cases} \frac{1}{2\varepsilon}, & \text{if } -\varepsilon < x < \varepsilon; \\ 0, & \text{if } |x| > \varepsilon. \end{cases}
$$

Corollary 7.4.2. *Let $f(x,t)=0$ and $g_\varepsilon(x)$ be defined as above. Then, for every $\varepsilon>0$,*

$$u(x,t)>0, \qquad (x,t)\in R^n\times(0,\infty).$$

7.4.2 Solution representation to the Black–Scholes equation

As an application we derive the explicit formula for the Black–Scholes equation.

In modern finance, the stock and bond options play an essential role for stability in the financial market. Option-price modeling becomes an important foundation for the market. There are several types of option contracts. The most active traded options are the European option and American option. We focus only on the European option that has an explicit formula.

Let $S(t)$ be a financial asset price such as a stock at time t. A fundamental assumption in the financial engineering is that a financial asset follows the geometric Brownian motion:

$$\frac{dS(t)}{S(t)}=\mu dt+\sigma dW(t),$$

where μ is the expected return rate and σ represents the volatility for the underlying asset $S(t)$. Let $P(s,t)$ be the option price for the stock s at time t and the value at the expiration date $t=T$ is known if a striking price is given. The task is to find the option price $P(s,0)$ at current time $t=0$.

Let r be the riskless interest rate. Moreover, the dividend for the stock S is neglected. Based on the no-arbitrage principle and Ito's lemma (see [17]), the option price $P(s,t)$ satisfies the following heat equation:

$$P_t+\frac{1}{2}\sigma^2 s^2 P_{ss}+sr P_s-rP=0, \qquad s>0, t>0. \tag{7.4.4}$$

The option price at expiration date T is fixed. Suppose the striking price of the stock is K. Then, the payoff value of the option at T is equal to

$$P(s,T)=\begin{cases} s-K, & \text{if } s>K,\\ 0, & \text{if } s\le K.\end{cases}$$

Define

$$N(x)=\frac{1}{\sqrt{2\pi}}\int_{-\infty}^{x} e^{-\frac{y^2}{2}}dy.$$

Theorem 7.4.2. *Suppose the volatility σ and the riskless interest rate $r>0$ are given. Then, the option price $P(s,t)$ has the following representation:*

$$P(s,t)=sN(d_1)-Ke^{-r(T-t)}N(d_2),$$

where T is the expiration date measured in year and K is the striking price, $s = s(t)$ represents the stock price at time t, d_1 and d_2 are defined as follows:

$$d_1 = \frac{\ln \frac{s}{K} + (r + \frac{1}{2}\sigma^2)(T-t)}{\sigma\sqrt{T-t}},$$
$$d_2 = d_1 - \sigma\sqrt{T-t}.$$

Proof. Introduce new variables

$$x = lns, \qquad \tau = T - t.$$

Then, $V(x,\tau) := P(s,t)$ satisfies

$$V_\tau - \frac{1}{2}\sigma^2 V_{xx} - (r - \frac{1}{2}\sigma^2)V_x + rV = 0, \; x \in (-\infty, \infty), 0 < \tau < T, \tag{7.4.5}$$
$$V(x,0) = (e^x - K)^+. \tag{7.4.6}$$

Set

$$V(x,\tau) = u(x,\tau)e^{\alpha\tau + \beta x},$$

where constant parameters α and β are chosen later.

The basic idea is that we choose α and β properly to eliminate lower-order terms in Eq. (7.4.5).

A direct calculation shows that $u(x,\tau)$ satisfies

$$u_\tau - \frac{1}{2}\sigma^2 u_{xx}$$
$$= [\beta\sigma^2 + r - \frac{1}{2}\sigma^2]u_x - [r - \beta(r - \frac{1}{2}\sigma^2) - \frac{1}{2}\sigma^2\beta^2 + \alpha]u.$$

It follows that

$$u_\tau - \frac{1}{2}\sigma^2 u_{xx} = 0,$$

provided that we choose α and β as follows:

$$\alpha = -r - \frac{1}{2\sigma^2}(r - \frac{1}{2}\sigma^2)^2, \qquad \beta = \frac{1}{2} - \frac{r}{\sigma^2}.$$

Moreover,

$$u(x,0) = e^{-\beta x}(e^x - K)^+, \qquad -\infty < x < \infty.$$

We use the representation for the Cauchy problem of the heat equation to obtain

$$u(x,\tau) = \int_{-\infty}^{\infty} \Phi(x - y, \tau)e^{-\beta y}(e^y - K)^+ dy,$$

where

$$\Phi(x-y,\tau)=\frac{1}{\sigma\sqrt{2\pi\tau}}e^{-\frac{(x-y)^2}{2\sigma^2\tau}}.$$

After some simplification and substituting back the original variable we find that

$$V(s,t)=sN(d_1)-Ke^{-r(T-t)}N(d_2),$$

where $s=s(t)$ represents the stock price at time t. □

The significance of the Black–Scholes formula is that the option price does not depend on the expectation of the stock return rate μ.

7.5 Application to the wave equation

In this section we will use the Fourier transform method to derive the solution representation for the Cauchy problem of the wave equation. The method is also used to prove an interesting split-energy property for the wave equation. The following assumption is needed throughout this section.

7.5.1 Solution representation for the wave equation

Consider the following wave equation subject to initial conditions:

$$u_{tt}-c^2\Delta u=f(x,t), \qquad (x,t)\in R^n\times(0,\infty), \tag{7.5.1}$$

$$u(x,0)=g(x),\ u_t(x,0)=h(x), \qquad x\in R^n. \tag{7.5.2}$$

H(7.5.1) Let $g(x),h(x)\in L^1(R^n)\bigcap L^2(R^n)$ and $f(x,t)\in L^1(R^n\times R^1_+)\bigcap L^2(R^n\times R^1_+)$.

Theorem 7.5.1. *Under the assumption H(7.5.1) the solution to the problem (7.5.1)–(7.5.2) has the following representation*

$$u(x,t)=g(x)*m_1(x,t)+h(x)*m_2(x,t)+\int_0^t f(x,\tau)*m_2(y,\tau)d\tau, \tag{7.5.3}$$

where

$$m_1(x,t)=(\cos(ct|y|)^\vee,\ m_2(x,t)=\left(\frac{\sin(ct|y|)}{c|y|}\right)^\vee,\ (x,t)\in R^n\times R^1_+. \tag{7.5.4}$$

Proof. We can decompose the problem into two separate problems as in Chapter 5. We consider the homogeneous equation first where $f(x,t)=0$. Let

$$\hat{u}:=\mathcal{F}[u].$$

We take the Fourier transform to Eq. (7.5.1) with respect to x to obtain

$$\hat{u}_{tt} + c^2|y|^2\hat{u} = 0, \qquad t > 0, \tag{7.5.5}$$

$$\hat{u}(y,0) = \hat{g}(y),\ \hat{u}_t(y,0) = \hat{h}(y), \qquad y \in R^n. \tag{7.5.6}$$

The general solution for Eq. (7.5.5) is equal to

$$\hat{u}(y,t) = c_1(|y|)\cos(ct|y|) + c_2(|y|)\sin(ct|y|),$$

where $c_1(|y|)$ and $c_2(|y|)$ are two arbitrary functions.

We use the initial conditions to obtain:

$$c_1(|y|) = \hat{g}(y), \qquad c_2(|y|) = \frac{\hat{h}(y)}{c|y|}.$$

It follows that

$$\begin{aligned} u(x,t) &= \left(\hat{g}\cos(ct|y|)\right)^{\vee} + \left(\hat{h}\frac{\sin(ct|y|)}{c|y|}\right)^{\vee} \\ &= g(x) * m_1(x,t) + h(x) * m_2(x,t), \end{aligned}$$

where $m_1(x,t)$ and $m_2(x,t)$ are defined as the inverse Fourier transform of $\cos(ct|y|)$ and $\frac{\sin(ct|y|)}{c|y|}$, respectively.

For the nonhomogeneous problem, we simply take $g(x) = 0$ and $h(x)$ to be replaced by $f(x,\tau)$ and use the Duhamel principle in Chapter 5 to obtain the solution representation. We combine two representations of solutions to obtain the desired result. □

7.5.2 Explicit representation for space dimensions $n = 1$ and $n = 3$

In this subsection we derive the explicit representation for the solution of the wave equation when the space dimension is equal to 1 and 3, which is equivalent to finding the inverse Fourier transform for $m_1(x,t)$ and $m_2(x,t)$. It turns out that there is an essential difference for even and odd dimensions.

For $n = 1$, we note that

$$\cos(ct|y|) = \frac{e^{-ict|y|} + e^{ict|y|}}{2}$$

and we use the property of the inverse Fourier transform to obtain

$$\left(\hat{g}\cos(ct|y|)\right)^{\vee} = \frac{1}{2}[g(x-ct) + g(x+ct)].$$

For $m_2(x,t)$, we note that

$$\left(\hat{h}\frac{\sin(ct|y|)}{c|y|}\right)^{\vee} = \frac{1}{2}\int_{-\infty}^{\infty}\hat{h}\frac{\sin(ct|y|)}{|y|}e^{iyx}dx$$

$$= \frac{1}{2} \int_{-\infty}^{\infty} \hat{h} \frac{e^{icyt} - e^{-icyt}}{iy} e^{iyx} dx$$
$$= \frac{1}{2} \int_{x-ct}^{x+ct} h(s) ds.$$

For $n = 3$, we claim the following identity:

$$\int_{B_R(0)} e^{-ix\cdot y} dy = \frac{4\pi}{|y|} \left(\frac{\sin(R|y|) - R|y|\cos(R|y|)}{|y|^2} \right).$$

Indeed, we choose spherical coordinates (r, θ, ψ) along the direction $y = (y_1, y_2, y_3)$. Then,

$$x \cdot y = |y| r \cos\theta; \qquad dy = r^2 \sin\theta dr d\theta d\psi.$$

It follows that

$$\int_{B_R(0)} e^{-ix\cdot y} dy = \int_0^{2\pi} \int_0^{\pi} \int_0^{R} e^{-ix\cdot y} r^2 sin\theta dr d\theta d\psi$$
$$= \frac{4\pi}{|y|} \int_0^R r \sin(r|y|) dr$$
$$= \frac{4\pi}{|y|} \left(\frac{\sin(R|y|) - R|y|\cos(R|y|)}{|y|^2} \right).$$

On the other hand,

$$\int_{B_R(0)} e^{ix\cdot y} dy = \int_0^R \left(\int_{|x|^2 = r} e^{ix\cdot y} dS \right) dr.$$

Consequently, we have

$$\int_{|x|^2 = r} e^{ix\cdot y} dS = \frac{d}{dR} \int_{B_R(0)} e^{ix\cdot y} dx$$
$$= \frac{4\pi}{|y|} \frac{d}{dR} [\sin(R|y|) - R|y|\cos(R|y|)]$$
$$= 4\pi R^2 \frac{\sin(R|y|)}{R|y|}.$$

Now, for $R = ct$, we have

$$e^{ix\cdot y} \frac{\sin(ct|y|)}{ct|y|} = \frac{1}{4c^2 \pi t} \int_{|\xi| = ct} e^{i(x+\xi)\cdot y} dS(\xi).$$

It follows that

$$\int_{R^3} \hat{h}(y) e^{ix\cdot y} \frac{\sin(ct|y|)}{ct|y|} dy = \frac{1}{4c^2 \pi t} \int_{|\xi| = ct} \left[\int_{R^3} e^{iy\cdot(x+\xi)} \hat{h}(y) dy \right] dS(\xi)$$

$$= \frac{1}{4c^2\pi t} \int_{|\xi|=ct} h(x+\xi) dS(\xi)$$
$$= t M_{ct}[h],$$

where $M_R[f]$ represents the average of the surface integral: for any f

$$M_R[f] := \frac{1}{4\pi R^2} \int_{|y|=R} f(x+y) dS(y).$$

It follows that

$$u(x,t) = \frac{d}{dt}(t M_{ct}[g]) + t M_{ct}[h].$$

When the dimension $n = 2$, we can use the same argument discussed in Chapter 5 to obtain the same representation as in Chapter 5.

From the solution representation we see that the value of $u(x,t)$ only depends on the surface $S = \{x = (x_1, x_2, x_3) : |x| = ct\}$. Hence, we have the following Huygen's principle that holds only for $n = 3$ (or n is odd).

Corollary 7.5.1. *(Huygens' principle) Let $g(x) = h(x) = 0$ on $R^3 \backslash B_R(0)$. Then,*

$$u(x,t) = 0, \qquad \forall |x|^2 < (ct - R)^2.$$

Proof. This is directly from the representation. □

7.5.3 Potential and kinetic energy

In this subsection we consider the homogeneous wave equation (7.5.1) in R^n. For simplicity we take $c = 1$.

H(7.5.2) Suppose

$$h(x), g(x), |\nabla g(x)| \in L^2(R^n).$$

Define the energy function as before:

$$E(t) = \frac{1}{2} \int_{R^n} \left[u_t^2 + |\nabla u|^2 \right] dx.$$

We see that

$$E(t) = E(0) = \frac{1}{2} \int_{R^n} \left[h(x)^2 + |\nabla g|^2 \right] dx.$$

A very interesting property for the wave equation is that the kinetic energy and potential energy are equal as $t \to \infty$.

Theorem 7.5.2. *Assume that the assumptions H(7.5.1)–H(7.5.2) hold. Let $u(x,t)$ be the solution of the homogeneous wave equation (7.5.1) with $f = 0$ in R^n. Then,*

$$\lim_{t\to\infty} \int_{R^n} |u_t|^2 dx = \lim_{t\to\infty} \int_{R^n} |\nabla u|^2 dx = E(0).$$

Proof. From the solution representation, we see that

$$\int_{R^n} |\nabla u|^2 dx = \int_{R^n} \left[|y|^2 |\hat{g}|^2 \cos^2(t|y|) + |\hat{h}|^2 \sin^2(t|y|) \right] dy$$
$$+ 2 \int_{R^n} |y| (\hat{h}\bar{\hat{g}} + \hat{g}\bar{\hat{h}}) \cos(t|y|) \sin(t|y|) dy$$
$$:= I_1 + I_2 + I_3.$$

Now, for any smooth function $f(x) \in C_0^\infty(R^n)$, we have

$$\begin{aligned} I_3 &= 2 \int_{R^n} f(y) \cos(t|y|) \sin(t|y|) dy \\ &= \int_0^\infty \sin(2tr) \int_{\partial B_r(0)} f(y) ds(y) dr \\ &= -\frac{1}{2t} \int_0^\infty \left[\frac{d}{dr} \cos(2tr) \right] \int_{\partial B_r(0)} f(y) ds(y) dr \\ &= \frac{1}{2t} \int_0^\infty \left\{ \cos(2tr) \left[\frac{d}{dr} \int_{\partial B_r(0)} f(y) ds(y) \right] \right\} dr. \end{aligned}$$

Since

$$\int_{\partial B_r(0)} f(y) ds(y) = \int_{\partial B_1(0)} f(ry) ds(y),$$

it follows that

$$\left| \int_{R^n} f(y) \cos(t|y|) \sin(t|y|) dy \right| \to 0, \qquad \text{as } t \to \infty.$$

If we use a compact smooth function to approximate $|y|(\hat{h}\bar{\hat{g}} + \hat{g}\bar{\hat{h}})$, we find that

$$I_3 \to 0 \qquad \text{as } t \to \infty.$$

To estimate I_1 and I_2, we use an elementary trigonometric identity:

$$\cos^2(t|y|) = \frac{1 + \cos(2t|y|)}{2}, \qquad \sin^2(2t|y|) = \frac{1 - \cos(2t|y|)}{2},$$

to obtain

$$I_1 \to \frac{1}{2} \int_{R^n} |y|^2 |\hat{g}|^2 dy, \qquad I_2 \to \frac{1}{2} \int_{R^n} |\hat{h}|^2 dy$$

as $t \to \infty$. Note that

$$\int_{R^n} |y|^2 |\hat{g}|^2 dy = \int_{R^n} |\nabla g|^2 dx, \qquad \int_{R^n} |\hat{h}|^2 dy = \int_{R^n} h^2(x) dx.$$

We conclude that

$$\int_{R^n} |\nabla u|^2 dx \to E(0), \qquad \text{as } t \to \infty.$$

The energy identity yields the same conclusion for the kinetic energy. □

7.6 Generalized Fourier transform

The Fourier transform defined in Section 7.1 is only valid for functions in $L^1(R^n)$. In some applications, one may extend the Fourier transform for a broader class of functions including the Dirac-delta function.

Recall the Dirac-delta function $\delta(x)$ is defined as follows:

$$\delta(x) = \begin{cases} 0, & \text{if } x \neq 0 \\ \infty, & \text{if } x = 0, \end{cases}$$

$$\int_{R^n} \delta(x) dx - 1.$$

The Dirac-delta function $\delta(x)$ is not a function in the classical sense. It is called a generalized function. There is a rich theory about the generalized function space (see [5]). Here, we present an introduction that is needed for the generalized Fourier transform.

7.6.1 The fundamental function space and generalized functions

Definition 7.6.1. A function $\phi(x) \in C^\infty(R^n)$ is said to be rapidly decreasing as $|x| \to +\infty$ if

$$(1+|x|)^k \frac{\partial^\alpha \phi(x)}{\partial x^\alpha} \to 0, \qquad \text{as } |x| \to +\infty,$$

for any multiindex $\alpha = (\alpha_1, \cdots, \alpha_n)$ with $k = \sum_{j=1}^n |\alpha_j| \geq 0$.

Clearly, the function

$$f(x) = e^{-|x|}, \qquad x \in R^n$$

is a rapidly decreasing function. Also, every function in $C_0^\infty(R^n)$ is a rapidly decreasing function.

Let

$$D(R^n) := \{\text{all rapidly decreasing functions}\}.$$

Then, it is easy to verify that $D(R^n)$ is a vector space and

$$C_0^\infty(R^n) \subset D(R^n).$$

We define a generalized function space

$$\mathcal{D}(R^n) := \{\text{all linear continuous transforms from } D(R^n) \text{ to } R^1\}.$$

The generalized function space extends the definition of a classical function. For any continuous function $f(x)$, we define a functional M_f from $C_0^\infty(R^n)$ as follows:

$$M_f[\psi] :=< f, \psi >:= \int_{R^n} f(x)\psi(x)dx, \forall\psi(x) \in C_0^\infty(R^n).$$

It is clear that $M_f \in \mathcal{D}(R^n)$. We can use f instead of M_f for brevity. It follows that

$$C(R^n) \subset \mathcal{D}(R^n).$$

Recall the Dirac-delta function $\delta(x)$, we know by the definition that for any $\psi(x) \in C_0^\infty$

$$M_\delta[\psi] :=< \delta, \psi >= \psi(0).$$

It is easy to see that M_δ is a linear continuous mapping from $C_0^\infty(R^n)$ into R^1. It follows that

$$\delta(x) \in \mathcal{D}(R^n).$$

7.6.2 The Fourier transform for generalized functions

Now, we extend the Fourier transform for functions in $\mathcal{D}(R^n)$ as follows.

Definition 7.6.2. We say the Fourier transform of a generalized function f

$$\hat{f} := \mathcal{F}[f],$$

if

$$< \hat{f}, \psi >=< f, \mathcal{F}[\psi] >, \qquad \forall\psi(x) \in C_0^\infty(R^n).$$

Its inverse Fourier transform is defined similarly.

It is easy to prove that all properties of the Fourier transform in Section 7.2 hold for the generalized Fourier transform.

The next proposition shows that the definition is the same when $f(x)$ is of class $L^1(R^n)$.

Proposition 7.6.1. *If $f(x) \in L^1(R^n)$, then*

$$< \mathcal{F}[f], \psi >=< f, \mathcal{F}[\psi] >, \qquad \forall\psi(x) \in C_0^\infty(R^n).$$

Proof. By the definition of the Fourier transform, we have

$$< \mathcal{F}[f], \psi > = \int_{R^n} \mathcal{F}[f]\psi(y)dy$$

$$= \frac{1}{(2\pi)^{n/2}} \int_{R^n} \int_{R^n} e^{-ix\cdot y} f(x)\psi(y)dydx$$
$$= \int_{R^n} f(x)\mathcal{F}[\psi]dx$$
$$=< f, \mathcal{F}[\psi] > . \qquad \square$$

The following property will be used in finding a fundamental solution to a partial differential equation.

Proposition 7.6.2. *Let $a \in R^n$:*

$$(a) \quad \mathcal{F}[\delta(x-a)] = \frac{1}{(2\pi)^{n/2}} e^{-ia\cdot y}.$$
$$(b) \quad \mathcal{F}[1] = \delta(x).$$

Proof. (a) By the definition, for any $\psi \in \mathcal{D}(R^n)$,

$$< \mathcal{F}[\delta(x-a)], \psi > =< \delta(x-a), \mathcal{F}[\psi] >$$
$$=< \delta(x-a), \frac{1}{(2\pi)^{n/2}} \int_{R^n} \psi(y)e^{-ix\cdot y}dy >$$
$$= \frac{1}{(2\pi)^{n/2}} \int_{R^n} \psi(y)e^{-ia\cdot y}dy$$
$$= \frac{1}{(2\pi)^{n/2}} < e^{-ia\cdot y}, \psi >,$$

which is the formula (a).

The formula (b) is similar. $\square$

Example 7.6.1. Find the Fourier transform for $f(x) = e^{x^2}\delta(x-2)$, $x \in R^1$.

Solution. By definition, we see that

$$\mathcal{F}[f(x)] =< f, \mathcal{F}[\psi] >= e^4\psi(2) = e^4 < \mathcal{F}[\delta(x-2)], \psi >, \qquad \forall\psi \in C_0^\infty(R^1).$$

It follows that

$$\mathcal{F}[f] = e^4.$$

Example 7.6.2. Let $a > 0$ be a constant. Find the Fourier transform for $f(r) = \delta(r-a)$ with $r = \sqrt{x_1^2 + x_2^2 + x_3^2}$.

Solution. By the definition, similar to Example 7.6.1 we find that

$$\mathcal{F}[f] = \frac{a\sin(a|y|)}{|y|},$$

where $|y| = \sqrt{y_1^2 + y_2^2 + y_3^2}$.

7.6.3 Applications to some classical equations

In this section we give a few examples for some partial differential equations with a δ-source. It will be seen in Chapter 8 that the solution for this type of PDE is called the fundamental solution.

Example 7.6.3. Find the solution for the Laplace equation with a δ-source.

Solution. Consider the Laplace equation with a delta source function in R^2:

$$-\Delta\Phi(x) = \delta(x), \qquad x \in R^2. \tag{7.6.1}$$

If we use polar coordinates, Eq. (7.6.1) is equivalent to

$$\Phi_{rr} + \frac{1}{r}\Phi_r + \frac{1}{r^2}\Phi_{\theta\theta} = \delta(r,\theta), \qquad (r,\theta) \in (0,\infty) \times (0, 2\pi). \tag{7.6.2}$$

It is clear that the Laplace operator and delta function are invariant under the orthonormal transformation of the coordinate. It follows that we may assume that the fundamental solution is just a function of $r = |x|$ and integrable in $L^1(R^n)$. Hence, Eq. (7.6.2) is equivalent to

$$\Phi_{rr} + \frac{1}{r}\Phi_r = \delta(r,\theta), \qquad (r,\theta) \in (0,\infty) \times (0, 2\pi). \tag{7.6.3}$$

For any $\psi(r,\theta) \in C_0^\infty(R^2)$, define

$$\hat{\psi}(r) := \frac{1}{2\pi}\int_0^{2\pi} \psi(r,\theta)d\theta.$$

Then,

$$\begin{aligned} < \Delta\Phi, \psi > &= \int_{R^2} \Phi\Delta\psi dx \\ &= 2\pi \int_0^\infty [\Phi(r)\Delta\hat{\psi}]rdr \\ &= 2\pi \int_0^\infty \Phi(r)[\hat{\psi}_{rr} + \frac{1}{r}\hat{\psi}_r]rdr \\ &= 2\pi \int_0^\infty \Phi(r)\frac{d}{dr}[r\frac{d}{dr}\hat{\psi}(r)]dr. \end{aligned}$$

If we assume

$$r\Phi(r) \to 0, \text{ as } r \to 0 \text{ and } r\Phi'(t) = -\frac{1}{2\pi},$$

then we see that

$$< -\Delta\Phi, \psi > = \hat{\psi}(0) = \psi(0,0) = < \delta, \psi > .$$

It follows that

$$r\Phi'(r) = -\frac{1}{2\pi},$$

which yields

$$\Phi(r) = -\frac{1}{2\pi}\ln r, \qquad r > 0.$$

We can use the same idea to derive the fundamental solution for $n > 2$. In Chapter 8 we will use a direct approach to derive the general fundamental solution.

Example 7.6.4. Find the solution for the heat equation with the δ-function as an initial value.

Solution. Consider the heat equation with a delta-initial value:

$$u_t - \Delta u = 0, \qquad x \in R^n, t > 0, \tag{7.6.4}$$

$$u(x,0) = \delta(x), \qquad x \in R^n. \tag{7.6.5}$$

We take the Fourier transform to obtain

$$\hat{u_1}' + |y|^2\hat{u_1} = 0,\ \hat{u_1}(0) = 1.$$

It follows that

$$\hat{u_1}(y,t) = e^{-|y|^2 t}.$$

If we take the inverse Fourier transform to obtain

$$\Phi(x,t) = \mathcal{F}^{-1}[e^{-|y|^2 t}] = \frac{1}{(2\pi t)^{\frac{n}{2}}}\int_{R^n} e^{ix\cdot y - \frac{|x|^2}{2t}}\,dy = \frac{1}{(4\pi t)^{\frac{n}{2}}}e^{-\frac{|x|^2}{4t}}.$$

Example 7.6.5. Find the solution for the wave equation with a δ-source and initial δ-velocity in R^3:

$$u_{tt} - c^2\Delta u = 0, \qquad (x,t) \in R^3 \times (0,\infty), \tag{7.6.6}$$

$$u(x,0) = 0,\ u_t(x,0) = \delta(x). \tag{7.6.7}$$

Solution. We take the Fourier transform to obtain

$$\hat{u}'' + c^2|y|^2\hat{u} = 0,$$
$$\hat{u}(y,0) = 0,\ \hat{u}'(y,0) = 1.$$

The unique solution of the above ODE is equal to

$$\hat{u}(y,t) = \frac{\sin(crt)}{cr}, \qquad r = |y|.$$

It follows that

$$u(x,t) = \mathcal{F}^{-1}[\hat{u}] = \left(\frac{\sin(crt)}{cr}\right)^{\vee}.$$

The inverse Fourier transform for the above function depends on the dimension that we already obtained in Section 7.5.

7.7 Notes and remarks

The Fourier transform is a convenient tool in deriving the solution representation for partial differential equations. However, it is a challenge to prove that the solution has sufficient regularity. Most of the derivation is formal. One needs advanced analysis tools to justify the derivation. In Chapter 8, we will use different methods to derive various solution representations rigorously. The material in Section 7.6 is just a brief introduction for the generalized function. The beginners may skip this section without any difficulty for the other materials in the book.

7.8 Exercises

1. Let $f_n(x) \to f(x)$ as $n \to \infty$ in $L^1(R^n)$. Prove

$$\mathcal{F}[f_n] \to \mathcal{F}[f], \qquad \text{as } n \to \infty.$$

2. Prove Property 7.2.1.
3. Let

$$f(x) = \frac{5}{1+x^2}, \qquad x \in R^1.$$

Find the Fourier transform $\hat{f}(y)$.

4. Let

$$f(x) = e^{-x^2+5x}.$$

Find the Fourier transform $\hat{f}(y)$ of $f(x)$.

5. Let $R_+^n = R^{n-1} \times R_+$. Consider

$$\begin{aligned} \Delta u &= 0, \qquad x \in R_+^n, \\ u(x)\big|_{x_n=0} &= f(x'), \end{aligned}$$

where $x' = (x_1, \cdots, x_{n-1})$.
Suppose

$$||u||_{L^\infty} < \infty.$$

Use the Fourier transform to find the solution representation.

6. Let $u(x,t)$ be the solution of the following heat equation

$$\begin{aligned} &u_t - \Delta u = \delta(x,t), \qquad &(x,t) \in R^n \times (0,\infty), \\ &u(x,0) = 0, \qquad &x \in R^n. \end{aligned}$$

Use the Fourier transform to find the explicit solution representation.

7. Let a be a constant and $f(x) \in L^1(R^n) \bigcap L^2(R^n)$. Let $u(x,t)$ be the solution of the following heat equation

$$\begin{aligned} &u_t - \Delta u + au = 0, \qquad &(x,t) \in R^n \times (0,\infty), \\ &u(x,0) = f(x), \qquad &x \in R^n. \end{aligned}$$

Use the Fourier transform to find the explicit solution representation.

8. Let $f(x,t) \in L^1(R^1 \times R_+)$. Let $u(x,t)$ be the solution of the following wave equation:

$$\begin{aligned} &u_{tt} - u_{xx} = f(x,t), \qquad &(x,t) \in R^1 \times (0,\infty), \\ &u(x,0) = u_t(x,0) = 0, \qquad &x \in R^1. \end{aligned}$$

Use the Fourier transform to find the explicit solution representation.

9. Let $a > 0$ and $\gamma > 0$ be constants. Let $u(x,t)$ be the solution of the following wave equation:

$$\begin{aligned} &u_{tt} - u_{xx} + \gamma u_t = a\cos(t), \qquad &(x,t) \in R^n \times (0,\infty), \\ &u(x,0) = u_t(x,0) = 0, \qquad &x \in R^n. \end{aligned}$$

Use the Fourier transform to find the explicit solution representation. Does the solution decay as $t \to \infty$?

10. Let $v(x) \in L^1(R^n) \bigcap L^2(R^n)$ and $g(x) \in L^1(R^n)$. Consider the following Schrödinger equation:

$$\begin{aligned} &iu_t + \Delta u + v(x)u = 0, \qquad &(x,t) \in R^n \times (0,\infty), \\ &u(x,0) = g(x), \qquad &x \in R^n. \end{aligned}$$

Prove the problem has a unique solution.

11. Let a be a constant and $g(x) \in L^1(R^n) \bigcap L^2(R^n)$. Find the solution representation for the following problem:

$$\begin{aligned} &u_t - \Delta u + au = 0, \qquad &(x,t) \in R^n \times (0,\infty), \\ &u(x,0) = g(x), \qquad &x \in R^n. \end{aligned}$$

Does the solution decay as $t \to \infty$?

CHAPTER

8 The fundamental solution and Green's representation

8.1 Introduction to the fundamental solution

To see a physical motivation, we consider a point source such as a charged particle located at the origin 0 in a vacuum. Then, the density of the electric field is equal to

$$\rho(x) = c\delta(x), \qquad x \in R^n,$$

where c is a physical constant.

The conservation of charge implies that the potential function $u(x)$ for an electric field $\mathbf{E}(x) = -\nabla u$ satisfies

$$-\Delta u = c\delta(x).$$

This leads us to study a partial differential equation with a point source. A fundamental solution for a partial differential equation is a smooth function with isolated singular point that satisfies the equation with a point source (Dirac-delta function). It serves as an explicit expression for the inverse differential operator.

8.1.1 The definition of the fundamental solution

We begin with a simple case with one variable. Let $f(x) \in L^1(R^1) \bigcap L^2(R^1)$.

Consider

$$L_0[u] := -u''(x) = f(x), \qquad x \in R^1. \tag{8.1.1}$$

Formally, if the inverse of the differential operator L_0 exists, then we have

$$u(x) = L_0^{-1} f(x), \qquad x \in R^1.$$

It will be interesting and convenient if we can find the explicit form of the inverse operator L_0^{-1}.

To find such an explicit inverse operator, we consider an auxiliary equation with a δ-source function with mass at any point $x \in R^1$:

$$-\Phi''(x, y) = \delta(y - x), \qquad y \neq x, y \in R^1.$$

Partial Differential Equations and Applications. https://doi.org/10.1016/B978-0-44-318705-6.00014-8

Then,

$$u(x) = \int_{R^1} \delta(x-y)u(y)dy = \int_{R^1} (-\Phi'')u(y)dy.$$

Suppose that

$$\Phi(x, y), \Phi'(x, y) \to 0 \qquad \text{as } |y| \to \infty.$$

After taking integration by parts twice, we obtain

$$u(x) = \int_{R^1} \Phi(x, y) f(y)dy.$$

If we can find the explicit form for $\Phi(x)$, then we obtain an explicit expression for the solution $u(x)$. The function $\Phi(x)$ is called a fundamental solution for the operator L_0. This leads to the following definition.

Definition 8.1.1. A function $\Phi(x)$ is called a fundamental solution to a partial differential equation

$$L[u] = 0, \qquad x \in R^n,$$

if

$$L[\Phi] = \delta(x), \qquad x \in R^n \backslash \{0\}.$$

Clearly, the fundamental solution may not be unique. We often require that a fundamental solution satisfies

$$\int_{R^n} \Phi(x)dx = 1.$$

However, this condition is not necessary.

If we replace the point source by $\delta(x-y)$ for any $y \in R^n$, then the functional solution depends on y, denoted by $\Phi(x-y)$. Clearly, we see that $\Phi(x-y)$ is symmetric with respect to x and y:

$$\Phi(x-y) = \Phi(y-x), \qquad \forall x, y \in R^n, x \neq y.$$

Moreover, $\Phi(x, y)$ is singular at $x = y$.

Theorem 8.1.1. *Let $f(x) \in L^2(R^n)$. Suppose L is a self-adjoint differential operator of second order and $\Phi(x)$ is the fundamental solution of the operator L. Let $u(x) \in H^2(R^n)$ be a continuous solution of the equation*

$$L[u] = f(x), \qquad x \in R^n.$$

Then,

$$u(x) = \int_{R^n} \Phi(x-y) f(y)dy, \qquad x \in R^n.$$

Proof. By the definition of the Dirac-delta function, we see that

$$\begin{aligned} u(x) &= \int_{R^n} \delta(x-y)u(y)dy =< L[\Phi], u > \\ &=< \Phi, L[u] >=< \Phi, f > \\ &= \int_{R^n} \Phi(x-y) f(y)dy. \end{aligned}$$

□

8.1.2 Some elementary examples

Example 8.1.1. Let $a > 0$. Find the fundamental solution for the following ODE:

$$u' + au = \delta(x), \qquad x \in R^1. \tag{8.1.2}$$

Solution. We recall that the Heaviside function $H(x)$ in one dimension satisfies in the sense of distribution:

$$\frac{dH(x)}{dx} = \delta(x), \qquad x \in R^1.$$

We rewrite this equation as follows:

$$\frac{d}{dx}(e^{ax}u) = e^{ax}\delta(x), \qquad x \in R^1.$$

Since

$$< e^{ax}\delta(x), f >= f(0) =< \delta, f >, \qquad \forall f \in C(R^1),$$

it follows that

$$\Phi(x) = e^{-ax}H(x), \qquad x \in R^1$$

is a fundamental solution to the ODE (8.1.2).

Example 8.1.2. Find a fundamental solution for the second-order ODE:

$$-u''(x) = \delta(x), \qquad x \in R^1.$$

Solution. We first have

$$\frac{d}{dx}[u'(x) + H(x)] = 0.$$

It follows that

$$u(x) = \begin{cases} a, & \text{if } x < 0, \\ -x, & \text{if } x > 0, \end{cases}$$

where a is a constant. It also follows that

$$\Phi(x) = a(1 - H(x)) - xH(x), \qquad x \in R^1.$$

One may choose $a = 0$ if $\Phi(x) \to 0$ as $|x| \to \infty$.

If we replace $\delta(x)$ by $\delta(x-y)$, then

$$\Phi(x, y) = a(1 - H(x-y)) - (x-y)H(x-y).$$

In the next few sections we will derive the fundamental solutions for some classical equations.

8.2 The fundamental solution of the Laplace equation

In this section we derive the explicit fundamental solution for the Laplace operator. Some properties will be derived from the expression.

8.2.1 The derivation of the fundamental solution

Definition 8.2.1. The solution of the Laplace equation

$$L_0[\Phi] := -\Delta\Phi = \delta(x), \qquad x \in R^n \setminus \{0\}, \tag{8.2.1}$$

is called a fundamental solution for the operator L_0 with

$$\int_{R^n} \Phi(x)dx = 1.$$

Let $f(x) \in L^1(R^n) \bigcap L^2(R^n)$. Consider the following Laplace equation:

$$L_0[u] := -\Delta u = f(x), \qquad x \in R^n.$$

Then, by Theorem 8.1.1,

$$u(x) = L_0^{-1} f(x) = \int_{R^n} \Phi(y-x) f(y)dy, \qquad x \in R^n,$$

provided that $u(x) \in H^2(R^n)$.

To find the fundamental solution of Eq. (8.2.1), we assume that the fundamental solution of the Laplace operator is radially symmetric in R^n. Let

$$\Phi(x) = \Phi(r),$$

where $r = |x| = \sqrt{x_2^2 + \cdots + x_n^2}$. Then,

$$\Phi_{x_i}(r) = \Phi'(r)\frac{x_i}{r}, \qquad \Phi_{x_i x_i}(r) = \Phi''(r)(\frac{x_i}{r})^2 + \Phi'(r)\left[\frac{1}{r} - \frac{x_i^2}{r^3}\right].$$

Consequently,

$$\Delta\Phi = \Phi''(r) + \frac{n-1}{r}\Phi'(r) = 0.$$

We assume $\Phi'(r) \neq 0$, the general solution is equal to

$$\Phi(r) = \begin{cases} c_1 \ln r + c_2, & \text{if } n = 2, \\ \frac{c_1}{r^{n-2}} + c_2, & \text{if } n \geq 3, \end{cases}$$

where c_1 and c_2 are arbitrary constants.

We choose $c_2 = 0$ and c_1 such that

$$\int_{R^n} \Phi(|x|)dx = 1.$$

By using polar coordinates, we find the constant c_1 easily.

Theorem 8.2.1. *The fundamental solution of the Laplace operator is equal to*

$$\Phi(x) := \begin{cases} -\frac{1}{2\pi} \ln |x|, & \text{if } n = 2, \\ \frac{1}{n(n-2)\omega(n)} \frac{1}{|x|^{n-2}}, & \text{if } n \geq 3, \end{cases}$$

where $\omega(n)$ is the area of the unit sphere in R^n.

Moreover, for any $f(x) \in L^1(R^n) \bigcap L^2(R^n)$ the solution of the Poisson equation

$$-\Delta u = f(x), \qquad x \in R^n, \tag{8.2.2}$$

is in $H^2(R^n)$, which can be expressed by

$$u(x) = \int_{R^n} \Phi(x-y) f(y) dy, \qquad x \in R^n.$$

Proof. We have seen the expression from Theorem 8.1.1. Here we present a direct proof. For any $x \neq y$, a direct calculation yields

$$\Phi_{x_i} = \frac{1}{n\omega(n)}(x_i - y_i)|x-y|^{-n},$$

$$\Phi_{x_i x_j} = \frac{1}{n\omega(n)}\{|x-y|^2\delta_{ij} - n(x_i - y_i)(x_j - y_j)\}|x-y|^{-n-2}.$$

It follows that

$$|\Phi_{x_i}| \leq \frac{1}{n\omega(n)}|x-y|^{-n+1},$$

$$|\Phi_{x_i x_j}| \leq \frac{1}{\omega(n)}|x-y|^{-n}.$$

Note that for any smooth function $u(x)$, $v(x)$ and any domain $\Omega \subset R^n$, we have the following Green's identity (3.4.3) in Chapter 3 with $p(x) = 1$ in Ω:

$$\int_\Omega (v\Delta u - u\Delta v)dx = \int_{\partial\Omega} (v\nabla_\nu u - u\nabla_\nu v)\, ds.$$

It follows that for $v = \Phi$:

$$\int_\Omega (\Phi(x-y)\Delta u - u\Delta\Phi(x-y))dy = \int_{\partial\Omega} (\Phi\nabla_\nu u - u\nabla_\nu \Phi)\, ds(y).$$

However, since Φ is singular at $y = x$, we must modify the identity. We choose a small ball $B_\rho(x) \subset \Omega$ and apply Green's identity in $\Omega \backslash B_\rho(x)$.

Then,

$$\begin{aligned}
&\int_{\Omega\backslash B_\rho(x)} (\Phi(x-y)\Delta u - u\Delta\Phi(x-y))dy \\
&= \int_{\Omega\backslash B_\rho(x)} (\Phi(x-y)\Delta u dy \\
&\to -\int_\Omega (\Phi(x-y)f(y)dy \qquad \text{as } \rho \to 0.
\end{aligned}$$

Let

$$I = \int_{\partial B_\rho(x)} (\Phi\nabla_\nu u - u\nabla_\nu \Phi)\, ds(y).$$

Now, assume $\nabla_\nu u$ is bounded, then

$$|\int_{\partial B_\rho(x)} \Phi\nabla_\nu u ds(y)| \le n w(n)\rho^{n-1}\Phi(\rho) \max_{B_\rho(x)} |\nabla u| \to 0 \quad \text{as } \rho \to 0.$$

$$\begin{aligned}
&\int_{\partial B_\rho(x)} u(y)\nabla_\nu \Phi ds(y) = -\Phi'(\rho) \int_{\partial B_\rho(x)} u(y)ds(y) \\
&= -\frac{1}{n\omega(n)\rho^{n-1}} \int_{\partial B_\rho(x)} u(y)ds(y) \\
&\to -u(x) \quad \text{as } \rho \to 0.
\end{aligned}$$

It follows that

$$u(x) = \int_{\partial\Omega} (u(y)\Phi_\nu(x-y) - \Phi(x-y)u_\nu)\, ds(y) + \int_\Omega \Phi(x-y)f(y)dy.$$

For $\Omega = R^n$ with $n \ge 3$, since

$$|\Phi(x-y)|, |\Phi_\nu(x-y)| \to 0, \qquad \text{as } |y| \to \infty,$$

we find

$$u(x) = \int_{R^n} \Phi(x-y) f(y) dy.$$

For $n = 2$, we use the fact that $u \in H^2(R^2)$ to obtain the same conclusion. The above derivation is rigorous as long as the solution is in $H^2(R^n)$. This is indeed valid from the regularity theory below if $f(x) \in C(R^n) \bigcap L^2(R^n)$. □

The next example shows that the fundamental solution may be found formally by using the Fourier-transform method if the inverse Fourier transform can be obtained explicitly.

Example 8.2.1. Let $a > 0$ be a constant. Find the fundamental solution for the following equation

$$-\Delta u + au = \delta(x), \qquad x \in R^n.$$

Solution. We use the Fourier transform to obtain

$$(a + |y|^2)\hat{\Phi}(y) = 1.$$

It follows from Chapter 7 that

$$\Phi(x) = \left(\frac{1}{a+|y|^2}\right)^{\vee} = \frac{1}{2^{n/2}} \int_0^{\infty} \int_{R^n} t^{-\frac{n}{2}} e^{-at-\frac{|x|^2}{4t}} dx dt.$$

To satisfy the condition

$$\int_{R^n} \Phi(x) dx = 1,$$

we obtain the fundamental solution

$$\Phi(x) = \frac{1}{(4\pi)^{n/2}} \int_0^{\infty} \int_{R^n} t^{-\frac{n}{2}} e^{-at-\frac{|x|^2}{4t}} dx dt.$$

When $a = 0$, if we assume that the fundamental solution is a function of $|x|$ only, then we can obtain the same fundamental solution as in Theorem 8.2.1 by using the generalized Fourier transform as in Section 7.6.

8.2.2 Regularity

Let $f(x) \in L^1(R^n) \bigcap L^2(R^n)$;

$$v(x) = \int_{R^n} \Phi(x, y) f(y) dy.$$

Theorem 8.2.2. *Let $\Phi(x)$ be the fundamental solution of the Laplace operator. Then,*

(a) If $f(x) \in L^1(R^n)$, then $v(x) \in C^1(R^n)$ and for all $i = 1, \cdots, n$,

$$v_{x_i}(x) = \int_{R^n} \Phi_{x_i}(x, y) f(y) dy.$$

(b) If $f(x) \in L^2(R^n) \bigcap C^\alpha(R^n)$ with $\alpha \in (0, 1]$, then for any domain $\Omega \subset R^n$ with C^2-boundary $\partial\Omega$ and any $x \in \Omega$,

$$\begin{aligned} v_{x_i x_j}(x) &= \int_{R^n} \Phi_{x_i x_j}(x, y)[f(y) - f(x)] dy \\ &\quad - f(x) \int_{\partial\Omega} \Phi_{x_i}(x, y) \nu_i(y) ds(y), \ \forall i, j = 1, 2, \cdots, n. \end{aligned}$$

Proof. From the expression of $\Phi(x, y)$, for any $\rho > 0$ we see that

$$\int_{B_\rho(x)} \frac{1}{|x - y|^{n-1}} dy < \infty,$$
$$\int_{B_\rho(x)} \frac{|x - y|^\alpha}{|x - y|^n} dy < \infty.$$

Hence, the conclusion follows if $f(x)$ satisfies the condition in (a). To prove (b), we note that for any open domain $\Omega \subset R^n$ with C^1-boundary and all $i, j = 1, 2, \cdots n$,

$$u(x) := \int_\Omega \Phi_{x_i x_j}(x, y)[f(y) - f(x)] dy - f(x) \int_{\partial\Omega} \Phi_{x_i}(x, y) \nu_i(y) ds(y), x \in \Omega$$

is well defined. Due to a singularity of $\Phi(x, y)$ at $x = y$, we use an approximation technique as in [12] to prove

$$u(x) = v_{x_i x_j}(x), \qquad x \in \Omega.$$

For any $\varepsilon > 0$, define

$$v_\varepsilon(x) = \int_\Omega \Phi_{x_i} \eta \left(\frac{|x - y|}{\varepsilon} \right) f(y) dy.$$

where $\eta(s)$ is the mollifier function defined in Section 2.2 of Chapter 2. Then

$$\begin{aligned} \frac{\partial v_\varepsilon}{\partial x_j} &= \int_\Omega \left\{ \Phi_{x_i}(x, y) \eta_\varepsilon \left(\frac{|x - y|}{\varepsilon} \right) \right\}_{x_j} [f(y) - f(x)] dy \\ &\quad + f(x) \int_\Omega \left\{ \Phi_{x_i}(x, y) \eta_\varepsilon \left(\frac{|x - y|}{\varepsilon} \right) \right\}_{x_j} dy \\ &= \int_\Omega \left\{ \Phi_{x_i}(x, y) \eta_\varepsilon \left(\frac{|x - y|}{\varepsilon} \right) \right\}_{x_j} [f(y) - f(x)] dy \end{aligned}$$

$$+ f(x)\int_{\partial\Omega}\Phi_{x_i}(x,y)\nu_j ds(y).$$

It follows that

$$|u(x) - v_\varepsilon(x)_{x_j}| = \left|\int_{B_{2\varepsilon}(x)}\left\{[1-\eta\left(\frac{|x-y|}{\varepsilon}\right)]\Phi_{x_i}\right\}(f(y)-f(x)dy\right|$$
$$\le C[f]_\alpha\int_{B_{2\varepsilon}(x)}[|\Phi_{x_i x_j}| + \frac{1}{\varepsilon}|\Phi_{x_i}]|x-y|^\alpha dy$$
$$\le c\varepsilon^\alpha \to 0, \qquad \text{as } \varepsilon \to 0.$$

□

Corollary 8.2.1. *Let $u(x)$ be the solution of the Laplace equation (8.2.2). Then,*
(a) If $f(x) \in L^2(R^n)$, then $u(x) \in H^2(R^n)$ and

$$||u||_{H^2(R^n)} \le C||f||_{L^2(R^n)}.$$

(b) If $f(x) \in C_0^\alpha(R^n)\bigcap L^2(R^n)$, then $u(x) \in C^{2+\alpha}(R^n)$ and

$$||u||_{C^{2+\alpha}(R^n)} \le C[||f||_{C_0^\alpha(R^n)} + ||f||_{L^2(R^n)}],$$

where C depends only on n.

Proof. For (a), we take a smooth approximation $f_n(x) \in C_0^\infty(R^n)$ with

$$||f_n - f||_{L^2(R^n)} \to 0$$

as $n \to \infty$. From the solution representation, we see that

$$||u||_{L^2(R^n)} + ||\nabla u||_{L^2(R^n)} \le C||f||_{L^2(R^n)},$$

where C is a constant depending only on n.

On the other hand, from the equation we see that

$$||\Delta u||_{L^2(R^n)} = ||f||_{L^2(R^n)}.$$

Note that from the theory of Sobolev space the norm of $H^2(R^n)$ is equivalent to

$$||u||_{L^2(R^n)} + ||\Delta u||_{L^2(R^n)}.$$

It follows that the conclusion (a) holds.

The proof of (b) needs delicate estimates for the fundamental solution $\Phi(x, y)$. We skip the detail. The reader can find its proof in [12]. □

8.3 Green's functions for the Laplace operator

In this section we extend the idea in the preceding section to find the solution representation of the Laplace equation in a general domain in R^n.

8.3.1 The definition of Green's functions

Let Ω be a domain in R^n. Consider the following boundary value problem:

$$-\Delta u = f(x), \qquad x \in \Omega, \tag{8.3.1}$$

$$u(x) = g(x), \qquad x \in \partial\Omega. \tag{8.3.2}$$

Let $x \in \Omega$. Suppose there exists a function $G(x, y)$ that solves the following problem:

$$-\Delta_y G = \delta(y - x), \qquad y \in \Omega, y \neq x, \tag{8.3.3}$$

$$G(x, y) = 0, \qquad y \in \partial\Omega. \tag{8.3.4}$$

From Green's identity (3.4.3) in Chapter 3, we have

$$u(x) = \int_\Omega G(x, y) f(y) dy - \int_{\partial\Omega} [g(y) G_\nu(x, y)] ds(y), \qquad x \in \Omega, \tag{8.3.5}$$

where $G_\nu(x, y) = \nabla_y G(x, y) \cdot \nu(y)$ and $\nu(y)$ is the outward unit normal on $\partial\Omega$.

Definition 8.3.1. For any $x \in \Omega$, the solution of the following boundary value problem:

$$-\Delta_y G = \delta(y - x), \qquad y \in \Omega, \; y \neq x, \tag{8.3.6}$$

$$G(x, y) = 0, \qquad y \in \partial\Omega, \tag{8.3.7}$$

is called a Green's function for the Laplace operator.

We need to answer two basic questions. The first question is what conditions are needed for Ω such that a Green's function exists. The second is to find the Green's function in Ω. We will discuss the basic method to construct Green's functions in the next section.

Theorem 8.3.1. *Let Ω be a domain with a C^1-boundary. Then, there exists a Green's function $G(x, y)$ for the Laplace operator.*

Proof. From the definition of Green's function, we construct $G(x, y)$ from the fundamental solution $\Phi(x, y)$. Since we know that the fundamental solution is harmonic in R^n except for $x \neq y$, $x \in \Omega$, we consider the following auxiliary problem:

$$-\Delta_y \psi(x, y) = 0, \qquad y \in \Omega, \tag{8.3.8}$$

$$\psi(x, y) = \Phi(x, y), \qquad y \in \partial\Omega. \tag{8.3.9}$$

Since $\Phi(x, y)$ is smooth on $\partial\Omega$, there exists a unique solution $\psi(x, y)$ (see Chapter 6). It follows that

$$G(x, y) = \Phi(x, y) - \psi(x, y), \qquad y \in \Omega, y \neq x$$

satisfies all conditions, which is a Green's function in Ω. □

8.3.2 Properties of Green's functions

Proposition 8.3.1. *The Green's function is symmetric:*

$$G(x, y) = G(y, x), \qquad x, y \in \Omega, x \neq y.$$

Proof. This is due to the symmetry of the delta function and the fundamental solution. Indeed, for any $z_1, z_2 \in \Omega$, set $u(x) = G(x, z_1)$ and $v(x) = G(x, z_2)$. Then, from Green's identity, we have

$$\begin{aligned}&\int_\Omega \big[G(x, z_1)\Delta G(x, z_2) - G(x, z_2)\Delta G(x, z_1)\big]dx \\ &= \int_{\partial\Omega} \big[G(x, z_1)G_\nu(x, z_2) - G(x, z_2)G_\nu(x, z_1)\big]ds = 0.\end{aligned}$$

Since

$$-\Delta G(x, z_1) = \delta(x - z_1), \qquad -\Delta G(x, z_2) = \delta(x - z_2), \qquad x \in \Omega,$$

it follows that

$$G(z_2, z_1) = G(z_1, z_2), \qquad z_1, z_2 \in \Omega. \qquad \square$$

Theorem 8.3.2. *Let $n \geq 3$ and $\partial\Omega \in C^2$. Then,*

$$0 < G(x, y) \leq \Phi(x, y), \qquad |\nabla_x G(x, y)| \leq \frac{C}{|x - y|^{n-1}}, \qquad x, y \in \Omega, x \neq y, \tag{8.3.10}$$

where C is a constant that depends only on n and $\partial\Omega$.

Proof. From Theorem 8.2.2, we know that

$$G(x, y) = \Phi(x, y) - \psi(x, y), \qquad y \in \Omega.$$

The maximum principle yields the first estimate. The second estimate comes directly from the fundamental solution. $\square$

The reader can find many additional properties for a Green's function in advanced PDE books.

8.3.3 Green's functions in special domains

In this subsection we introduce a powerful method called the "mirror-reflection method" to construct a Green's function when a domain has a special shape. The basic idea is that we use the fundamental solution to construct a Green's function in which a harmonic function with a value along with the boundary is equal to the fundamental solution. We use several concrete examples to illustrate the method.

Example 8.3.1. Construct the Green's function for the Laplace operator L_0 in

$$\Omega = R^n_+ := \{x = (x_1, x_2, \cdots, x_n) \in R^n : x_n > 0\}.$$

Solution. We start with the fundamental solution $\Phi(x, y)$. What we need is a harmonic function with 0 value at $x_n = 0$. By the reflection of $\Phi(x, y)$ along $y_n = 0$, we see that

$$G(x, y) = \Phi(x_1, \cdots, x_n, y_1, \cdots, y_n) - \Phi(x_1, \cdots, x_n, y_1, \cdots, -y_n), \qquad x, y \in R^n.$$

It is clear that, for any $x \in R^n$, $G(x, y)$ satisfies

$$\begin{aligned} &-\Delta_y G(x, y) = 0, \text{ in } R^n, \\ &G(x, y)|_{y_n=0} = 0. \end{aligned}$$

It follows that $G(x, y)$ is the Green's function.

As an application, we consider the following boundary value problem:

$$-\Delta u = f(x), \qquad x \in R^n_+, \tag{8.3.11}$$

$$u(x) = g(x), \qquad \text{on } x_n = 0. \tag{8.3.12}$$

Now, the unit outward normal direction $\nu = (0, \cdots, 0, -1)$ for the boundary of R^n_+. It follows that

$$G_\nu(x, y) = -G_{y_n}(x, y) = -\frac{1}{n\omega(n)}\left[\frac{x_n - y_n}{|x - y|^n} + \frac{x_n + y_n}{|x - y|^n}\right].$$

On $y_n = 0$, we see that

$$G_\nu(x, y) = -G_{y_n}(x, t) = -\frac{2x_n}{n\omega(n)}\left[\frac{1}{|x - y|^n}\right].$$

It follows that the solution of the Poisson equation (8.3.11)–(8.3.12) is given by

$$u(x) = \frac{2x_n}{n\omega(n)} \int_{\partial R^n_+} \frac{g(y)}{|x - y|^n} dy + \int_{R^n_+} f(y) G(x, y) dy.$$

Example 8.3.2. Construct the Green's function for the Laplace operator L_0 in

$$\Omega = \{(x_1, x_2) \in R^2 : x_1 \geq 0, x_2 \geq 0\}.$$

Solution. We start with the fundamental solution $\Phi(x, y) = G(x_1, x_2, y_1, y_2)$ in R^2. If we reflect $\Phi(x_1, x_2, y_1, y_2)$ along the axis $y_2 = 0$ and set

$$\Phi_1(x_1, x_2, y_1, y_2) = \Phi(x_1, x_2, y_1, y_2) - \Phi(x_1, x_2, y_1, -y_2),$$

then, we see that

$$-\Delta_y \Phi_1 = 0, \in \Omega, \qquad \Phi_1|_{y_2=0} = 0, \qquad y_1 \in R^1.$$

From $\Phi_1(x_1, x_2, y_1, y_2)$, we reflect Φ_1 along $y_1 = 0$ and set

$$\Phi_2(x_1, x_2, y_1, y_2) = \Phi_1(x_1, x_2, y_1, y_2) - \Phi_1(x_1, x_2, -y_1, y_2).$$

Then, we see $\Phi_2(x_1, x_2, y_1, y_2)$ satisfies

$$-\Delta_y \Phi_2 = 0,\ y \in \Omega, \qquad \Phi_2|_{y_1=0} = \Phi_2|_{y_2=0} = 0, \qquad y_1 \in R^1.$$

It follows that $\Phi_2(x_1, x_2, y_1, y_2)$ is the Green's function for Ω.

Example 8.3.3. Construct the Green's function for a ball in R^n associated with the Laplace operator,

$$\Omega = \{x = (x_1, x_2, \cdots, x_n) : |x| < 1\}.$$

Solution. For any point $x \in B_1(0)$, we define a refection of $x \in B_1(0)$ about $\partial B_1(0)$, denoted by $\tilde{x}$, as follows:

$$\tilde{x} = \frac{x}{|x|^2}, \qquad x \in B_1(0).$$

It is clear that $\tilde{x} \in R^n \backslash B_1(0)$ and $\tilde{x} \to 1$ as $x \to \partial B_1(0)$. For any $x \in B_1(0)$, let $\psi(x, y)$ be the solution of the following problem:

$$\begin{aligned} -\Delta_y \psi(x, y) &= 0, \qquad && y \in B_1(0), \\ \psi(x, y) &= \Phi(x, y), && y \subset \partial B_1(0). \end{aligned}$$

Now, define

$$G(x, y) = \Phi(x, y) - \psi(x, y).$$

Then, it is clear that $G(x, y)$ is harmonic in $B_1(0)$ and $G(x, y) = 0$ on $\partial B_1(0)$. We claim that

$$\psi(x, y) = \Phi(|x|(y - \tilde{x})), \qquad x, y \in B_1(0).$$

Indeed, for $n \geq 3$ and any fixed $x \in B_1(0)$, $\Phi(y - \tilde{x})$ is harmonic in $B_1(0)$ since $\tilde{x} \in R^n \backslash B_1(0)$. From the expression of $\Phi(y - x)$, we see that

$$|x|^{2-n} \Phi(|x|(y - \tilde{x}))$$

is harmonic in $B_1(0)$. Moreover, for $y \in \partial B_1(0)$,

$$\begin{aligned} (|x||y - \tilde{x}|)^2 &= |x|^2 \big(|y|^2 - \frac{2y \cdot x}{|x|^2} + \frac{1}{|x|^2}\big) \\ &= |x|^2 - 2y \cdot x + 1 = |x - y|^2. \end{aligned}$$

Hence, on $\partial B_1(0)$,

$$(|x||y - \tilde{x}|)^{-(n-2)} = |x - y|^{-(n-2)}.$$

Consequently,

$$\psi(x, y) = \Phi(x, y), \qquad \text{on } \partial B_1(0).$$

Therefore

$$G(x, y) = \Phi(x, y) - \Phi(|x|(y - \tilde{x})), \qquad x, y \in B_1(0), x \neq y.$$

As an application, we have the following theorem.

Theorem 8.3.3. *Let $g(x) \in C(\partial B_r(0))$. Let $u(x)$ be the solution of the following Poisson problem:*

$$\begin{aligned} -\Delta u &= f(x), & x \in B_r(0), \\ u(x) &= g(x), & x \in \partial B_r(0). \end{aligned}$$

Then,

$$u(x) = -\frac{r^2 - |x|^2}{n\omega(n)r} \int_{\partial B_r(0)} \frac{g(y)}{|x - y|^n} ds(y) + \int_{B_r(0)} f(y)G(x, y)dy, \qquad x \in B_r(0).$$

Proof. We first calculate $G_\nu(x, y)$ on $\partial B_1(0)$.

$$\begin{aligned} G_{y_i}(x, y) &= \Phi_{y_i}(y - x) - \Phi(|x|(y - x))_{y_i} \\ &= \frac{1}{n\omega(n)} \frac{x_i - y_i}{|x - y|^n} + \frac{-1}{n\omega(n)} \frac{y_i|x|^2 - x_i}{(|x||y - \tilde{x}|)^n} \\ &= \frac{1}{n\omega(n)} \frac{x_i - y_i}{|x - y|^n} + \frac{-1}{n\omega(n)} \frac{y_i|x|^2 - x_i}{(|x - y|)^n}. \end{aligned}$$

Here at the final step, we have used the fact $|x||y - \tilde{x}| = |x - y|$ on $\partial B_1(0)$.

Note that on $\partial B_1(0)$, the outward normal $\nu = y$. Hence,

$$\begin{aligned} G_\nu(x, y) &= \sum_{i=1}^{n} y_i G_{y_i}(x, y) \\ &= -\frac{1}{n\omega(n)} \frac{1}{|x - y|^n} \sum_{i=1}^{n} y_i[(y_i - x_i) - y_i|x|^2 + x_i] \\ &= -\frac{1}{n\omega(n)} \frac{1 - |x|^2}{|x - y|^n}. \end{aligned}$$

If we set $u^*(x) = u(rx)$, then we immediately obtain the formula. □

In the next subsection we will introduce a different method to find a Green's function for the rectangular domain

$$\Omega = R = \{(x, y) : 0 < x < L, 0 < y < H\}.$$

8.3.4 Green's functions via eigenfunctions

In this subsection we introduce a new method to construct a Green's function based on eigenvalues and eigenfunctions.

Let $p(x) \in C^1(\bar{\Omega})$ and $p(x) \geq p_0 > 0$ on Ω for some $p_0 > 0$. and $q(x) \in L^\infty(\Omega)$. For any $f(x) \in L^2(\Omega)$, consider the following problem:

$$L[u] := -\nabla[p(x)\nabla u] + q(x)u = f(x), \qquad x \in \Omega, \tag{8.3.13}$$

$$B[u] := 0, \qquad x \in \partial\Omega, \tag{8.3.14}$$

where the boundary operator

$$B[u] = u \qquad \text{or} \qquad B[u] = \nabla_\nu u.$$

We can define a fundamental solution as well as a Green's function for the operator L in R^n associated with the operator B.

Let $G(x, y)$ be the Green's function associated with the operators L and B in Ω. Then,

$$u(x) = \int_\Omega G(x, y) f(y) dy, \qquad x \in \bar{\Omega}.$$

Consider the eigenvalue problem:

$$L[\phi] = \lambda\phi, \qquad x \in \Omega, \tag{8.3.15}$$

$$B[\phi] = 0, \qquad x \in \partial\Omega. \tag{8.3.16}$$

Suppose $M = \{\lambda_n\}_{n=1}^\infty$ are all eigenvalues and $N = \{\phi_n(x)\}_{n=1}^\infty$ are all orthogonal eigenfunctions (which forms a basis for $L^2(\Omega)$).

If $u(x)$ is a solution of the problem (8.3.13)–(8.3.14), then we can express $u(x)$ as a series solution:

$$u(x) = \sum_{n=1}^\infty a_n \phi_n(x), \qquad x \in \Omega.$$

It follows that

$$L[u] = \sum_{n=1}^\infty a_n L[\phi_n] = \sum_{n=1}^\infty a_n \lambda_n \phi_n.$$

On the other hand, $f(x)$ can be expressed as a series of eigenfunctions:

$$f(x) = \sum_{n=1}^\infty b_n \phi_n(x), \qquad x \in \Omega,$$

where

$$b_n = \frac{< f, \phi_n >}{< \phi_n, \phi_n >}, \qquad n = 1, 2, \cdots.$$

Suppose that $\lambda_n \neq 0$ for all $n \geq 1$, we see that

$$a_n = \frac{< f, \phi_n >}{\lambda_n < \phi_n, \phi_n >}, \qquad n = 1, 2, \cdots.$$

Consequently, due to the uniqueness of the solution we obtain

$$u(x) = \sum_{n=1}^{\infty} a_n \phi_n(x) = \int_{\Omega} f(y) \sum_{n=1}^{\infty} \frac{\phi_n(x)\phi_n(y)}{\lambda_n \int_{\Omega} \phi_n^2(z)dz} dy.$$

It follows that

$$\begin{aligned} G(x, y) &= \sum_{n=1}^{\infty} \frac{\phi_n(x)\phi_n(y)}{\lambda_n \int_{\Omega} \phi_n^2(z)dz} \\ &= \sum_{n=1}^{\infty} \frac{1}{\lambda_n \int_{\Omega} \phi_n^2(z)dz} [\phi_n(x)\phi_n(y)], \qquad x, y \in \Omega, x \neq y. \end{aligned}$$

We summarize the analysis to obtain the following theorem.

Theorem 8.3.4. *Suppose the eigenvalues and eigenfunctions for the problem (8.3.15)–(8.3.16) are given by*

$$\{\lambda_n\}_{n=1}^{\infty}, \qquad \{\phi_n\}_{n=1}^{\infty}.$$

If $\lambda_n \neq 0$, $\forall n \geq 1$, then the Green's function associated with the operators L and B is equal to

$$G(x, y) = \sum_{n=1}^{\infty} \frac{1}{\lambda_n \int_{\Omega} \phi_n^2(z)dz} [\phi_n(x)\phi_n(y)]. \quad \square$$

Example 8.3.4. For the Laplace operator associated with a Dirichlet boundary condition, find the Green's function for the domain

$$R = \{(x_1, x_2) \in R^2 : 0 < x_1 < L, 0 < x_2 < H\}.$$

Solution. For the Laplace operator associated with Dirichlet boundary conditions in R, we know that the eigenvalues and corresponding eigenfunctions are

$$\begin{aligned} \lambda_{nm} &= \left(\frac{n\pi}{L}\right)^2 + \left(\frac{m\pi}{H}\right)^2, \\ \phi_{nm}(x_1, x_2) &= \sin\left(\frac{n\pi x_1}{L}\right) \sin\left(\frac{m\pi x_2}{H}\right), \ n, m = 1, 2, \cdots. \end{aligned}$$

It follows that the Green's function for R is equal to

$$G(x_1, x_2, y_1, y_2) = \frac{4}{HL} \sum_{n,m=1}^{\infty} \frac{1}{\lambda_{nm}} \sin\left(\frac{n\pi x_1}{L}\right) \sin\left(\frac{m\pi x_2}{H}\right) \sin\left(\frac{n\pi y_1}{L}\right) \sin\left(\frac{m\pi y_2}{H}\right).$$

8.3.5 Generalized Green's function for the Laplace operator

We have seen in the previous subsection that there is no Green's function if $\lambda = 0$ is one of the eigenvalues for the eigenvalue problem (8.3.15)–(8.3.16). However, we may modify the definition to define a generalized Green's function when there exists an eigenvalue that is equal to 0.

Suppose there exists an eigenvalue $\lambda_m = 0$ for some m. Let $N_m := \{\phi_m(x)\}$ be the corresponding subspace of eigenfunctions corresponding to λ_m. Then, by the Fredholm Alternative, the problem (8.3.13)–(8.3.14) has a unique solution if and only if the nonhomogeneous term $f(x)$ satisfies

$$< f, \phi_m >= 0, \qquad \forall \phi_m \in N_m.$$

To illustrate the basic idea, we assume $dim N_m = 1$.

Suppose $G(x, y)$ is the Green's function associated with operators L and B:

$$\begin{aligned} -L_y G(x, y) &= \delta(x - y), && x, y \in \Omega, x \neq y, \\ \nabla_\nu G(x, y) &= 0, && y \in \partial\Omega. \end{aligned}$$

If λ_m is an eigenvalue and $\phi_m \subset N_m$ is the corresponding eigenfunction for the operator L associated with the boundary operator B, then the Fredholm Alternative implies

$$0 =< \delta(x - y), \phi_m(y) >= \int_\Omega \delta(x - y)\phi_m(y)dy = \phi_m(x), \qquad \forall x \in \Omega,$$

which is a contradiction.

To overcome the issue we introduce a forcing term $c(x)\phi_m(y)$ to replace $\delta(x - y)$, where $c(x)$ will be chosen later. Then, the Fredholm Alternative implies

$$0 =< \delta(x - y) + c(x)\phi_m(y), \phi_m(y) >= \phi_m(x) + c(x) \int_\Omega \phi_m^2 dy.$$

It follows that

$$c(x) = -\frac{\phi_m(x)}{\int_\Omega \phi_m^2 dy}.$$

This leads to the following definition.

Definition 8.3.2. Let $\lambda_m = 0$ be an eigenvalue of the problem (8.3.15)–(8.3.16) and $\phi_m(x)$ be the corresponding eigenfunction. Then, the solution of the following problem

$$\begin{aligned} -L_y G(x, y) &= \delta(x - y) - \frac{\phi_m(x)\phi_m(y)}{\int_\Omega \phi_m^2 dy}, && x, y \in \Omega, x \neq y, \\ \nabla_\nu G(x, y) &= 0, && y \in \partial\Omega, \end{aligned}$$

is called a generalized Green's function for the operator L subject to the Neumann boundary operator B.

With the modified Green's function, we can have a similar representation. Consider the problem

$$-L[u] = f(x), \qquad x \in \Omega, \tag{8.3.17}$$

$$\nabla_\nu u(x) = 0, \qquad x \in \partial\Omega. \tag{8.3.18}$$

Theorem 8.3.5. *Let $G_m(x, y)$ be a generalized Green's function associated with the operator L subject to the boundary condition $\nabla_\nu G = 0$. Then, the solution $u(x)$ of problem (8.3.17)–(8.3.18) can be expressed in terms of a Green's function:*

$$u(x) = \int_\Omega f(y) G_m(x, y) dy + \frac{\phi_m(x)}{|\Omega|} \int_\Omega u(y) \phi_m(y) dy,$$

where $\phi_m(x)$ is the eigenfunction corresponding to the eigenvalue $\lambda_m = 0$. □

We consider the Laplace operator subject to the Neumann boundary condition:

$$-\Delta u = f(x), \qquad x \in \Omega, \tag{8.3.19}$$

$$\frac{\partial u}{\partial \nu} = g(x), \qquad x \in \partial\Omega. \tag{8.3.20}$$

First, the problem (8.3.19)–(8.3.20) has a unique solution if and only if

$$\int_\Omega f(x) dx + \int_{\partial\Omega} g(x) ds = 0.$$

We know $\lambda_1 = 0$ is an eigenvalue and we choose $\phi_1(x) = \frac{1}{\sqrt{|\Omega|}}$ as the corresponding eigenfunction. Define the generalized Green's function as follows:

$$-\Delta_y G_m(x, y) = \delta(x - y) - \frac{1}{|\Omega|}, \qquad x, y \in \Omega, x \neq y,$$

$$\frac{\partial G_m}{\partial \nu} = 0, \qquad y \in \partial\Omega.$$

Theorem 8.3.6. *(a) The generalized Green's function $G_m(x, y)$ is equal to*

$$G_m(x, y) = \frac{1}{|\Omega|} + \sum_{n=2}^{\infty} \phi_n(x) \phi_n(y).$$

(b) The solution of the problem (8.3.19)–(8.3.20) is equal to

$$u(x) = \int_\Omega G_m(x, y) f(y) dy + \int_{\partial\Omega} G_m(x, y) g(y) ds(y) + \bar{u},$$

where

$$\bar{u} = \frac{1}{|\Omega|} \int_\Omega u(x) dx.$$

Proof. First, suppose

$$\delta(x-y)=\sum_{n=1}^{\infty}a_n(y)\phi_n(x)=\frac{1}{|\Omega|}+\sum_{n=2}^{\infty}a_n(y)\phi_n(x).$$

Then,

$$\phi_m(y)=\int_\Omega \delta(x-y)\phi_m(x)dx=a_m(y).$$

Hence,

$$\delta(x-y)=\sum_{n=1}^{\infty}\phi_n(x)\phi_m(y), \qquad x,y\in\Omega,\ x\neq y.$$

Next, set

$$G_m(x,y)=\sum_{n=2}^{\infty}b_n(y)\phi_n(x).$$

We claim that

$$-\Delta_x G_m(x,y)=\sum_{n=2}^{\infty}b_n(y)\lambda_n\phi_n(x).$$

Indeed, if we choose

$$b_n(y)=\frac{\phi_n(y)}{\lambda_n}, \qquad n=2,3,\cdots,$$

then,

$$\frac{1}{|\Omega|}+\sum_{n=2}^{\infty}b_n(y)\lambda_n\phi_n(x)=\frac{1}{|\Omega|}+\sum_{n=2}^{\infty}\phi_n(y)\phi_n(x)=\delta(x-y),$$

which implied the claim since $\phi_1(x)=\frac{1}{\sqrt{|\Omega|}}$.

(b) By using Green's identity, we find that

$$u(x)=\int_\Omega G_m(x,y)f(y)dy+\int_{\partial\Omega}G_m(x,y)g(y)ds(y)+\bar{u}, \qquad x\in\Omega. \quad \square$$

8.4 The fundamental solution of the heat equation

In this section we will derive the fundamental solution for the heat operator. Some elementary properties are also deduced. With the help of the fundamental solution we derive the explicit solution representation for the Cauchy problem associated with the heat operator.

8.4.1 The definition of the fundamental solution

Let $f(x,t) \in L^1(R^n \times R^1_+) \bigcap L^2(R^n \times R^1_+)$, $g(x) \in L^2(R^n)$. Consider a Cauchy problem for the heat equation in $R^n \times R^1_+$:

$$u_t - \Delta u = f(x,t), \qquad (x,t) \in R^n \times R^1_+, \tag{8.4.1}$$

$$u(x,0) = g(x), \qquad x \in R^n. \tag{8.4.2}$$

We are seeking an explicit solution representation.

Similar to the Laplace equation, naturally for any (y,τ) we need to find a function that satisfies

$$\Phi_t - \Delta_x \Phi = \delta(x-y, t-\tau), \qquad (x,t) \neq (y,\tau),$$

where $\delta(x-y,t-\tau)$ is the Dirac-delta function with

$$\int_0^t \int_{R^n} \delta(x-y,t-\tau) f(y,\tau) dy d\tau = f(x,t), \qquad \forall f(x,t) \in C(R^n \times R^1_+).$$

Definition 8.4.1. The solution of the following heat equation

$$\Phi_t - \Delta_x \Phi = \delta(x-y, t-\tau), \qquad x, y \in R^n, t > \tau, \tag{8.4.3}$$

and

$$\Phi(x-y, t-\tau) = 0, \ t < \tau, \qquad \int_{R^n} \Phi(x-y, t-\tau) dy = 1,$$

is called a fundamental solution for the heat equation.

Obviously, the fundamental solution $\Phi(x,t)$ is singular at $(x,t) = (0,0)$. To find the fundamental solution, we note that if $u(x,t)$ is a solution of the heat equation, then for any real number λ, $u(\lambda x, \lambda^2 t)$ is also a solution. This motivates us to find an explicit solution by choosing $\lambda = \frac{1}{\sqrt{t}}$ to eliminate the variable t. Moreover, we seek the solution that is radially symmetric with respect to the space variables. This leads us to set

$$\Phi(x,t) = v(y), \qquad (x,t) \in R^n \times R^1_+,$$

where $y = \frac{|x|^2}{t}$.

However, it will be seen that $v(y)$ does not satisfy the following condition:

$$\int_{R^n} \Phi(x,t) dx < \infty, \qquad \forall t \in R^1_+.$$

Hence, $\Phi(x,t)$ needs a more complicated form.

Now, we use the method of separation of variables to seek a solution in the following form:

$$\Phi(x,t) = w(t) v(y), \qquad (x,t) \in R^n \times R^1_+.$$

Then, the heat equation for Φ is equivalent to

$$tw'v - 2nwv' - wy[4v'' + v'] = 0, \qquad (x,t) \in R^n \times R^1_+.$$

Consequently, we obtain

$$\frac{tw'(t)}{w(t)} = \frac{2nv'(y) + y[4v''(y) + v'(y)]}{v} = k,$$

where k is an unknown constant.

It follows that the general solution for $w(t)$ is equal to

$$w(t) = ct^k, \qquad t > 0.$$

It is difficult to find $v(y)$ for any constant k. However, if we choose $k = -\frac{n}{2}$, we have

$$2nv'(y) + y[4v''(y) + v'(y)] = -\frac{n}{2}v(y).$$

We can rewrite the equation in the following form:

$$(y^{\frac{n}{2}}v')' + \frac{1}{4}(y^{\frac{n}{2}}v)' = 0,$$

which implies

$$y^{\frac{n}{2}}[v'(y) + \frac{1}{4}v(y)] = \text{constant}.$$

The above constant must be equal to 0 if we require that $v(y), v'(y) \to 0$ as $y \to \infty$.

It follows that

$$v(y) = e^{-\frac{y}{4}}.$$

Therefore the fundamental solution has the following form:

$$\Phi(x,t) = \frac{c}{t^{\frac{n}{2}}}e^{-\frac{|x|^2}{4t}}.$$

We choose c such that

$$\int_{R^n} \Phi(x,t)dy = 1.$$

It follows that

$$\Phi(x,t) = \frac{1}{(4\pi t)^{\frac{n}{2}}}e^{-\frac{|x|^2}{4t}}, \qquad x \in R^n, t > 0.$$

For convenience, we often extend the fundamental solution to $(-\infty, 0)$ by setting

$$\Phi(x-y, t-\tau) = 0, \qquad \forall t < \tau.$$

Theorem 8.4.1. *The fundamental solution of the heat equation is given by*

$$\Phi(x-y,t-\tau)=\frac{1}{(4\pi(t-\tau))^{\frac{n}{2}}}e^{-\frac{|x-y|^2}{4(t-\tau)}},\qquad x,y\in R^n,t>\tau.\quad\square$$

8.4.2 Some properties of the fundamental solution

From the expression of $\Phi(x,t)$, we know that $\Phi(x,t)$ is analytic with respect to x and smooth with respect to t if $t>0$.

Property 8.4.1. *The fundamental solution $\Phi(x-y,t-\tau)$ is symmetric with respect to (x,t) and (y,τ):*

$$\Phi(x-y,t-\tau)=\Phi(y-x,\tau-t).$$

Proof. This is clear from the explicit expression. $\square$

Proposition 8.4.1. *For any $(x,t)\neq(y,\tau)$ in $R^n\times R^1_+$,*

$$(a)\quad 0<\Phi(x-y,t-\tau)\le C(t-\tau)^{-\frac{n}{2}}e^{-\frac{|x-y|^2}{4(t-\tau)}};$$

$$(b)\quad |\nabla_x\Phi(x-y,t-\tau)|\le C(t-\tau)^{-\frac{n+1}{2}}e^{-\frac{|x-y|^2}{4(t-\tau)}},$$

where C is a constant.

Proof. (a) is obvious from the construction.

For (b) we note that for any $\mu\in(0,1)$ and any $z\in[0,\infty)$

$$z^{\frac{n}{2}-\mu}e^{-z}\le C,$$

where C depends only on n and μ. $\square$

For convenience, for any $T>0$ we set

$$Q_T=R^n\times(0,T],\qquad Q=R^n\times(0,\infty).$$

Let $f(x,t)$ be measurable in Q_T. Define

$$V(x,t)=\int_0^t\int_{R^n}\Phi(x-y,t-\tau)f(y,\tau)dyd\tau.$$

Theorem 8.4.2. *Let $\Phi(x-y,t-\tau)$ be the fundamental solution of the heat operator.*

(a) If $f(x,t)$ is in $L^1(R^n\times R^1_+)$, then for all $1\le i\le n$

$$\frac{\partial V(x,t)}{\partial x_i}=\int_0^t\int_{R^n}\Phi_{x_i}(x-y,t-\tau)f(y,\tau)dyd\tau.$$

(b) If $f(x,t)$ is Hölder continuous with respect to (x,t) with exponent $\alpha \in (0,1)$ in $\bar{Q}_T$, then, for all $1 \le i, j \le n$,

$$\frac{\partial V(x,t)}{\partial t} = \int_0^t \int_{R^n} \Phi_t(x-y,t-\tau)[f(y,\tau)-f(x,\tau)]dyd\tau + f(x,t),$$
$$\frac{\partial^2 V(x,t)}{\partial x_i x_j} = \int_0^t \int_{R^n} \Phi_{x_i x_j}(x-y,t-\tau)[f(y,\tau)-f(x,\tau)]dyd\tau.$$

Proof. (a) is directly from differentiation from the expression, since the function is integrable in Q.

To prove (b), we note that $f(x,t)$ is Hölder continuous in $\bar{Q}_T$, then

$$|f(x,t)-f(y,\tau)| \le C[|x-y|^\alpha + |t-\tau|^{\frac{\alpha}{2}}], \forall x, y, \in R^n, t, \tau \in [0,T].$$

It follows that

$$|\Phi_t(x-y,t-\tau)[f(y,\tau)-f(y,t)]| \le C\frac{ze^{-z^2}}{|t-\tau|^{1-\frac{\alpha}{2}|}}, \tag{8.4.4}$$

$$|\Phi_{x_i x_j}(x-y,t-\tau)[f(y,\tau)-f(x,\tau)]| \le C\frac{ze^{-z^2}}{|t-\tau|^{1-\alpha}}, \tag{8.4.5}$$

where $z = \frac{|x-y|}{\sqrt{|t-\tau|}}$.

Since both functions on the right-hand side of the above inequalities (8.4.4)–(8.4.5) are integrable in Q_T, we see the desired derivatives are justified. □

8.4.3 The Cauchy problem for the heat equation

In this subsection we will derive the solution representation for the Cauchy problem in terms of the fundamental solution. The main challenge is that the fundamental solution $\Phi(x-y,t-\tau)$ is singular at the point (y,τ). However, the difficulty can be overcame by using some similar ideas to the case when we deal with the fundamental solution associated with the Laplace operator.

H(8.4.1)

(a) Let $f(x,t) \in L^1(Q) \bigcap L^2(Q_T)$.

(b) Let $g(x) \in L^2_{loc}(R^n)$ and

$$|g(x)| \le Ce^{h|x|^2},$$

where

$$0 < h < \frac{1}{4T}.$$

Consider the following Cauchy problem in Q_T:

$$u_t - \Delta u = f(x,t), \qquad (x,t) \in Q_T, \tag{8.4.6}$$

$$u(x,0)=g(x), \qquad x\in R^n. \tag{8.4.7}$$

Theorem 8.4.3. *Under the assumption H(8.4.1), the Cauchy problem (8.4.6)–(8.4.7) has a unique solution*

$$u(x,t)\in L^2((0,T);H^2(R^n))\bigcap C([0,T];L^2(R^n)),$$

which can be expressed by

$$u(x,t)=\int_{R^n}\Phi(x,y,t,0)g(y)dy+\int_0^t\int_{R^n}\Phi(y-x,t-\tau)f(y,\tau)dyd\tau.$$

Moreover, if $f(x,t)\in C^{\alpha,\frac{\alpha}{2}}(Q_T)$*, then*

$$u(x,t)\in C^{2+\alpha,1+\frac{\alpha}{2}}(R^n\times(0,\infty)).$$

Proof. Let

$$\begin{aligned}
u_1(x,t)&=\int_{R^n}\Phi(x,y,t,0)g(y)dy\\
u_2(x,t)&=\int_0^t\int_{R^n}\Phi(y-x,t-\tau)f(y,\tau)dyd\tau\\
&=\int_0^t\int_{R^n}\Phi(y,\tau)f(x-y,t-\tau)dyd\tau.
\end{aligned}$$

For $t>0$, $u_1(x,t)$ is differentiable with respect to x and t. Moreover,

$$u_{1t}-\Delta u_1=\int_{R^n}[\Phi_t-\Delta\Phi(x,t,y,0)]g(y)dy=0.$$

Furthermore, since $g(x)\in L^2(R^n)$,

$$\begin{aligned}
\lim_{t\to0+}||u_1(x,t)-g(x)||^2_{L^2(R^n)}&=\lim_{t\to0+}||\int_{R^n}\Phi(x,y,t,0)(g(y)-g(x))dy||^2_{L^2(R^n)}\\
&\le C\lim_{t\to0+}||g(x-z\sqrt{t})-g(x)||_{L^2(R^n)}=0.
\end{aligned}$$

For $u_2(x,t)$, we rewrite the expression as follows:

$$u_2(x,t)=\int_0^t\int_{R^n}\Phi(y-x,t-\tau)f(y,\tau)dyd\tau.$$

By Theorem 8.4.2, we have

$$u_{2t}=\int_0^t\int_{R^n}\Phi_t(y-x,t-\tau)[f(y,\tau)-f(x,\tau)]dyd\tau+f(x,t)$$

$$= \int_0^t \int_{R^n} \Phi_t(y - x, t - \tau)[f(y, \tau) - f(x, \tau)]dyd\tau + f(x, t)$$

$$\Delta u_2 = \int_0^t \int_{R^n} \Delta_x \Phi(y - x, t - \tau)[f(y, \tau) - f(x, \tau)]dyd\tau.$$

It follows that

$$\int\int_Q [\Phi_t - \Delta_x \Phi] f(y, \tau)dx = \int\int_Q \delta(x - y, t - \tau) f(y, \tau)dyd\tau = f(x, t).$$

However, since $\Phi(x - y, t - \tau)$ is singular at (x, t), we need to take away a small parabolic cylinder from $R^n \times R^1_+$. Let $r > 0$ be small and $Q_r(x, t) = B_r(x) \times (t - r^2, t]$. Moreover, we first assume that $f(x, t)$ is smooth with a compact support in $R^n \times R^1_+$. Then,

$$\begin{aligned}
u_{2t} - \Delta u_2 &= \int_0^t \int_{R^n} \Phi(y, \tau)[(\frac{\partial}{\partial t} - \Delta_x) f(x - y, t - \tau)dyd\tau \\
&+ \int_{R^n} \Phi(y, t) f(x - y, 0)dy. \\
&= \int_0^t \int_{R^n} \Phi(y, \tau)[(-\frac{\partial}{\partial \tau} - \Delta_y) f(x - y, t - \tau)dyd\tau \\
&+ \int_{R^n} \Phi(y, t) f(x - y, 0)dy \\
&= \int_\varepsilon^t \int_{R^n} \Phi(y, \tau)[(-\frac{\partial}{\partial \tau} - \Delta_y) f(x - y, t - \tau)dyd\tau \\
&+ \int_\varepsilon^t \int_{R^n} \Phi(y, \tau)[(-\frac{\partial}{\partial \tau} - \Delta_y) f(x - y, t - \tau)dyd\tau \\
&+ \int_{R^n} \Phi(y, t) f(x - y, 0)dy \\
&= J_1 + J_2 + J_3.
\end{aligned}$$

Clearly, since $\Phi(y, s)$ is a fundamental solution, we see that

$$J_1 = \int_{R^n} \Phi(y, \varepsilon) f(x - y, t - \varepsilon)dy - \int_{R^n} \Phi(x - y, t) f(x - y, 0)dy.$$

Also,

$$|J_2| \le [||f_t||_0 + ||\Delta f||_0]\varepsilon \to 0, \qquad \text{as } \varepsilon \to 0.$$

It follows that

$$u_{2t} - \Delta u_2 = f(x, t), \qquad (x, t) \in R^n \times R^1_+.$$

When $f(x,t) \in L^\infty(R^n \times R^1_+) \bigcap C^{\alpha,\frac{\alpha}{2}}(Q)$, we use a smooth approximation $f_n(x,t) \in C_0^\infty(R^n \times R^1_+)$ to obtain

$$u_n(x,t) = \int_{R^n} \Phi(x,y,t,0)g(y)dy + \int_0^t \int_{R^n} \Phi(y-x,t-\tau)f_n(y,\tau)dyd\tau.$$

Now, for any $T > 0$,

$$||u_n||_{L^\infty(R^n \times R^1_+)} \le ||g||_{L^\infty(R^n)} + T||f||_{L^\infty(R^n \times R^1_+)}.$$

Similarly, we see $u_n(x,t)$ is locally Hölder continuous and

$$||u_n||_{C^{\alpha,\frac{\alpha}{2}}(R^n \times (0,T])} \le C.$$

By the compactness theorem and the regularity theory ([8,21]), we see that $u(x,t)$ satisfies

$$u(x,t) = \int_{R^n} \Phi(x,y,t,0)g(y)dy + \int_0^t \int_{R^n} \Phi(y-x,t-\tau)f(y,\tau)dyd\tau.$$

From Theorem 8.4.2, we see that

$$u(x,t) \in C^{2+\alpha,1+\frac{\alpha}{2}}(R^n \times (0,\infty))$$

and it satisfies Eq. (8.4.6)–(8.4.7). The uniqueness follows from the maximum principle that will be proved in the next theorem. □

8.4.4 The maximum principle for the Cauchy problem

In this subsection we prove a maximum principle for the Cauchy problem. Unlike the maximum principle for a bounded domain, we need a condition that controls the growth of the solution as $|x| \to \infty$. Roughly speaking, we should require that a solution $u(x,t)$ grows at most like $e^{C|x|^2}$ for some constant $C > 0$.

Theorem 8.4.4. *(The maximum principle for the Cauchy problem) Let $u(x,t) \in C^{2,1}(Q_T) \bigcap C(R^n \times [0,T))$ satisfy*

$$\begin{aligned} &u_t - \Delta u \ge 0, \qquad (x,t) \in Q_T, \\ &u(x,0) \ge 0. \end{aligned}$$

Moreover, there exists a constant $\alpha > 0$ such that

$$\lim_{R\to\infty} inf_{B_R(0)\times[0,T]} u(x,t)e^{\alpha R^2} \ge 0. \tag{8.4.8}$$

Then,

$$u(x,t) \ge 0, \qquad (x,t) \in Q_T.$$

Proof. We choose $\beta > \alpha$ and set $T_1 = \frac{1}{8\beta}$. Introduce an auxiliary function

$$v(x,t) = (1-4\beta t)^{-\frac{n}{2}} e^{\frac{\beta|x|^2}{(1-4\beta t)}}.$$

Then, a direct calculation shows that

$$v_t - \Delta v = 0, \qquad (x,t) \in Q_{T_1}.$$

For any $\varepsilon > 0$, set

$$w(x,t) = u(x,t) + \varepsilon v(x,t).$$

Then,

$$\begin{aligned} &w_t - \Delta w \geq 0, && (x,t) \in Q_{T_1}, \\ &w(x,0) \geq 0, && x \in R^n. \end{aligned}$$

Moreover, by the assumption,

$$\lim_{|R|\to\infty} inf_{B_R(0)} w(x,t) \geq 0, \qquad \text{uniformly for } t \in [0,T].$$

For any fixed point $(x_0, t_0) \in Q_{T_1}$ and any small constant $a > 0$, we can choose R sufficiently large such that

$$w(x,t) \geq -a, \qquad \forall |x| \geq R, t \in [0, T_1].$$

On $Q_R - B_R(0) \times (0, T_1]$, $V(x,t) := w(x,t) + a$ satisfies

$$\begin{aligned} &V_t - \Delta V \geq 0, && (x,t) \in Q_R, \\ &V(x,t) \geq 0, && (x,t) \in \partial B_R(0) \times (0, T_1], \\ &V(x,0) \geq 0, && x \in R^n. \end{aligned}$$

We use the maximum principle in a bounded domain to obtain

$$V(x,t) \geq 0, \qquad (x,t) \in Q_R.$$

Therefore

$$w(x_0, t_0) \geq -a.$$

Since a can be arbitrarily small and $(x_0, t_0) \in Q_{T_1}$ is arbitrary, we conclude $u(x,t) \geq 0$ in Q_{T_1}. We can continue this process to obtain the nonnegativity of $u(x,t)$ in the interval $[T_1, T_1 + \frac{1}{8\beta}]$. After a finite number of steps, we see that

$$u(x,t) \geq 0, \qquad (x,t) \in R^n \times [0,T].$$

□

A direct consequence of Theorem 8.2.1 is that the solution of the Cauchy problem (8.4.1)–(8.4.2) is unique for a small $T > 0$ if for any large $R > 0$,

$$|u(x,t)| \leq Ce^{\alpha|x|^2}, \qquad \forall |x| \geq R,$$

where C and α are independent of R but depend on T.

Another consequence of Theorem 8.4.4 is that the propagation speed of the heat conduction is infinite.

Corollary 8.4.1. *Let $g(x) \in C(R^1)$ and $f(x,t) \geq 0$ on $R^n \times R^1_+$ be nonnegative and $g(x) \neq 0$ on R^1. Then, the solution $u(x,t)$ of the Cauchy problem must be positive:*

$$u(x,t) > 0, \qquad \forall (x,t) \in Q. \quad \square$$

Corollary 8.4.2. *(The maximum norm estimate) Let $g(x) \in L^\infty(R^n)$ and $f(x,t) \in L^\infty(Q)$. Then, the solution $u(x,t)$ of the Cauchy problem (8.4.6)–(8.4.7) must have the following estimate:*

$$\sup_{Q_T} u(x,t) \leq C(T)[||g||_{L^\infty(R^n)} + ||f||_{L^\infty(Q)}], \qquad \forall (x,t) \in Q. \quad \square$$

Remark 8.4.1. The growth condition (8.4.8) is optimal for the uniqueness of the Cauchy problem (8.4.6)–(8.4.7). One can construct a counterexample that there exists an infinite number of solutions $u(x,t)$ for the Cauchy problem if

$$\int_0^t \int_{R^1} |u(y,\tau)| e^{-K|y|^{2+\varepsilon}} dy d\tau < \infty$$

for any $\varepsilon > 0$.

8.5 Green's function of the heat equation

In this section we define a Green's function associated with the heat equation. The basic goal is to find the explicit solution representation for the heat equation in a general domain. Many ideas are parallel to the case associated with the Laplace equation in a general domain.

8.5.1 The definition of a Green's function

We have seen in Section 8.4 that a Green's function for the Laplace operator in a bounded domain is constructed from the fundamental solution, except with zero value on the boundary. It is natural to define a Green's function associated with the heat operator as follows.

Let Ω be a domain in R^n and $Q_T = \Omega \times (0,T]$. For any $\tau \in (0,T]$, Let $Q_T(\tau) = \Omega \times (\tau, T]$.

Definition 8.5.1. We call $G(x-y,t-\tau)$ a Green's function for the heat operator associated with a Dirichlet boundary condition, if for any function $f(x) \in L^2(\Omega)$,

$$G_t - \Delta_x G = \delta(x-y,t-\tau), \qquad (x,t) \in Q_T(\tau), \tag{8.5.1}$$

$$G(x-y,t-\tau) = 0, \qquad (x,t) \in \partial\Omega \times (\tau, T], \tag{8.5.2}$$

$$\lim_{t\to\tau+} \| \int_\Omega G(x-y,t-\tau) f(y) dy - f \|_{L^2(\Omega)} = 0. \tag{8.5.3}$$

To construct a Green's function, we only need to use the fundamental solution Φ with a correction term, denoted by $\psi(x-y,t-\tau)$, which satisfies the heat equation with the boundary value Φ. For any fixed $y \in \Omega$, $\tau \in [0,T)$, we consider

$$\psi_t - \Delta_x \psi = 0, \qquad x \in \Omega, \tau < t < T, \tag{8.5.4}$$

$$\psi(x-y,t-\tau) = \Phi(x-y,t-\tau), \qquad x \in \partial\Omega, \tau < t < T, \tag{8.5.5}$$

$$\psi(x-y,0) = 0, \qquad x \in \Omega \times \{t-\tau\}. \tag{8.5.6}$$

The existence of $\psi(x-y,t-\tau)$ is known, since $\Phi(x-y,t-\tau)$ is smooth on $\partial\Omega$ for $t > \tau$. Then, by definition,

$$G(x-y,t-\tau) = \Phi(x-y,t-\tau) - \psi(x-y,t-\tau)$$

is the Green's function for the heat equation associated with the Dirichlet boundary condition.

8.5.2 Green's representation

From the definition of the Green's function, we have the following elementary properties.

Property 8.5.1. *Let $G(x-y,t-\tau)$ be the Green's function for the heat operator associated with a Dirichlet boundary condition. Then,*

(a) $G(x-y,t-\tau) = G(y-x,t-\tau)$.

(b)

$$0 < G(x-y,t-\tau) \le \frac{C}{(t-\tau)^{n/2}} e^{-\frac{|x-y|^2}{4(t-\tau)}}, \qquad \forall x,y \in Q_T, t > \tau.$$

Consider the following problem:

$$u_t - \Delta u = f(x,t), \qquad (x,t) \in Q_T, \tag{8.5.7}$$

$$u(x,t) = 0, \qquad (x,t) \in \partial\Omega \times (0,T], \tag{8.5.8}$$

$$u(x,0) = g(x), \qquad x \in \Omega. \tag{8.5.9}$$

From the construction of the Green's function, we have the following theorem.

Theorem 8.5.1. *Let $f(x,t)$ be $L^2(Q_T)$ and $g(x) \in L^2(\Omega)$. Then, the solution $u(x,t)$ of Eq. (8.5.7)–(8.5.9) can be expressed by*

$$u(x,t) = \int_\Omega G(x-y,t)g(y)dy + \int_0^t \int_\Omega G(x-y,t-\tau)f(y,\tau)dyd\tau.$$

Moreover,

(a) if $f(x,t) \in C^{\alpha,\frac{\alpha}{2}}(Q_T)$ for some $\alpha \in (0,1)$, then $u(x,t) \in C^{2+\alpha,1+\frac{\alpha}{2}}(Q_T)$;
(b) if $f(x,t) \in C^{\alpha,\frac{\alpha}{2}}(\bar{Q}_T)$ and $g(x) \in C^\alpha(\bar{\Omega})$, then $u(x,t) \in C^{2+\alpha,1+\frac{\alpha}{2}}(\bar{Q}_T)$.

Proof. The argument is similar to the case for the Laplace operator. $G(y-x,\tau-t)$ has a singular point (x,t). For the definition of the heat operator, we have

$$\begin{aligned}
u(x,t) &= \int_0^t \int_\Omega u(y,\tau)\delta(y-x,\tau-t)dyd\tau \\
&= \int_0^t \int_\Omega u(y,\tau)[G_\tau(y-x,\tau-t) - \Delta_y G(y-x,\tau-t)]dyd\tau \\
&= \int_\Omega G(y,t)g(y)dy + \int_0^t \int_\Omega [u_\tau - \Delta_y u]G(y-x,\tau-t)dyd\tau \\
&= \int_\Omega G(y-x,t)g(y)dy + \int_0^t \int_\Omega f(y,\tau)G(y-x,\tau-t)dyd\tau.
\end{aligned}$$

This formula is the same as what we derived in Chapter 7 by using the Fourier transform. The regularity results are based on the estimates for the fundamental solution Φ. We omit the proof. The reader may find the detailed proof in [9,19,21]. □

8.5.3 Green's functions for some special domains

Similar to the Laplace equation, the method of a mirror-image reflection is an effective way to construct a Green's function. We give a few examples here.

Example 8.5.1. Find the Green's function for the heat operator in the half-space:

$$\Omega = R^n_+ := \{x = (x_1, x_2, \cdots, x_n) \in R^n, x_n > 0\}.$$

Solution. For $y = (y_1, y_2, \cdots, y_n) \in R^n_+$, we define

$$\tilde{y} = (y_1, y_2, \cdots, -y_n).$$

Then, by using the symmetry we see that the Green's function in $R^n_+ \times (0, T]$ is equal to

$$G(x-y,t-\tau) = \Phi(x-y,t-\tau) - \Phi(x-\tilde{y}),t-\tau).$$

In particular, in one-space dimension we have

$$G(x-y,t-\tau) = \frac{1}{\sqrt{4\pi(t-\tau)}}\left(e^{-\frac{(x-y)^2}{4(t-\tau)}} - e^{-\frac{(x+y)^2}{4(t-\tau)}}\right).$$

Example 8.5.2. Find the Green's function for the heat operator in the domain $\Omega = \{x \in R^1 : 0 < x < L\}$, where $L > 0$.

Solution. To satisfy the boundary condition $x = 0$, we use the reflection method to have

$$G_1(x-y,t-\tau) := \frac{1}{\sqrt{4\pi(t-\tau)}}\left(e^{-\frac{(x-y)^2}{4(t-\tau)}} - e^{-\frac{(x+y)^2}{4(t-\tau)}}\right).$$

However, G_1 does not satisfy the boundary condition at $x = L$, so we begin with G_1 and subtract a new function such that $G_1 = 0$ at $y = L$:

$$G_2(x-y,t-\tau) := G_1(x+L-y,t-\tau) - G_1(x-y,t-\tau).$$

G_2 satisfies the heat equation and the boundary value at $y = L$. However, G_2 does not satisfy the boundary condition at $y = 0$. We begin with G_2 and use the mirror-image reflection to obtain $G_3(x-y,t-\tau)$. Then, we continue this process to obtain

$$G(x-y,t-\tau) = \sum_{m=-\infty}^{\infty} \Phi(x+mL-y,t-\tau) - \sum_{m=-\infty}^{\infty} \Phi(x+mL+y,t-\tau).$$

Justification of the convergence for the above series relies on the exponential decay property from the fundamental solution. One can extend the method into a rectangle in n-space dimension.

8.6 Finite-time blowup and extinction

In this section we study a semilinear heat equation to demonstrate some phenomena that are quite different from a linear problem. We will show that the solution may blow up in finite time when a nonlinear source $f(s)$ grows fast enough as long as the initial datum satisfies certain conditions. On the other hand, we show that the solution may become zero in finite time if $f(s)$ sinks fast enough.

8.6.1 Finite-time blowup in a bounded domain

Let Ω be a bounded domain in R^n and $Q = \Omega \times (0,\infty)$. To ensure the existence of a solution in a short time, we assume the following conditions.

H(6.1) Let $f(s)$ be a locally Lipschitz continuous function on $[0,\infty)$.
H(6.2) Let $g(x) \in L^\infty(\Omega)$.

Consider the following semilinear problem:

$$u_t - \Delta u = f(u), \qquad (x,t) \in Q, \tag{8.6.1}$$

$$u(x,t) = 0, \qquad (x,t) \in \partial\Omega \times (0,\infty), \tag{8.6.2}$$

$$u(x,0) = g(x), \qquad x \in \Omega. \tag{8.6.3}$$

Our goal is to find sufficient conditions on $f(s)$ such that the problem (8.6.1)–(8.6.3) has a solution in Q or the solution blows up (unbounded in $L^p(\Omega)$ for some $p \geq 1$) in finite time. We first prove a local existence result.

Lemma 8.6.1. *Let the assumptions H(6.1)–H(6.2) hold. Then, there exists a maximal time $T^* \in (0, \infty)$ such that the problem (8.6.1)–(8.6.3) has a unique solution in Q_{T^*}.*

Moreover, if $T^* < \infty$, then

$$\lim_{T \to T^*-} \max_{0 \leq t \leq T} ||u||_{L^\infty(\Omega)} = \infty.$$

Proof. We use the contraction mapping principle to prove the result. Choose $V = L^\infty(Q_T)$ for some $T > 0$. Set $K_0 = 1 + ||g||_{L^\infty(\Omega)}$ and

$$L(K_0) := \max\{L : |f(s_1) - f(s_2)| \leq L|s_1 - s_2|, 0 \leq s_1, s_2 \leq K_0\}.$$

For any $v(x,t) \in V$, we define a mapping $M : V \to V$ with $M[v] = u(x,t)$, where $u(x,t)$ is the solution to the following linear problem:

$$u_t - \Delta u = f(v), \qquad (x,t) \in Q_T, \tag{8.6.4}$$

$$u(x,t) = 0, \qquad (x,t) \in \partial\Omega \times (0, T], \tag{8.6.5}$$

$$u(x,0) = g(x), \qquad x \in \Omega. \tag{8.6.6}$$

Clearly, the mapping M is well defined, since the linear problem (8.6.4)–(8.6.6) has a unique solution.

Since $g(x) \in L^\infty(\Omega)$, we see from the Green's representation that

$$||u||_{L^\infty(\Omega)} \leq ||g||_{L^\infty(\Omega)} + [|f(0)| + L(K_0)|\Omega|]t \leq K_0, \qquad t \in [0, T_0],$$

provided that $T_0 < \frac{1}{|f(0)|+L(K_0)|\Omega|}$.

By using Green's representation, we see that

$$||M[v_1] - M[v_2]||_{L^\infty(\Omega)} \leq L(K_0)t||v_1 - v_2||_{L^\infty(\Omega)}, \qquad t \in (0, T_0].$$

Therefore if we choose T_1 with $T_1 < \frac{1}{2}\min\{T_0, \frac{1}{L(K_0)}\}$, then the mapping M is a contraction mapping from V into V. The contraction-mapping principle implies that the mapping M has a unique fixed point. This fixed point of the mapping M is the unique solution of the semilinear problem (8.6.1)–(8.6.3). Define

$$g_1(x) = u(x, T_1), \qquad x \in \Omega.$$

We can consider the problem (8.6.1)–(8.6.3) in $\Omega \times [T_1, T]$ with a new initial value $g_1(x)$ to obtain a local solution on $[T_1, T_2]$. As long as $u(x, T_2)$ is bounded, we can repeat the process to obtain a local solution for the problem (8.6.1)–(8.6.3) over $\Omega \times [T_2, T_3]$ for some $T_3 > T_2$. Let

$$T^* = \sup\{T : \text{ the problem (8.6.1)–(8.6.3) has a unique solution on } Q_T\}.$$

If $T^* < \infty$, then

$$\lim_{T \to T^*-} ||u(x,T)||_{L^\infty(\Omega)} = \infty. \qquad \square$$

Next, we show that the local solution $u(x,t)$ will blow up in finite time if $f(s)$ grows fast enough. From now on we assume that $g(x) \geq 0$ and $f(0) \geq 0$. The maximum principle implies that

$$u(x,t) \geq 0, \qquad \forall (x,t) \in Q.$$

Theorem 8.6.1. *Let $f(s)$ be monotone increasing and convex on $[a, \infty)$ for some large $a > 0$. Moreover, $f(s)$ satisfies*

$$\int_a^\infty \frac{ds}{f(s)} < \infty.$$

Then, the solution $u(x,t)$ of the problem (8.6.1)–(8.6.3) blows up in finite time as long as

$$y_0 := \int_\Omega g(x)\psi(x)dx$$

is suitably large, where $\psi(x)$ is the principal eigenfunction corresponding to the first eigenvalue of the Laplace operator associated with a Dirichlet boundary condition.

Namely, there exists a constant $T^* > 0$ such that

$$\lim_{t \to T^*-} ||u||_{L^2(\Omega)} = \infty,$$

provided that y_0 is suitably large.

Proof. Let $\psi(x)$ be the eigenfunction corresponding to the first eigenvalue λ_1 for the Laplace equation associated with a Dirichlet boundary condition. We further choose $\psi(x) \geq 0$ such that

$$\int_\Omega \psi(x)dx = 1.$$

Define

$$y(t) = \int_\Omega u(x,t)\psi(x)dx.$$

Since $f(s)$ is convex, by Jensen's inequality we have

$$\int_\Omega f(u)\psi(x)dx \geq f(y(t)), \qquad t \geq 0.$$

It follows from Eq. (8.6.1) that

$$y'(t) + \lambda_1 y(t) \geq f(y(t)), \qquad t \geq 0.$$

As long as y_0 is suitably large such that

$$f(y_0) - \lambda_1 y_0 > 0,$$

then

$$T^* \le \int_0^{T^*} \frac{dy(t)}{f(y(t)) - \lambda_1 y(t)} = \int_{y_0}^{y(T^*)} \frac{ds}{f(s) - \lambda_1 s} < \infty,$$

i.e., $y(t)$ must blow up in finite time. □

A classical example for Theorem 8.6.1 is that

$$f(u) = |u|^{p-1}u, \qquad p \in (0, \infty).$$

The solution to the problem (8.6.1)–(8.6.3) blows up in finite time if $p > 1$ and the initial datum is suitably large, while the solution exists globally if $p \in [0, 1]$. An interesting observation is that the size condition for the initial datum is necessary.

Corollary 8.6.1. *Let $f(s) = s^p$ with $p > 1$. Then, the problem (8.6.1)–(8.6.3) has a global solution if $g(x)$ is sufficiently small.*

Proof. Let $n \ge 2$. Set

$$\phi(x,t) = \delta(t+1)\Phi(x,t+1), \qquad x \in R^n, t \ge 0,$$

where δ is a positive constant to be chosen later.

Then,

$$\begin{aligned}
&\phi_t - \Delta\phi - \phi^p \\
&= \delta\Phi(x,t+1) - \delta^p(t+1)^p\Phi(x,t+1) \\
&= \frac{\delta}{(4\pi)^{n/2}}(t+1)^{-\frac{n}{2}}e^{-\frac{|x-y|^2}{4(t+1)}} - \frac{\delta^p}{(4\pi)^{pn/2}}(t+1)^{-\frac{pn}{2}}e^{-p\frac{|x-y|^2}{4(t+1)}} \\
&> 0 \qquad \text{(since } n \ge 2 \text{ and } \Omega \text{ is bounded)}, \forall t \ge 0,
\end{aligned}$$

provided that δ is suitably small.

We choose $g(x)$ sufficiently small such that

$$0 \le g(x) \le \phi(x,0), \qquad \forall x \in \Omega.$$

Let $u(x,t)$ be the solution of problem (8.6.1)–(8.6.3) in Q_{T^*}. Define

$$w(x,t) = \phi(x,t) - u(x,t), \qquad \forall (x,t) \in Q_{T^*}.$$

Then, $w(x,t)$ satisfies

$$w_t - \Delta w = f'(\theta(x,t))w, \qquad (x,t) \in Q_{T^*},$$

$$w(x,t) \geq 0, \qquad (x,t) \in \partial_p Q_{T^*},$$

where $\theta(x,t)$ is the mean value between $\phi(x,t)$ and $u(x,t)$.

The maximum principle implies that

$$0 \leq u(x,t) \leq \phi(x,t), \qquad \forall (x,t) \in Q_{T^*}.$$

Note that $\phi(x,t)$ is uniformly bounded for all $t \in [0,\infty)$. Therefore $u(x,t)$ has a global upper bound, which implies $T^* = \infty$. □

If the Dirichlet boundary condition (8.6.2) is replaced by a Neumann type, a very different situation occurs. From the physical point of view, it is expected that the solution will blow up in finite time for any initial datum since there is no heat loss across the boundary.

Theorem 8.6.2. *Let $f(s)$ be monotone increasing and convex on $[a,\infty)$ for some large $a > 0$. Moreover, $f(s)$ satisfies*

$$\int_a^\infty \frac{ds}{f(s)} < \infty.$$

Then, the solution $u(x,t)$ of the problem (8.6.1), (8.6.3) along with the Neumann boundary condition

$$\nabla_\nu u(x,t) = 0, \qquad (x,t) \in \partial\Omega \times (0,\infty), \tag{8.6.7}$$

blows up in finite time as long as

$$y_0 := \int_\Omega g(x)dx > 0.$$

Proof. The idea is the same as for Theorem 8.6.1. Define

$$y(t) = \frac{1}{|\Omega|}\int_\Omega u(x,t)dx \geq 0.$$

Then, integrating Eq. (8.6.1) over Ω and using Jensen's inequality, we obtain

$$y'(t) = \frac{1}{|\Omega|}\int_\Omega f(u)dx \geq f(y(t)).$$

It follows that

$$T \leq \int_{y_0}^{y(T)} \frac{ds}{f(s)} \leq \int_{y_0}^\infty \frac{ds}{f(s)} < \infty.$$ □

Next, we consider the following problem with a nonlinear Neumann condition:

$$u_t - \Delta u = 0, \qquad (x,t) \in Q, \tag{8.6.8}$$

$$\nabla_\nu u(x,t) = f(u), \qquad (x,t) \in \partial\Omega \times (0,\infty), \tag{8.6.9}$$

$$u(x,0) = g(x) \geq 0, \qquad x \in \Omega. \tag{8.6.10}$$

Theorem 8.6.3. *Let $f(s) \in C^1[0,\infty)$ with $f(0) \geq 0$ be positive on $(0,\infty)$ and satisfy*

$$\lim_{s\to\infty} sup \frac{f(s)}{s} = \infty, \quad \int_a^\infty \frac{ds}{f(s)} < \infty.$$

If

$$g(x) \geq 0, \int_\Omega g(x)dx > 0,$$

then the solution $u(x,t)$ to the problem (8.6.8)–(8.6.10) must blow up in finite time.

We leave the proof as an exercise. A much more detailed analysis can be found in [18].

8.6.2 Finite-time blow up in R^n

There is a very interesting result for the heat equation (8.6.1) when Ω is replaced by the whole space R^n. Let $p \in (1,\infty)$ and $Q = R^n \times R^1_+$. Consider

$$u_t - \Delta u = u^p, \qquad (x,t) \in Q, \tag{8.6.11}$$

$$u(x,0) = g(x) \geq 0, \qquad x \in R^n. \tag{8.6.12}$$

We always assume that $g(x) \in L^1(Q)$ is nonnegative and

$$G_0 := \int_{R^n} g(x)dx > 0.$$

Then, the problem (8.6.11)–(8.6.12) has a nonnegative solution in Q_T for some $T > 0$. Moreover, we know that the solution will blow up in finite time if $g(x)$ is suitably large.

Define

$$p_0 = 1 + \frac{2}{n}.$$

p_0 is called Fujita's critical number (see [11]).

Theorem 8.6.4. *Let $G_0 > 0$. Then,*

(a) There exists a solution to the problem (8.6.11)–(8.6.12) globally if $p > p_0$ and $g(x)$ is sufficiently small.

(b) Every solution to the problem (8.6.11)–(8.6.12) will blow up in finite time if $1 < p \leq p_0$.

Proof. By the solution representation, we have

$$u(x,t) = \int_{R^n} \Phi(x-y,t-\tau)g(y)dy + \int_0^t \int_{R^n} \Phi(x-y,t-\tau)u^p(y,\tau)dyd\tau$$

$$:= u_1(x,t) + u_2(x,t).$$

To prove (a), we construct an upper solution that is global in time. Let $\varepsilon > 0$ be small. Define

$$\psi(x,t) = \varepsilon t^{\eta}\Phi(x,t), \qquad t \geq 1.$$

A direct computation shows that

$$\begin{aligned}\psi_t - \Delta\psi - \phi^p &= \varepsilon\eta t^{\eta-1}\Phi(x,t) - \varepsilon^p t^{\eta p}\Phi(x,t)^p \\ &= \frac{\varepsilon\eta}{(4\pi)^{\frac{n}{2}}} t^{-[1+\frac{n}{2}-\eta]} exp\{-\frac{|x|^2}{4t}\} - \frac{\varepsilon^p\eta}{(4\pi)^{\frac{np}{2}}} t^{-p[1+\frac{n}{2}-\eta]} exp\{-\frac{p|x|^2}{4t}\} \\ &> 0, \qquad \forall t \geq 1,\end{aligned}$$

provided that η is chosen to be sufficiently small such that

$$1 + \frac{n}{2} - \eta < p(\frac{n}{2} - \eta)$$

and then choose ε sufficiently small such that

$$\frac{\varepsilon\eta}{(4\pi)^{\frac{n}{2}}} > \frac{\varepsilon^p\eta}{(4\pi)^{\frac{np}{2}}}.$$

It follows that

$$u(x,t) \leq \psi(x,t+1), \qquad \forall (x,t) \in Q.$$

Now, if $g(x) \leq \psi(x,1)$, we use the maximum principle to obtain

$$u(x,t) \leq \psi(x,t+1), \qquad \forall (x,t) \subset Q.$$

Thus $u(x,t)$ is uniformly bounded in Q. Hence, $u(x,t)$ must exist globally.

To prove (b), we assume that the solution $u(x,t)$ exists for all $t \geq 0$.

For any fixed $s > 0$, a direct calculation shows that the fundamental solution

$$\Phi(x,s) = \frac{1}{(4ns)^{\frac{n}{2}}} e^{-\frac{|x|^2}{4s}}$$

satisfies

$$\Delta\Phi(x,s) + \frac{n}{2s}\Phi(x,s) \geq 0, \qquad x \in R^n.$$

Define

$$y(t) := \int_{R^n} u(x,t)\Phi(x,s)dx.$$

Then,

$$y'(t) = \int_{R^n} u_t\Phi(x,s)dx$$

$$
\begin{aligned}
&= \int_{R^n} [(\Delta u + u^p)\Phi(x,s)]dx \\
&= \int_{R^n} [u\Delta\Phi + u^p\Phi(x,s)]dx \\
&\geq -\frac{n}{2s}y(t) + \int_{R^n} u^p\Phi(x,s)dx \\
&\geq -\frac{n}{2s}y(t) + y(t)^p,
\end{aligned}
$$

where at the final step, Jensen's inequality is used.

Let

$$\lambda = \frac{n}{2s}.$$

Then,

$$\frac{d}{dt}(e^{\lambda t}y(t)) \geq y(t)^p e^{\lambda t} = [y(t)e^{\lambda t}]^p e^{\lambda(1-p)t},$$

which is equivalent to

$$y(t)^{p-1} \geq \frac{\lambda}{1 + (\lambda y(0)^{1-p} - 1)e^{\lambda(p-1)t}}.$$

This implies that $y(t)$ blows up in finite time as long as

$$\lambda y(0)^{1-p} - 1 < 0.$$

On the other hand,

$$y(0) = \frac{1}{(4\pi s)^{\frac{n}{2}}} \int_{R^n} g(x)e^{-\frac{|x|^2}{4s}} dx.$$

Hence,

$$\lambda y(0)^{1-p} - 1 < 0$$

is equivalent to

$$\frac{1}{(4\pi)^{\frac{n}{2}}} \int_{R^n} g(x)e^{-\frac{|x|^2}{4s}} dx > \left(\frac{n}{2}\right)^{p-1} s^{\frac{n}{2}+1-p}.$$

Note that $1 < p < p_0 = 1 + \frac{n}{2}$, if

$$\int_{R^n} g(x)e^{-\frac{|x|^2}{4s}} dx > 0,$$

we can choose s sufficiently small such that the condition

$$\lambda y(0)^{1-p} - 1 < 0$$

holds.

For the critical case $p = p_0 = 1 + \frac{n}{2}$, we show that $u(x,t)$ will become sufficiently large if t is large. Indeed, first, since $g(x) \geq 0$,

$$\int_{R^n} \Phi(x-y,t)g(y)dy > 0,$$

we may assume that there exists a small constant $c_0 > 0$ such that

$$g(x) \geq c_0, \qquad x \in B_1(0).$$

Hence, for $t > 1$,

$$\begin{aligned} u(x,t) &\geq u_1(x,t) \\ &\geq \frac{c_0}{(2\pi t)^{\frac{n}{2}}} \int_{B_1(0)} exp\{-\frac{|x|^2+|y|^2}{2}\}dy \\ &= \frac{c_1}{(2\pi t)^{\frac{n}{2}}} exp\{-\frac{|x|^2}{2}\}, \end{aligned}$$

where $c_1 > 0$ is a constant depending only on c_0 and n.

Now, for $t > 2$

$$\begin{aligned} u(x,t) &\geq u_2(x,t) \\ &\geq \int_1^{t/2} \int_{R^n} \Phi(x-y,t-\tau)u_1^p(y,\tau)dyd\tau \\ &\geq \frac{c_3}{t^{\frac{n}{2}}} exp\{-\frac{|x|^2}{2}\} \int_1^{t/2} \int_1^t \frac{t^{\frac{n}{2}}}{\tau^{1+\frac{n}{2}}(t-\tau)^{n/2}} \\ &\quad \times \int_{R^n} exp\{\frac{|x|^2}{t} - \frac{|x|^2+|y|^2}{t} - \frac{p|y|^2}{2\tau}\}dyd\tau \\ &\geq \frac{c_3}{t^{\frac{n}{2}}} exp\{-\frac{|x|^2}{2}\} log(t/2), \end{aligned}$$

where c_3 is a constant.

If we choose t_0 sufficiently large and use $u(x,t_0)$ as an initial value, then $y(t_0)$ will be sufficiently large. Consequently, since $p > 1$,

$$\lambda y(t_0)^{1-p} - 1 < 0,$$

which implies that $y(t)$ blows up in finite time. □

8.6.3 Finite-time extinction and dead core

In this subsection we investigate a different phenomenon in which a nonnegative solution of (8.6.1)–(8.6.3) becomes zero in the whole domain (called extinction) in finite time. Namely, when a cooling process in the domain is fast enough, then it will become the extinction in finite time. Let $Q = \Omega \times (0,\infty)$ where $\Omega \subset R^n$ with a smooth boundary.

H(6.3) Let $g(x) \in L^2(\Omega)$ be nonnegative. Let $f(s) \in C^1(0,\infty)$, $f(0) = 0$ and $f(s) \le 0$ be monotone decreasing on $[0,\infty)$ and

$$\sup_{s\ge 0} \frac{f(s)}{s^p} \le -a,$$

where $p \in (0,1)$ and $a > 0$.

A typical example is that

$$f(u) = -u^p, \qquad p \in (0,1).$$

With the assumption H(6.3), the comparison principle yields that

$$0 \le u(x,t) \le ||g||_{L^\infty(\Omega)}, \qquad (x,t) \in Q.$$

It follows that the problem (8.6.1)–(8.6.3) has a global solution in Q.

Theorem 8.6.5. *Let H(6.3) hold. Then, there exists a time T such that the solution (8.6.1)–(8.6.3) will become zero in $\Omega \times (T,\infty)$.*

Proof. By the assumption for f, we see that $f(u) \le -au^p$ for $u \ge 0$. Define

$$y(t) := \frac{1}{2}\int_\Omega u(x,t)^2 dx, \qquad t \ge 0.$$

Then, we use Eq. (8.6.1) and the condition for $f(s)$ to obtain

$$y'(t) + ay(t)^\beta \le 0,$$

where $\beta = \frac{1+p}{2}$.

It follows that

$$\frac{1}{1-\beta}[y(t)^{1-\beta} - y(0)^{1-\beta}] + t \le 0.$$

Since $\beta < 1$ and $y(t) \ge 0$, we see $y(t)$ must be equal to 0 at a finite time T. □

A typical example for the extinction phenomenon is that

$$f(u) = -u^p, \qquad p \in (0,1).$$

Corollary 8.6.2. *Let $h(x,t) \in L^\infty(\partial\Omega \times (0,\infty))$ be nonnegative. Suppose the boundary condition (8.6.2) is replaced by*

$$u(x,t) = h(x,t), \qquad (x,t) \in \partial\Omega \times (0,\infty). \tag{8.6.13}$$

Then, there exists a subregion $\Omega(t) \subset \Omega \times (T, \infty)$ for some $T > 0$ such that

$$u(x,t) = 0, \qquad (x,t) \in \Omega(t).$$

Proof. Let

$$H_0 := ||h||_{L^\infty(\partial\Omega)\times(0,\infty)} + ||g||_{L^\infty(\Omega)}.$$

Let $w(x,t)$ be the solution of the following auxiliary problem:

$$\begin{aligned} w_t - \Delta w &= f(w), && (x,t) \in Q, \\ w(x,t) &= H_0, && (x,t) \in \partial\Omega \times (0,\infty), \\ w(x,t) &= g(x), && x \in \Omega. \end{aligned}$$

The comparison theorem implies that

$$0 \le u(x,t) \le w(x,t), \qquad (x,t) \in Q.$$

Note that $w(x,t)$ attains a maximum at every lateral point $\partial\Omega \times (0,\infty)$, Hopf's Lemma implies that

$$\frac{\partial w(x,t)}{\partial \nu} > 0, \qquad (x,t) \in \partial\Omega \times (0,\infty).$$

The rest of the proof follows the same steps as in the proof of Theorem 8.6.5. □

Corollary 8.6.2 shows that the chemical concentration will become zero as long as there exists a strong absorption even if the concentration on the boundary is very large. The subregion $\Omega(t)$ is called a *dead core* in the literature.

8.7 Notes and remarks

The fundamental solution for a differential operator plays an important role in advanced analysis. Many interesting results such as DiGiorgi–Nash's estimate for elliptic and parabolic equations are derived from these fundamental solutions. Schauder's estimate for a general elliptic and parabolic equation can be established via various estimates for the fundamental solution ([9,12,19,21,32,34]). We focused only on the Laplace and the heat operators to illustrate the basic ideas. Green's function is an interesting topic that can be used to deduce many properties of the solution for the Laplace and the heat equations. The boundary-element method in numerical analysis is based on Green's functions, which is an important topic in advanced PDEs ([9,21]).

The materials in Sections 8.1–8.5 are fairly straightforward. Section 8.6 is an advanced topic that is usually not covered in the elementary PDEs. Blowup and extinction in finite time for the solution of a semilinear heat equation are established. The interested readers may find much more material for topics in [14] and [15].

8.8 Exercises

1. Find a constant A such that the following problem has a unique solution:

$$L[u] = -u'' - (\frac{5\pi}{L})^2 u = A + x, \qquad 0 < x < L,$$
$$u(0) = u(L) = 0.$$

2. Let

$$\Omega = \{(x, y) : 0 < x < L, -\infty < y < \infty\}.$$

Find the Green's function in Ω associated with the Laplace operator.

3. Let

$$\Omega = B_a(0)^+ = \{(x, y) : x^2 + y^2 < a^2, y > 0\}.$$

Find the Green's function in $B_a(0)^+$ associated with the Laplace operator subject to the Dirichlet boundary condition.

4. Let $L > 0$, $H > 0$, $K > 0$ and

$$\Omega = \{(x, y, z) \in R^3 : 0 < x < L, 0 < y < H, 0 < z < K\}.$$

Find the Green's function in Ω associated with the Laplace operator subject to the Dirichlet boundary condition.

5.

$$\Omega = B_a(0)^+ = \{(x, y) : x^2 + y^2 < a^2, y > 0\}.$$

Find the Green's function in $Q = B_a(0) \times (0, \infty)$ associated with the heat operator subject to the Dirichlet boundary condition.

6. Let $p \in (0, \infty)$ and $u(x, t)$ be the solution of the following problem:

$$u_t - \Delta u = -|u|^{p-1} u, \qquad (x, t) \in Q,$$
$$u(x, t) = 0, \qquad (x, t) \in \partial\Omega \times (0, \infty),$$
$$u(x, 0) = g(x) \geq 0, \qquad x \in \Omega.$$

Prove the solution must exist and must be unique.

7. Let $u(x, t)$ be the solution of the following problem with $c > 0$:

$$u_t - \Delta u = \frac{1}{1 - u}, \qquad (x, t) \in Q,$$
$$u(x, t) = 1, \qquad (x, t) \in \partial\Omega \times (0, \infty),$$
$$u(x, 0) = g(x) \geq 0, \qquad x \in \Omega,$$

where $0 \leq g(x) < 1$ on Ω. Prove (a) the solution must be bounded:

$$0 \leq u(x, t) \leq 1.$$

(b) $||\nabla u||_{L^\infty(\Omega)}$ blows up in finite time.

8. Consider the following problem:

$$u_t - c^2 u_{xx} = u(1-u), \qquad (x,t) \in R^1 \times (0,\infty),$$
$$u(x,0) = g(x) \geq 0, \qquad x \in \Omega.$$

Suppose

$$\lim_{x \to -\infty} g(x) = 0, \ \lim_{x \to \infty} g(x) = 1.$$

Prove the problem has a traveling-wave solution $u(x,t) = v(x - ct)$ with

$$v(-\infty) = 0, \qquad v(\infty) = 1.$$

9. Let Ω be a bounded domain in R^n with C^1-boundary and $p > 1$. Consider the following problem:

$$u_t - \Delta u = 0, \qquad (x,t) \in \Omega \times (0,\infty),$$
$$\nabla_\nu u = u^p, \qquad (x,t) \in \partial\Omega \times (0,\infty),$$
$$u(x,0) = g(x) \geq 0, \qquad x \in \Omega.$$

Prove $u(x,t)$ blows up in finite time if

$$\int_\Omega g(x)dx > 0.$$

10. Let Ω be a bounded domain and $p > 1$. Consider the following problem:

$$u_t - \Delta u = \lambda u - u^p, \qquad (x,t) \in \Omega \times (0,\infty),$$
$$u(x,t) = 0, \qquad (x,t) \in \partial\Omega \times (0,\infty),$$
$$u(x,0) = g(x) \geq 0, \qquad x \in \Omega.$$

Prove that $u(x,t)$ is uniformly bounded if $\lambda < \lambda_1$, where λ_1 is the principal eigenvalue of the Laplace operator associated with Dirichlet boundary conditions.

11. Let Ω be a bounded domain in R^n with C^1-boundary and $p > 1$. Consider the following problem:

$$u_t - \Delta u = 0, \qquad (x,t) \in \Omega \times (0,\infty),$$
$$\nabla_\nu u = -u^p, \qquad (x,t) \in \partial\Omega \times (0,\infty),$$
$$u(x,0) = g(x) \geq 0, \qquad x \in \Omega.$$

Prove $u(x,t)$ must be nonnegative. Does the solution develop a dead core if $p \in (0,1)$?

CHAPTER

9 Systems of first-order partial differential equations

9.1 Some first-order PDE models in physical sciences

In this section we describe some physical models governed by systems of first-order PDEs. These models are derived by using several physical laws that are different from previous chapters.

9.1.1 The mathematical model of traffic flow and Hamilton–Jacobi equations

We begin with a simple model that describes the traffic flow in highways. Let $\rho(x,t)$ represent the number of cars per unit distance at time t and location x. $\rho(x,t)$ is often referred as the traffic density. Let $q(x,t)$ represent the traffic flow that measures the number of cars per unit time passing at place x and time t.

If we consider an arbitrary section of the highway between $x=a$ and $x=b$, then the total number of cars between $x=a$ and $x=b$ is equal to

$$\int_a^b \rho(x,t)dx.$$

From the conservation of cars moving into the roadway at $x=a$ and moving out at $x=b$, we have

$$\frac{d}{dt}\int_a^b \rho(x,t)dx = q(a,t)-q(b,t) = -\int_a^b q_x(x,t)dx.$$

It follows that

$$\int_a^b [\rho_t + q_x]dx = 0.$$

Since the interval $[a,b]\subset R^1$ is arbitrary, we see that

$$\rho_t + q_x = 0, \qquad x\in R^1, t\geq 0,$$

which is essentially the same as the continuity equation.

Partial Differential Equations and Applications. https://doi.org/10.1016/B978-0-44-318705-6.00015-X

On the other hand, the traffic flow q must be proportional to the car velocity, denoted by u. It is clear that the velocity u must depend on the density ρ. Therefore it is reasonable to assume that the car velocity u typically is a function of the density ρ. It follows that

$$q = q(x, t, \rho).$$

If there exist branch roads that merge to the main highway, one may add an additional number of cars as a source term, denoted by b. This leads to the following first-order partial differential equation:

$$\rho_t + a(x, t, \rho)\rho_x = b(x, t, \rho), \tag{9.1.1}$$

where $a(x, t, \rho)$ and $b(x, t, \rho)$ are known functions.

It is expected that one should be able to find the traffic density $\rho(x, t)$ at any time if an initial density is known.

A more general class of the first-order nonlinear equation in R^n is given by

$$u_t + H(x, t, u_{x_1}, \cdots, u_{x_n}) = 0,$$

which is often called the Hamilton–Jacobi equation.

9.1.2 The fundamental equations in fluid mechanics

Fluid dynamics is an important field in modern sciences and engineering fields. The fundamental equations in fluid mechanics are based on two conservation laws. The first one is called the continuity equation that is based on the conservation of mass. The second equation is called Euler's equation for a compressible fluid or the Navier–Stokes equation for an incompressible fluid, which is based on the conservation of momentum (or Newton's second law).

Consider a fluid occupying a region $\Omega \subset R^3$. Let $\rho(x, t)$ be the density of the fluid at location $x \in R^3$ and time t. Let $\mathbf{V}(x, t)$ be the velocity field of the fluid. Suppose the trajectory of a particle at location x and time t in the fluid is denoted by $X(t) =< x_1(t), x_2(t), x_3(t) >$ for $t \geq 0$.

Let D be any subdomain of Ω. Then, the change of total mass of the fluid in D must be equal to the total fluid flowing out through the boundary of D in the normal direction:

$$\frac{d}{dt}\int_D \rho(x, t)dx = -\int_{\partial D} [\rho\mathbf{V} \cdot \nu(x)]ds,$$

where $\nu(x)$ is the outward unit normal at $x \in \partial D$.

Gauss's divergence theorem implies that

$$\int_D [\rho_t + div(\rho\mathbf{V})]dx = 0, t > 0.$$

Since D is arbitrary in Ω, we find that

$$\rho_t + div(\rho\mathbf{V}) = 0, \qquad x \in \Omega, t > 0, \tag{9.1.2}$$

Eq. (9.1.2) is called the continuity equation that holds for all kinds of fluids. For an incompressible fluid with constant density such as water or crude oil, we see that

$$div\mathbf{V} = 0.$$

Next, we derive an equation based on the conservation of momentum (or Newton's second law). By definition, the velocity $\mathbf{V}$ is equal to

$$X'(t) = \mathbf{V}(X(t), t), \qquad t \geq 0.$$

The acceleration of the fluid particle is equal to

$$X''(t) = \frac{\partial \mathbf{V}}{\partial t} + (\mathbf{V} \cdot \nabla)\mathbf{V}.$$

For any subdomain $D \subset \Omega$, the total momentum of the fluid over the domain D is equal to

$$\int_D \rho \left[\frac{\partial \mathbf{V}}{\partial t} + (\mathbf{V} \cdot \nabla)\mathbf{V} \right] dx.$$

There are two types of forces acting on the fluid. The first one is the Kelvin force or potential force due to the pressure $p(x, t)$. The other type of force is the external force such as the gravitational force and resistance force, denoted by $\mathbf{F}$. Suppose we neglect the force due to the internal friction of fluid (viscosity). Therefore the total momentum produced by the Kelvin force and external force in D are equal to

$$-\int_{\partial D} p\nu ds + \int_D \mathbf{F} dx,$$

where ν represents the outward unit normal on ∂D.

The divergence theorem implies that

$$\int_{\partial D} p\nu ds = \int_D (\nabla p) dx.$$

Since D is arbitrarily in Ω, we obtain

$$\rho[\mathbf{V}_t + (\mathbf{V} \cdot \nabla)\mathbf{V}] = -\nabla p + \mathbf{F}. \tag{9.1.3}$$

Eq. (9.1.3) is called Euler's equation.

For a fluid such as a gas, the pressure p is a function of density ρ. Therefore we obtain the fundamental equation for the gas dynamics:

$$\rho_t + div(\rho\mathbf{V}) = 0, \qquad x \in \Omega, t > 0, \tag{9.1.4}$$

$$\rho[\mathbf{V}_t + (\mathbf{V} \cdot \nabla)\mathbf{V}] = -\nabla p(\rho) + \mathbf{F}, \qquad x \in \Omega, t > 0. \tag{9.1.5}$$

For a fluid with viscosity, there is an additional internal force due to the friction (viscosity), which can be expressed by (see [10]):

$$\mu \Delta \mathbf{V},$$

where μ is referred to as the kinematic viscosity.

Hence, Euler's equation becomes

$$\rho[\mathbf{V}_t + (\mathbf{V} \cdot \nabla)\mathbf{V}] = \mu \Delta \mathbf{V} - \nabla p + \mathbf{F}. \tag{9.1.6}$$

For an incompressible fluid with a constant density, Eq. (9.1.6) along with the continuity equation (9.1.2) is called the Navier–Stokes equation.

9.1.3 Maxwell's equations in electric and magnetic fields

Suppose that charged particles occupy a medium $\Omega \subset R^3$. From the physical experiments, we know that charged particles produce an electric field and a magnetic field in Ω. Let the electric and magnetic densities be denoted by $\mathbf{E}(x,t)$ and $\mathbf{H}(x,t)$, respectively.

By using the conservation laws for current (similar to the continuity equation) and charged particles (similar to the conservation of energy) as well as experimental laws (Ampere's and Faraday's laws) and constitutive assumptions, we obtain the following Maxwell's system (see a more detailed derivation in [36]):

$$\varepsilon \mathbf{E}_t + \sigma \mathbf{E} = \nabla \times \mathbf{H} + \mathbf{J}_0, \qquad x \in \Omega, t > 0, \tag{9.1.7}$$

$$\mu \mathbf{H}_t + \nabla \times \mathbf{E} = \mathbf{M}_0, \qquad x \in \Omega, t > 0, \tag{9.1.8}$$

where $\mathbf{J}_0$ and $\mathbf{M}_0$ represent known electric and magnetic currents, respectively, generated from the external system, ε and μ are called the electric permittivity and magnetic permeability, respectively. Moreover, σ is the electric conductivity.

With appropriate initial and boundary conditions, we obtain the mathematical model for the motion of electric and magnetic fields in a medium Ω.

We know that an nth-order ODE can be expressed as a system of first-order ODEs. In principle, by introducing suitable new variables, many high-order PDEs can be written as a system of first-order PDEs. However, it will be seen that the well-posedness for a system of first-order PDEs is much harder to deal with in comparison to those classical second-order PDEs. It is worthy noting that for a first-order PDE or system how to impose a boundary condition is a complicated task. It turns out that a boundary condition depends on the characteristic curves for the system.

9.2 The characteristic method

In this section we study an initial-value problem for a single first-order partial differential equation. We will introduce the characteristic method to solve the traffic-flow equation globally. The method is then extended to the transport model in higher space dimension.

9.2.1 The characteristic method with a single space variable

We begin with a linear traffic-flow equation. Let $a(x,t)$ be a continuous function for all $(x,t) \in R^1 \times R^1_+$ and $b(x,t) \in L^\infty(R^1 \times R^1_+)$. Moreover, $a(x,t)$ is Lipschitz continuous with respect to x.

Consider the traffic-flow equation with an initial condition:

$$u_t + a(x,t)u_x = b(x,t), \qquad (x,t) \in R^1 \times R^1_+, \tag{9.2.1}$$

$$u(x,0) = f(x), \qquad x \in R^1, \tag{9.2.2}$$

where $f(x) \in C(R^1)$.

Let $x_0 \in R^1$ be a fixed point. Define a characteristic curve $L : x = x(t)$ as follows:

$$\frac{dx}{dt} = a(x,t), \qquad t \in R^1_+, \tag{9.2.3}$$

$$x(0) = x_0, \qquad x_0 \in R^1. \tag{9.2.4}$$

From the theory of ODEs, there exists a unique global solution $x = x(t)$ to Eq. (9.2.3)–(9.2.4) for $t \geq 0$. Moreover, since the tangent at every point of the characteristic curve L is not equal to 0, the characteristic curve L is monotone increasing for $t > 0$.

Now, along the characteristic curve L we define a function

$$U(t) := u(x(t),t), t \geq 0.$$

We use Eqs. (9.2.1) and (9.2.3) to find

$$\begin{aligned}\frac{dU(t)}{dt} &= u_t(x(t),t) + u_x(x(t),t)\frac{dx(t)}{dt} \\ &= u_t(x(t),t) + a(x(t),t)u_x(x(t),t) \\ &= b(x(t),t).\end{aligned}$$

It follows that

$$\begin{aligned}U(t) &= U(0) + \int_0^t b(x(\tau),\tau)d\tau = u(x_0,0) + \int_0^t b(x(\tau),\tau)d\tau \\ &= f(x_0) + \int_0^t b(x(\tau),\tau)d\tau,\end{aligned}$$

i.e., the solution $u(x,t)$ can be found explicitly along the characteristic curve L. This method is called the *characteristic method.*

We summarize the above analysis as the following theorem.

Theorem 9.2.1. *Let $a(x,t) \in C(R^1 \times R^1_+)$ be Lipschitz continuous w.r.t. x and $b(x,t) \in L^\infty(R^1 \times R^1_+)$. Then, for any $f(x) \in C(R^1)$, the solution of the problem (9.2.1)–(9.2.2) is equal to*

$$u(x(t),t) = f(x_0) + \int_0^t b(x(\tau),\tau)d\tau,$$

where $x(t)$ is the characteristic curve obtained by the solution of the ODE:

$$\frac{dx}{dt} = a(x,t), \qquad (x,t) \in R^1 \times R^1_+,$$
$$x(0) = x_0, \qquad x_0 \in R^1.$$

Moreover, $u(x,t)$ continuously depends on the known data. □

Let us take a look at a special case where $a(x,t) = a$ is a constant. In this case, the characteristic curve L is a line $x = at + x_0$, $t \geq 0$. We can easily see the solution $u(x,t)$ along the line L:

$$u(x,t) = f(x-at) + \int_0^t b(x - a\tau, \tau)d\tau, \qquad (x,t) \in R^1 \times R^1_+.$$

From the characteristic method, unlike the second-order PDE we may not be able to prescribe a boundary condition for a first-order PDE. It depends on whether or not a boundary point can be determined along the characteristic curve. We use the following example to illustrate the idea.

Example 9.2.1. Consider the following initial-boundary value problem:

$$u_t + au_x = 0, \qquad (x,t) \in (0,\infty) \times (0,\infty), \tag{9.2.5}$$
$$u(0,t) = g(t), \qquad t > 0, \tag{9.2.6}$$
$$u(x,0) = f(x), \qquad x \in R^1_+, \tag{9.2.7}$$

where a is a constant, $f(x)$ and $g(t)$ are given functions.

A characteristic curve L starting with $x = x_0 \in R^1_+$ is defined by

$$\frac{dx(t)}{dt} = a, \qquad x(0) = x_0, \qquad x_0 \in [0,\infty).$$

It follows that every characteristic curve $L : x = at + x_0$, which is a line starting with x_0 and slope $\frac{1}{a}$ on the xt-plane.

When $a < 0$, then every characteristic line will hit every point on the t-axis for $t \geq 0$. Namely, the value of $u(0,t)$ is uniquely determined by the initial value $f(x)$, see Fig. 9.1 (b). In this case, the boundary condition (9.2.6) is not needed. On the other hand, when $a > 0$, every characteristic line does not intersect with the upper half of the t-axis, see Fig. 9.1 (a). Therefore we need a boundary condition at $x = 0$. Otherwise, the solution can not be determined uniquely in the region between $x = 0$ and $t = \frac{1}{a}x$, $t \geq 0$.

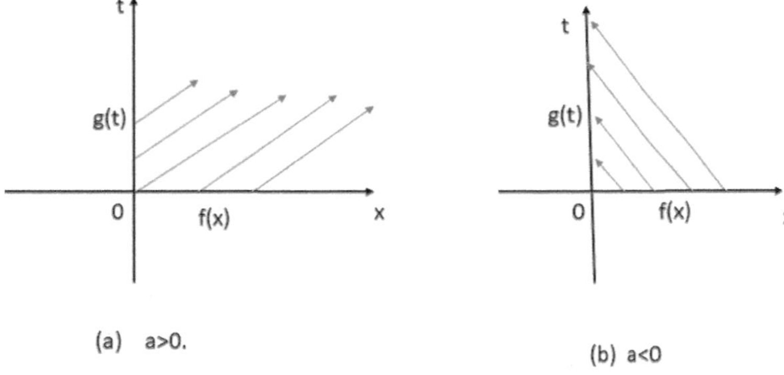

FIGURE 9.1

9.2.2 The solution of the traffic-flow equation in higher space dimension

We extend the above characteristic method to several space dimensions.

Let $f(x) \in C(R^n)$. Consider the traffic-flow equation in R^n with $n \geq 1$:

$$u_t + \sum_{i=1}^{n} a_i(x,t)u_{x_i} = b(x,t), \qquad (x,t) \in R^n \times R^1_+, \tag{9.2.8}$$

$$u(x,0) = f(x), \qquad x \in R^n. \tag{9.2.9}$$

For any $x_0 = (x_{01}, \cdots, x_{0n}) \in R^n$ we define the characteristic curve in R^n:

$$\frac{dx_i}{dt} = a_i(x,t), \qquad (x,t) \in R^n \times R^1_+, i = 1, 2, \cdots, n, \tag{9.2.10}$$

$$x_i(0) = x_{0i}. \tag{9.2.11}$$

Suppose $a_i(x,t)$ is locally Lipschitz continuous with respect to x in R^n. Then, the ODE system (9.2.10)–(9.2.11) has a unique solution

$$x_1 = x_1(t), \cdots, x_n = x_n(t), \qquad t \geq 0,$$

which represents a curve in R^n

Define

$$U(t) := u(x_1(t), \cdots, x_n(t), t), \qquad t \geq 0.$$

Then,

$$U'(t) = u_t + \sum_{i=1}^{n} u_{x_i} x_i'(t) = b(x(t), t).$$

It follows that

$$U(t) = f(x_0) + \int_0^t b(x(t), \tau)d\tau, \qquad (x,t) \in R^n \times R^1_+.$$

We summarize the above discussion to obtain the following theorem.

Theorem 9.2.2. *Let $a_i(x,t) \in C(R^1 \times R^1_+)$ be locally Lipschitz continuous with respect to x in $R^n \times R^1_+$ for all $1 \le i \le m$. Then, for every $f(x) \in C(R^n)$, the linear problem (9.2.8)–(9.2.9) has a unique solution as long as a solution of the ODE system (9.2.10)–(9.2.11) exists. Moreover, the solution continuously depends on the known data if $f(x)$ and $b(x,t)$ is locally Lipschitz continuous with respect to x.* □

9.3 Well-posedness for a system of first-order PDEs

We begin with a system of a first-order linear PDE with two variables $(x,t) \in R^1 \times R^1_+$. Let $\mathbf{U}(x,t)$ be an m-dimensional vector-valued function. Consider the following system of PDE with an initial condition:

$$\mathbf{U}_t + B(x,t)\mathbf{U}_x = A(x,t)\mathbf{U} + F(x,t), \qquad (x,t) \in R^1 \times (0,\infty), \tag{9.3.1}$$

$$\mathbf{U}(x,0) = \mathbf{U}_0(x), \qquad x \in R^1, \tag{9.3.2}$$

where $A(x,t) = (a_{ij}(x,t))_{m\times m}$ and $B(x,t) = (b_{ij}(x,t))_{m\times m}$ are $m \times m$ matrices with

$$a_{ij}(x,t) \in L^\infty(R^1 \times R^1_+), \qquad b_{ij}(x,t) \in L^\infty(R^1 \times R^1_+).$$

In general, we do not expect that the system (9.3.1)–(9.3.2) has a global solution in the classical sense since the characteristics may intersect each other at some time $t_0 > 0$. However, we can prove the global existence of a solution if the eigenvalues of $B(x,t)$ satisfy certain conditions.

9.3.1 Global solution for the diagonal system

We first consider a special case where the coefficient matrix B is in diagonal form $B(x,t) = diag\{b_1(x,t), \cdots, b_m(x,t)\}$ with

$$b_1(x,t) < b_2(x,t) < \cdots < b_n(x,t), \qquad (x,t) \in R^1 \times R^1_+.$$

Moreover, $b_i(x,t) \in C(R^1 \times R^1_+)$ and is Lipschitz continuous w.r.t. x.

For any $x_0 \in R^1$, define the characteristic curve $L_i : x = x_i(t)$, $t \ge 0$ by the following system of ODEs:

$$\frac{dx_i(t)}{dt} = b_i(x,t), \qquad t > 0, \tag{9.3.3}$$

$$x_i(0) = x_0, \qquad i = 1, 2, \cdots, m. \tag{9.3.4}$$

From the ODE theory, the system (9.3.3)–(9.3.4) has a unique solution $x_i(t)$ globally in $[0, \infty)$. Moreover, due to the assumption for $b_i(x,t)$ these characteristic curves do not intersect each other for $t > 0$.

Next, we follow the same method as for a single equation and define

$$v_i(t) = u_i(x_i(t), t), \qquad t > 0, i = 1 \cdots, m.$$

Then,

$$\begin{aligned}\frac{dv_i}{dt} &= u_{it} + u_{ix}x_i'(t)\\ &= u_{it} + b_i(x_i(t), t)u_{ix}\\ &= \sum_{j=1}^{m} a_{ij}v_j(x_j(t), t) + f_i(x_i(t), t), \qquad i = 1, \cdots, m.\end{aligned}$$

We perform the integration to obtain

$$v_i(t) = v_i(0) + \int_0^t \sum_{j=1}^{n} a_{ij}v_j(x_i(t), t)dt + \int_0^t f_i(x_i(t), t)dt, \qquad i - 1, \cdots, m. \tag{9.3.5}$$

Since a_{ij} is uniformly bounded in $R^1 \times R^1_+$, we know that the system (9.3.5) has a unique solution globally. We summarize the above analysis as the following theorem.

Theorem 9.3.1. *Let $A = (a_{ij})_{n\times n}$ and $F = (f_i)_{1\times n}$ with $a_{ij}(x,t), f_i(x,t) \in C(R^1 \times R^1_+)$. Let $B(x,t)$ be a diagonal matrix with every diagonal entry b_i satisfying*

$$b_1(x,t) < b_2(x,t) < \cdots < b_m(x,t), \qquad (x,t) \in R^1 \times R^1_+.$$

Moreover, $b_i(x,t) \in C(R^1 \times R^1_+)$ and is Lipschitz continuous w.r.t. x. Then, the system (9.3.1)–(9.3.2) has a unique global solution $\mathbf{U}(x,t) \in C^{1,1}(R^1 \times R^1_+)$. Moreover, the solution continuously depends on the known data if $a_{ij}(x,t)$, $b_{ij}(x,t)$, and $f_i(x,t)$ are Lipschitz continuous with respect to x. □

9.3.2 Global solution for the diagonizable system

For a general matrix $B(x,t)$, the system (9.3.1)–(9.3.2) may not have a global solution. However, we can prove that the system has a global solution if the matrix $B(x,t)$ is symmetric and has distinct eigenvalues. The idea is to introduce a linear transform to reduce the matrix $B(x,t)$ into a diagonal form.

Suppose the matrix $B(x,t)$ is symmetric and has m distinct eigenvalues $\lambda_i(x,t)$ with

$$\lambda_1(x,t) < \lambda_2(x,t) < \cdots < \lambda_m(x,t), \qquad (x,t) \in R^1 \times R^1_+.$$

Let $\mathbf{K}_i(x,t)$ be the eigenvector corresponding to $\lambda_i(x,t)$,

$$B\mathbf{K}_i = \lambda_i \mathbf{K}_i, \qquad (x,t) \in R^1 \times R^1_+.$$

We choose a unitary matrix Q such that

$$Q^T = Q^{-1}, \qquad QBQ^T = diag\{\lambda_1, \cdots, \lambda_m\} := \hat{B}(x,t).$$

Introduce a new vector function $\mathbf{V}$:

$$\mathbf{V}(x,t) := Q\mathbf{U}(x,t), \qquad (x,t) \in R^1 \times R^1_+.$$

Then, the system (9.3.1) can be written as follows:

$$\mathbf{V}_t + \hat{B}\mathbf{V}_x = \hat{A}\mathbf{V} + \hat{F}(x,t), \qquad (x,t) \in R^1 \times R^1_+, \tag{9.3.6}$$

$$\mathbf{V}(x,0) = \mathbf{V}_0(x), \qquad x \in R^1, \tag{9.3.7}$$

where

$$\hat{A} = Q_t Q^T + QAQ^T + \hat{B}Q_x Q^T, \qquad \hat{F} = Q\mathbf{F}.$$

From Theorem 9.3.1, we know that the system (9.3.1)–(9.3.2) has a unique solution globally by Theorem 9.3.1.

Theorem 9.3.2. *Let B be symmetric and C^2-differentiable. Moreover, B has n distinct eigenvalues $\{\lambda_i(x,t), i = 1, 2, \cdots, m\}$ that satisfy the following property:*

$$\lambda_1(x,t) < \cdots < \lambda_m(x,t), \qquad (x,t) \in R^1 \times R^1_+.$$

Let $A(x,t) \in C^1(R^1 \times R^1_+)$ and $\mathbf{F}(x,t) \in C(R^1 \times R^1_+)$. Then, the system (9.3.1)–(9.3.2) has a unique global solution $\mathbf{U}(x,t) \in C^{1,1}(R^1 \times R^1_+)$. □

9.4 The solution of Maxwell's equations

In this section we study Maxwell's equations. Due to the curl operator, the system has a distinguishing feature that is different from the systems of the first-order PDE considered in previous sections.

9.4.1 Formulation of Maxwell's system

Let Ω be a bounded domain in R^n. From Section 9.1, we know that the electric field $\mathbf{E}$ and magnetic field $\mathbf{H}$ satisfy the following system:

$$\varepsilon \mathbf{E}_t + \sigma(x,t)\mathbf{E} = \nabla \times \mathbf{H} + \mathbf{J}_0(x,t), \qquad x \in \Omega, t > 0, \tag{9.4.1}$$

$$\mu \mathbf{H}_t + \nabla \times \mathbf{E} = \mathbf{M}_0(x,t), \qquad x \in \Omega, t > 0. \tag{9.4.2}$$

To complete the mathematical model, we must impose appropriate initial and boundary conditions. The initial conditions are standard:

$$\mathbf{E}(x,0)=\mathbf{E}_0(x),\ \mathbf{H}(x,0)=\mathbf{H}_0(x), \qquad x\in\Omega. \tag{9.4.3}$$

The boundary condition is very different from previous boundary conditions considered in previous chapters. From the physics, on the surface of the medium Ω, one can impose that an electric field is given in the tangential direction or normal direction. Mathematically, we prescribe one of the following boundary conditions: either

$$\nu\times\mathbf{E}(x,t)=\mathbf{G}(x,t), \qquad (x,t)\in\partial\Omega\times(0,\infty), \tag{9.4.4}$$

or

$$\nu\cdot\mathbf{E}(x,t)=g(x,t), \qquad (x,t)\in\partial\Omega\times(0,\infty), \tag{9.4.5}$$

where ν represents the outward unit normal on $\partial\Omega$.

Note that the boundary condition (9.4.4) or (9.4.5) is very different from those boundary conditions studied in previous chapters. One can prescribe a similar condition for the magnetic field $\mathbf{H}$ instead of $\mathbf{E}$. We focus on the boundary condition (9.4.4) as an example.

9.4.2 The well-posedness of Maxwell's system

The local existence for the system (9.4.1)–(9.4.4) follows from the general theorem in the previous section. We are interested in the global solution. We give the following conditions for the known data.

For simplicity, we set $\varepsilon=\mu=1$. We only consider the boundary condition (9.4.4) with $\mathbf{G}(x,t)=0$ and $\mathbf{M}_0(x,t)=0$.

Introduce a new vector function

$$\mathbf{W}(x,t)=\int_0^t \mathbf{E}(x,\tau)d\tau, \qquad (x,t)\in Q.$$

Then, it is easy to see from Eq. (9.4.2) that

$$\mathbf{H}(x,t)=\mathbf{H}_0(x)-\nabla\times\mathbf{W}, \qquad (x,t)\in Q.$$

It follows that Maxwell's system is equivalent to the following system for $\mathbf{W}(x,t)$:

$$\mathbf{W}_{tt}+\nabla\times(\nabla\times\mathbf{W})+\sigma\mathbf{W}_t=\mathbf{J}_0+\nabla\times\mathbf{H}_0, \qquad (x,t)\in Q, \tag{9.4.6}$$

$$\nu\times\mathbf{W}(x,t)=0, \qquad (x,t)\in\partial\Omega\times(0,\infty), \tag{9.4.7}$$

$$\mathbf{W}(x,0)=0,\ \mathbf{W}_t(x,0)=\mathbf{E}_0(x), \qquad x\in\Omega. \tag{9.4.8}$$

Similar to a wave equation, we can define a weak solution to the system (9.4.6)–(9.4.8). Define

$$H_0^1(curl, div0, \Omega) := \{\mathbf{U}(x) \in L^2(\Omega) : curl\mathbf{U} \in L^2(\Omega),\ div\mathbf{U}(x) = 0 \text{ in } \Omega,\\ \nu \times \mathbf{U} = 0 \text{ on } \partial\Omega\}.$$

Definition 9.4.1. $\mathbf{W}(x,t)$ is called a weak solution to the system (9.4.6)–(9.4.8) if

$$\mathbf{W}(x,t) \in L^\infty((0,T]; L^2(\Omega)) \bigcap L^2((0,T]; H_0^1(curl, div0, \Omega)),\ \mathbf{W}_t \in L^2(Q_T),$$

and $\mathbf{W}$ satisfies the following integral equation:

$$\int\int_{Q_T} [-\mathbf{W}_t \cdot \mathbf{V}_t + (\nabla \times \mathbf{W}) \cdot (\nabla \times \mathbf{V}) + \sigma \mathbf{W}_t \cdot \mathbf{V}]dxdt$$
$$= \int\int_{Q_T} \mathbf{E}_0(x) \cdot \mathbf{V}(x,0)dx + \int\int_{Q_T} [\mathbf{J}_0 \cdot \mathbf{V} + \mathbf{H}_0 \cdot (\nabla \times \mathbf{V})]dxdt,$$

for every test function

$$\mathbf{V} \in H^1((0,T]; L^2(\Omega)) \bigcap L^2((0,T]; H_0^1(curl, div0, \Omega)), \mathbf{V}(x,T) = 0, \forall x \in \Omega.$$

The following basic conditions are assumed.

H(9.4.1) (a) Let $\sigma(x,t) \geq 0$ be bounded in Q. Let $\mathbf{J}_0(x,t)$ be in $L^2(Q)$.
(b) Let $\mathbf{E}_0(x) \in L^2(\Omega)$, $\mathbf{H}_0(x) \in H^1(\Omega)$ and the following consistency condition holds:

$$\nabla \cdot \mathbf{H}_0(x) = 0, \qquad x \in \Omega.$$

Theorem 9.4.1. *Under the hypothesis H(9.4.1), the problem (9.4.6)–(9.4.8) has a unique global weak solution* $\mathbf{W}(x,t) \in L^2((0,T]; H_0^1(curl, div0, \Omega))$ *with* $\mathbf{W}_t \in L^2(Q_T)$. *Thus there exists a unique weak solution* $(\mathbf{E}, \mathbf{H})$ *to the system (9.4.1)–(9.4.4) with*

$$\mathbf{E}(x,t), \mathbf{H}(x,t) \in L^\infty((0,\infty); L^2(\Omega))$$

with the following energy estimate:

$$\sup_{0<t<T} [||\mathbf{E}||_{L^2(\Omega)} + ||\mathbf{H}||_{L^2(\Omega)}] \leq C(T)\left[||\mathbf{E}_0||_{L^2(\Omega)} + ||\mathbf{H}_0||_{L^2(\Omega)} + ||\mathbf{J}_0||_{L^2(Q_T)}\right].$$

Moreover, if $\sigma(x,t) \geq \sigma_0 > 0$ *and* $\mathbf{J}_0(x,t)$ *decays to* 0 *as* $t \to \infty$, *then* $||\mathbf{E}||_{L^2(\Omega)}$ *decays to* 0. *Moreover, if* $\mathbf{J}_0 = 0$, *then,*

$$||\mathbf{E}||_{L^2(\Omega)} + ||\mathbf{H}||_{L^2(\Omega)} \leq Ce^{-\sigma_0 t}, \qquad t \geq 0.$$

Proof. We define an inner production for vector-valued functions $\mathbf{U}, \mathbf{V}$ as follows:

$$< \mathbf{U}, \mathbf{V} >= \int_\Omega (\mathbf{U} \cdot \mathbf{W})dx.$$

Let

$$M := H_0^1(curl, div0, \Omega).$$

Then, $M = H_0^1(curl, div0, \Omega)$ is a Hilbert space (see [5,36]). Moreover, the norm of M is equivalent to

$$||\mathbf{U}||_M = ||\nabla \times \mathbf{U}||_{L^2(\Omega)}.$$

Define a bilinear form:

$$B[\mathbf{W}, \mathbf{U}] = \int_\Omega [(\nabla \times \mathbf{W}) \cdot (\nabla \times \mathbf{U}) + \mathbf{W} \cdot \mathbf{U}]dx.$$

Then, one can easily verify that $B[\mathbf{W}, \mathbf{U}]$ satisfies the Lax–Milgram conditions. Now, we can use the same argument as for a wave equation in Chapter 5 to obtain a global weak solution for $\mathbf{W}(x, t)$.

To show the energy estimate, we note the following identity: if $\nu \times \mathbf{E} = 0$ on $\partial\Omega$, then

$$\int_\Omega (\nabla \times \mathbf{E}) \cdot \mathbf{H} dx = \int_\Omega \mathbf{E} \cdot (\nabla \times \mathbf{H}) dx.$$

We take the inner production for Eq. (9.4.1) with $\mathbf{E}$ and Eq. (9.4.2) with $\mathbf{H}$ to obtain

$$\begin{aligned}
&\frac{d}{dt}\frac{1}{2}\int_\Omega [|\mathbf{E}|^2 + |\mathbf{H}|^2]dx + \int_\Omega \sigma |\mathbf{E}|^2 dx \\
&\leq \int_\Omega [\mathbf{J}_0 \cdot \mathbf{E}]dx \\
&\leq \varepsilon \int_\Omega [|\mathbf{E}|^2]dx + \frac{1}{4\varepsilon}\int_\Omega [|\mathbf{J}_0|^2]dx.
\end{aligned}$$

Gronwall's inequality yields the desired estimate.

Next, we show that the solution decays if $\sigma(x, t) \geq \sigma_0 > 0$. By a similar estimate as above, we have for any $\varepsilon > 0$

$$\begin{aligned}
&\frac{d}{dt}\frac{1}{2}\int_\Omega [|\mathbf{E}|^2 + |\mathbf{H}|^2]dx + (\sigma_0 - \varepsilon)\int_\Omega |\mathbf{E}|^2 dx \\
&\leq C(\varepsilon)\int_\Omega |\mathbf{J}_0|^2 dx.
\end{aligned}$$

It follows that $||\mathbf{E}||_{L^2(\Omega)}$ decays to 0 if $||\mathbf{J}_0||_{L^2(\Omega)}$ decays to 0. Moreover, if $\mathbf{J}_0 = 0$, then

$$\int_\Omega [|\mathbf{E}|^2 + |\mathbf{H}|^2 dx \leq Ce^{-\sigma_0 t}. \qquad \square$$

9.5 The conservation law and shock waves

In this section we present an introduction to the conservation law with the following form

$$u_t + f(x,t,u)_x = 0.$$

The characteristic method introduced in the previous section can be used to find a solution of the conservation law.

9.5.1 A simple example: Solution of Burgers' equation

We begin with Burgers' equation in which $f(x,t,u) = \frac{1}{2}u^2$ to illustrate the characteristic method for a nonlinear equation.

Example 9.5.1. Consider an initial-value problem:

$$u_t + uu_x = 0, \qquad x \in R^1, t > 0, \tag{9.5.1}$$

$$u(x,0) = g(x), \qquad x \in R^1, \tag{9.5.2}$$

where $g(x)$ is a continuous function.

To find the solution to (9.5.1)–(9.5.2) we define a characteristic curve $L : x = x(t), t \geq 0$ as follows:

$$\frac{dx(t)}{dt} = u(x(t),t), \qquad t \geq 0, \tag{9.5.3}$$

$$x(0) = x_0, \tag{9.5.4}$$

where $x_0 \in R^1$ is arbitrary.

Then, along the characteristic curve L,

$$\frac{du(x(t),t)}{dt} = u_t + u_x x'(t) = u_t + uu_x = 0,$$

which implies that $u(x(t),t)$ must be a constant for $t \geq 0$.

It follows that every characteristic curve L must be a straight line with a slope $\frac{1}{u(x(t),t)}$. As long as we know $g(x)$, we can find the solution $u(x,t)$ in the classical sense for small $t > 0$. On the other hand, if there exist two points $x_1 < x_2$ such that

$$g(x_1) > g(x_2),$$

then the slope for the characteristic line L_1 starting with x_1 is less than the characteristic line L_2 starting with x_2. This implies that the solution $u(x,t)$ will not exist in the classical sense even if $g(x) \in C^\infty(R^1)$. See Fig. 9.2.

Example 9.5.2. Let

$$g(x) = \begin{cases} 0 & \text{if } x < 0, \\ 1 & \text{if } x > 0. \end{cases} \tag{9.5.5}$$

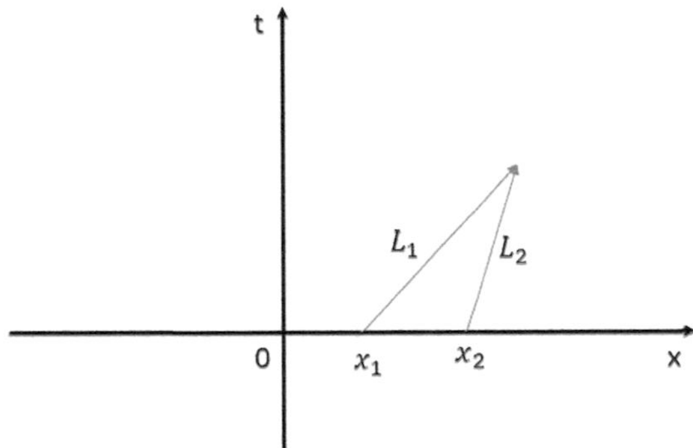

FIGURE 9.2

Find the solution of (9.5.1)–(9.5.2).

Solution. Define regions A, B, and C as follows:

$$A := \{(x,t) : -\infty < x < 0, t > 0\}.$$
$$B :== \{(x,t) : 0 < x < t, t > 0\}.$$
$$C := \{(x,t) : t < x < \infty, t > 0\}.$$

Based on the characteristic method, we easily see that $u(x,t) = 0$ if $(x,t) \in A$ and $u(x,t) = 1$ if $(x,t) \in C$. $u(x,t)$ cannot be determined in the region B.

Suppose we set

$$u_1(x,t) = \begin{cases} 0, & \text{if } (x,t) \in B, 0 < x < \frac{t}{2}; \\ 1, & \text{if } (x,t) \in B, \frac{t}{2} < x < t. \end{cases}$$

Then, we can easily see that $u_1(x,t)$ satisfies Eq. (9.5.1) except on the curve $L : x = \frac{t}{2}, t \geq 0$ (see Fig. 9.3). We will see that the curve L can be derived from the structure of the solution.

On the other hand, we define

$$u_2(x,t) = \frac{x}{t}, \qquad (x,t) \in B.$$

Then, u_2 is also a solution of (9.5.1)–(9.5.2). See Fig. 9.4.

From this example, we see that the solution of (9.5.1)–(9.5.2) is not unique in general. The curve $L : x = \frac{t}{2}, t \geq 0$ is called a shock wave. However, since $u(x,t)$ is bounded, this shock wave is nonphysical. The solution $u_2(x,t)$ is called a rarefaction wave.

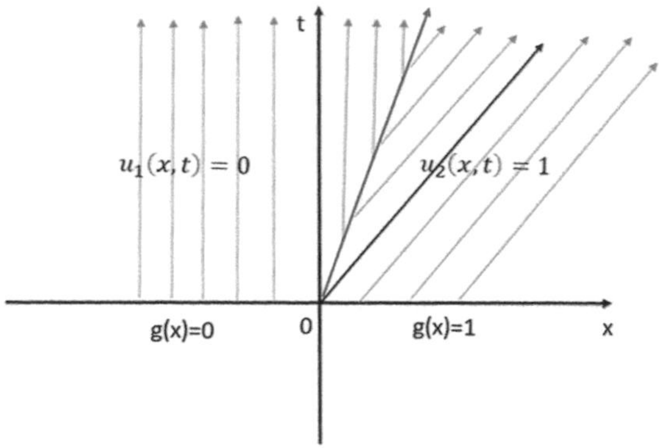

FIGURE 9.3

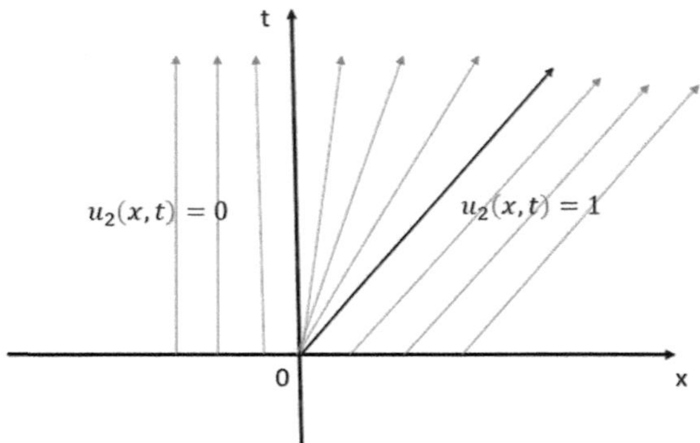

FIGURE 9.4

Example 9.5.3. Find $u(x,t)$ of (9.5.1)–(9.5.2) if $g(x)$ is given by

$$g(x) = \begin{cases} 0, & \text{if } x \leq 0, \\ x, & \text{if } 0 < x < 1, \\ 1, & \text{if } x \geq 1. \end{cases} \tag{9.5.6}$$

Solution. Note that the characteristic curve L is a straight line with slope 1 for $x > 0$, while the characteristic line is vertical for $x < 0$. Now, we define three regions A, B, and C as follows:

$$A := \{(x,t) : -\infty < x < 0, t > 0\}.$$
$$B := \{(x,t) : 0 < x < 1, t > 0\}.$$
$$C := \{(x,t) : 1 < x < \infty, t > 0\}.$$

Clearly, $u(x,t) = 0$ in A and $u(x,t) = 1$ in C. For any point $(x,t) \in B$, the slope of the characteristic line L passing any point $x_0 \in (0,1)$ at $t = 0$ is equal to

$$\frac{t}{x - x_0} = \frac{1}{u(x(t),t)} = \frac{1}{x_0}.$$

Since $u(x,t)$ is a constant along the line L, it follows that

$$u(x,t) = x_0 = \frac{x}{1+t}.$$

Hence,

$$u(x,t) = \begin{cases} 0, & \text{if } x < 0, \\ \frac{x}{1+t}, & \text{if } 0 < x < 1,\, t > 0, \\ 1, & \text{if } x \geq 1,\, t > 0. \end{cases}$$

Note that $u(x,t)$ is continuous in $R^1_+ \times R^1$. The trajectory of the solution $u(x,t)$ is the rarefaction wave, see Fig. 9.5.

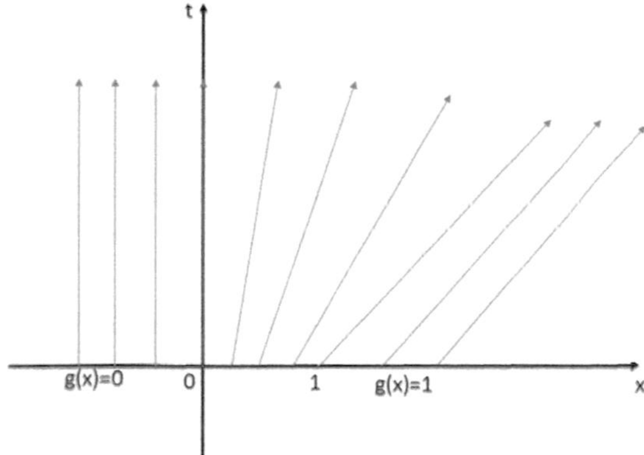

FIGURE 9.5

Example 9.5.4. Find $u(x,t)$ of (9.5.1)–(9.5.2) if $g(x)$ is given by

$$g(x) = \begin{cases} 1, & \text{if } x \leq -1, \\ 1 - x, & \text{if } -1 < x < 0, \\ 0, & \text{if } x \geq 0. \end{cases} \tag{9.5.7}$$

Solution. Note that the characteristic curve L is a straight line with slope 1 for $x < -1$, while the characteristic line is vertical for $x > 0$. It follows that two lines have an intersection point at $x = 0$, $t = 1$. Now, we define three regions A, B, and C as follows:

$$A := \{(x,t) : -\infty < x < t, 0 < t < 1\}.$$

$$B := \{(x,t) : t < x < 0, 0 < t < 1\}.$$
$$C := \{(x,t) : 0 < x < \infty, 0 < t < 1\}.$$

It is easy to see that $u(x,t) = 1$ in A and $u(x,t) = 0$ in C. For any point $(x,t) \in B$, the slope of the characteristic line L passing any point x_0 at $t = 0$ is equal to

$$\frac{t}{x - x_0} = \frac{1}{u(x(t),t)} = \frac{1}{1 - x_0}.$$

Since $u(x,t)$ is a constant along the line L, it follows that

$$u(x,t) = 1 - x_0 = 1 - \frac{x - t}{1 - t} = \frac{1 - x}{1 - t}.$$

Hence,

$$u(x,t) = \begin{cases} 1, & \text{if } x \le t, 0 \le t \le 1, \\ \frac{1-x}{1-t}, & \text{if } t < x < 1, 0 \le t \le 1, \\ 0, & \text{if } x \ge 1, 0 \le t \le 1. \end{cases}$$

For $t > 1$, if we define a curve $L : s(t) = \frac{1+t}{2}$, we find that

$$u(x,t) = \begin{cases} 1, & \text{if } x < s(t), t > 1, \\ 0. & \text{if } x > s(t)1, t > 1, \end{cases}$$

satisfies Eq. (9.5.2) except on the curve $L : x = \frac{1+t}{2}, t \ge 1$.

The solution $u(x,t)$ is called the shock-wave solution. The curve $L : x = \frac{1+t}{2}$, $t \ge 1$ is called a shock wave. See Fig. 9.6.

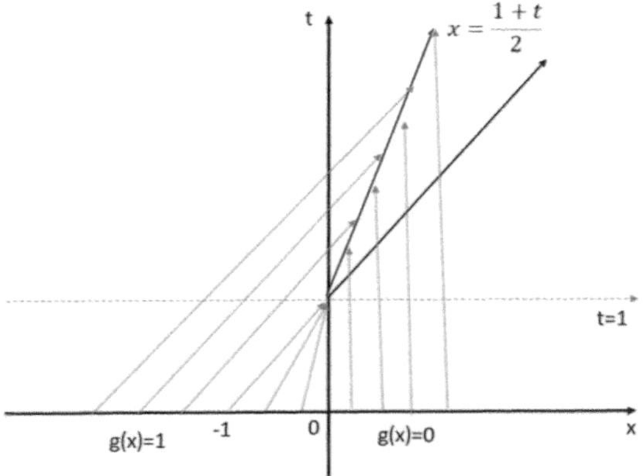

FIGURE 9.6

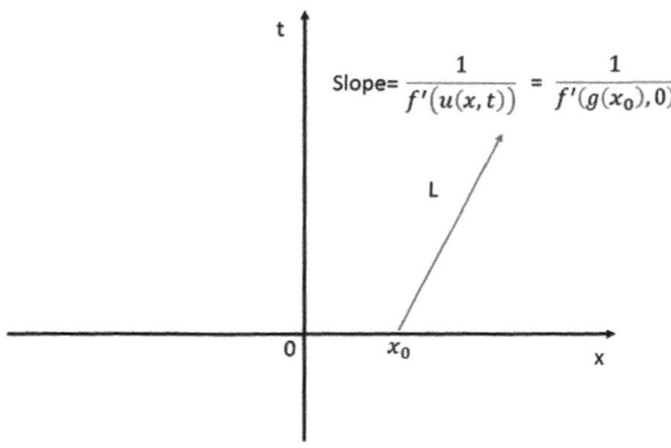

FIGURE 9.7

9.5.2 The solution for a general conservation law

Now, we can extend the above idea to a general nonlinear equation (9.5.1). Consider the following initial value problem:

$$u_t + f(u)_x = 0, \qquad (x,t) \in R^1 \times R^1_+, \tag{9.5.8}$$

$$u(x,0) = g(x), \qquad x \in R^1. \tag{9.5.9}$$

Theorem 9.5.1. *Let $f(u) \in C^2(R^1)$ and $f''(u) \geq 0$. Let $g(x)$ be in $C^1(R^1)$.*

(a) If $g(x)$ is monotone increasing in R^1, then the problem (9.5.8)–(9.5.9) has at least a classical solution globally in time.

(b) If there exist two points $x_1 < x_2$ with $g(x_1) > g(x_2)$, then the problem (9.5.8)–(9.5.9) has no global solution in the classical sense.

Proof. As in Example 9.5.1, for any $x_0 \in R^1$ we can define the characteristic curve as follows:

$$\frac{dx(t)}{dt} = f'(u(x(t),t)),$$
$$x(0) = x_0.$$

These characteristic curves are straight lines with slope $\frac{1}{f'(u(x(t),t))}$ (see Fig. 9.7). Moreover, all these lines do not intersect each other if $g(x)$ is monotone increasing in R^1. This implies that a solution as rarefaction waves exists globally. On the other hand, when $g(x)$ does not satisfy the monotone condition in (a), at least two lines will intersect at some $t = t_0 > 0$. Therefore the solution has a singularity when $t \to t_0$, which implies that $u(x,t)$ does not exist in the classical sense when $t \geq t_0$. □

From Example 9.5.2, we know that a classical solution for Eq. (9.5.8)–(9.5.9) may not exist, so it is natural to seek a solution in a weak sense.

For any $v(x,t) \in C_0^\infty(R^1 \times R_+^1)$, we find that

$$\int_0^\infty \int_{-\infty}^\infty [uv_t + f(u)v_x]dxdt + \int_{-\infty}^\infty g(x)v(x,0)dx = 0. \tag{9.5.10}$$

Definition 9.5.1. A function $u(x,t) \in L^1(R \times R_+^1)$ is called an integral solution to (9.5.8)–(9.5.9) if the integral equation (9.5.10) holds.

We have seen from Example 9.5.2 that the integral solution may not be unique. The problem (9.5.8)–(9.5.9) may have an infinite number of nontrivial solutions for some $g(x)$. The interested reader may consult [29]. To find a solution that can fit the physical model, we need to specify additional conditions, called the *entropy conditions*. We shall not investigate this general issue in this book. However, we will discuss a special case in the following subsection.

9.5.3 Rankine–Hugoniot condition

In this subsection we assume that $u(x,t)$ takes only two values u_l and u_r in $R^1 \times R_+^1$ except a curve $L : x = s(t)$, $t \geq 0$. Moreover, we assume that $s(t)$ is differentiable in $(0,\infty)$. Define

$$[[u]] = u_r - u_l := \text{jump of } u(x,t) \text{ across the curve } L,$$
$$[[f]] := f(u_r) - f(u_l) := \text{ jump of } f(u) \text{ across the curve } L,$$
$$\sigma := s'(t) := \text{ speed the curve } C \text{ moves as } t \text{ increases.}$$

Since the curve L is differentiable in $(0,\infty)$, we see the unit normal to the curve L is equal to

$$\nu = \left\{ \frac{1}{\sqrt{1+s'(t)^2}}, -\frac{s'(t)}{\sqrt{1+s'(t)^2}} \right\}, \qquad t > 0.$$

If we use Green's theorem, from the integral identity (9.5.10) we see that

$$\int_C \{[[[f]]\nu_1 + [[u]]\nu_2]v(x,t)\}ds = 0, \qquad \forall v(x,t) \in C_0^\infty(R^1 \times R_+^1).$$

It follows that

$$s'(t) = \frac{[[f(u)]]}{[[u]]}, \qquad t > 0. \tag{9.5.11}$$

Eq. (9.5.11) is called the Rankine–Hugoniot condition and the curve $C : x = s(t)$, $t \geq 0$ is called a *shock wave*.

For Example 9.5.2, we have

$$f(u) = \frac{u^2}{2}, \ [[f(u)]] = \frac{1}{2}, \ [[u]] = 1.$$

The Rankine–Hugoniot condition implies

$$s'(t) = \frac{1}{2}, \qquad t > 0.$$

Hence,

$$s(t) = \frac{t}{2}, \qquad t > 0.$$

Similarly, we can deduce the curve $x = s(t)$ in other examples.

9.6 A class of semilinear wave equations

In this section we study the global solvability for a class of semilinear wave equations. It will be seen that the problem has very different features in comparison with the problems for semilinear heat equations. The regularity of the weak solution for a wave equation may develop a singularity even if the solution is uniformly bounded. It is still a very active research topic in the study of nonlinear hyperbolic partial differential equations (see [30]). We introduce different methods to prove the solution of a semilinear wave equation blows up in finite time.

9.6.1 The global existence

Let Ω be a bounded domain in R^n with C^1-boundary and $T > 0$,

$$Q_T = \Omega \times (0, T], \qquad Q := \Omega \times (0, \infty).$$

Suppose $f(s)$ is locally Lipschitz in R^1.

Consider the following problem:

$$u_{tt} - \Delta u = f(u), \qquad (x,t) \in Q, \tag{9.6.1}$$

$$u(x,t) = 0, \qquad (x,t) \in \partial\Omega \times (0, \infty), \tag{9.6.2}$$

$$u(x,0) = g_1(x), u_t(x,0) = g_2(x), \qquad x \in R^n, \tag{9.6.3}$$

where $g_1(x) \in H^1(\Omega)$, $g_2(x) \in L^2(\Omega)$.

Define

$$F(u) = \int_0^u f(s)ds, \qquad u \in R^1.$$

Let

$$E(t) := \frac{1}{2}\int_\Omega [u_t^2 + |\nabla u|^2]dx - \int_\Omega F(u)dx.$$

It is clear that

$$E(t) = E(0), \qquad \forall t \in [0, \infty).$$

If

$$\int_{\Omega} F(u)dx \leq 0, \qquad \forall t \in [0, \infty),$$

then

$$\int_{\Omega} [|u_t|^2 + |\nabla u|^2]dx \leq \int_{\Omega} [|g_2|^2 + |\nabla g_1|^2]dx,$$

provided that the initial data $g_1(x) \in H^1(\Omega)$ and $g_2(x) \in L^2(\Omega)$.

Theorem 9.6.1. *Let the following sign condition hold:*

$$uf(u) \leq 0, \qquad \forall u \in R^1.$$

Suppose $g_1(x) \in H^1(\Omega)$, $g_2(x) \in L^2(\Omega)$ and $F(g_1) \in L^1(\Omega)$. Then, the problem (9.6.1)–(9.6.3) has a global weak solution $u(x,t)$ with

$$u_t(x,t), \nabla u \in L^\infty((0,\infty); L^2(\Omega)).$$

Moreover,

$$\sup_{t \geq 0} E(t) \leq E(0).$$

Proof. First, we note that the sign condition implies that

$$\int_0^u f(s)ds \leq 0, u \in R^1.$$

We use a similar argument for the linear wave equation to establish the global existence. As a first step we construct an approximation solution by using the finite-element method. Choose all orthogonal eigenfunctions for the Laplace operator associated with the homogeneous Dirichlet boundary condition, denoted by $\{\psi_k(x)\}_{k=1}^\infty$. Define

$$u_N(x,t) = \sum_{k=1}^{N} d_k^N(t)\psi_k(x),$$

where $d_k^N(t)$ is the solution of the nonlinear ODE:

$$d_k^N(t)'' + \lambda_k d_k^N(t) =< f(u_N), \psi_k >, \ k = 1, 2, \cdots, N, \tag{9.6.4}$$

$$d_k^N(0) =< g_1^N, \psi_k >, \ d_k^N(t)' =< g_2^N, \psi_k >, \tag{9.6.5}$$

where $g_1^N = \sum_{k=1}^N < g_1, \psi_k > \psi_k$, $g_2^N = \sum_{k=1}^N < g_2, \psi_k > \psi_k$. Since $f(s)$ is locally Lipschitz continuous, the above ODE system (9.6.4)–(9.6.5) has a local solution

$\{d_k^N(t), k = 1, 2 \cdots, N, \}$ on $[0, T_0]$ for $T_0 > 0$. To prove the global existence, we note that

$$\sum_{k=1}^{N} < f(u_N), d_k^N(t)'\psi_k > = \frac{d}{dt}\int_\Omega \int_0^{u_N} f(s)dsdx.$$

It follows that

$$\frac{d}{dt}\frac{1}{2}\sum_{k=1}^{N}\left[(d_k^N(t)')^2 + \lambda_k(d_k^N)^2\right] - \frac{d}{dt}\int_\Omega \int_0^{u_N} f(s)dsdx = 0, \qquad t \geq 0.$$

Hence,

$$\begin{aligned}
&\frac{1}{2}\sum_{k=1}^{N}\left[(d_k^N(t)')^2 + \lambda_k(d_k^N(t))^2\right] - \int_\Omega \int_0^{u_N(x,t)} f(s)dsdx,\\
&= \frac{1}{2}\sum_{k=1}^{N}\left[(d_k^N(0)')^2 + \lambda_k(d_k^N(0))^2\right] - \int_\Omega \int_0^{g_1^N(x,0)} f(s)dsdx,
\end{aligned}$$

which is uniformly bounded on $[0, \infty)$. By using the size condition for $f(s)$, we see that

$$\sum_{k=1}^{N}\left[(d_k^N(t)')^2 + \lambda_k(d_k^N)^2\right] \leq C, \qquad t \geq 0,$$

where C depends only on known data.

Consequently, the local solution $\{d_k^N(t)\}$ can be extended into $[0, \infty)$. Next, from the construction of the approximation solution $u_N(x, t)$, we have

$$\frac{d}{dt}E(u_N) \leq 0.$$

It follows that

$$\sup_{t\geq 0}\left[||(u_N)_t||_{L^2(\Omega)} + ||\nabla u_N||_{L^2(\Omega)}\right] \leq C,$$

where C is a constant independent of N.

If we use the weak compactness of $L^2(Q_T)$ and the compact embedding from $H^1(Q_T)$ into $L^2(Q_T)$, we can extract a subsequence from $u_N(x, t)$ (still denoted by u_N) such that $u_N(x, t)$ converges to a limit function $u(x, t)$ strongly in $L^2(Q_T)$ as well as a.e. in Q_T. Moreover, $(u_N)_t$, $\nabla u_N(x, t)$ converge, respectively, to $u_t(x, t)$, $\nabla u(x, t)$ weakly in $L^2(Q_T)$. Since $f(s)$ is continuous in R^1, we see that $u(x, t)$ is a weak solution for the problem (9.6.1)–(9.6.3). □

If a bounded domain Ω is replaced by R^n, we can obtain the same result in Theorem 9.6.1 for Eqs. (9.6.1) and (9.6.3) in R^n.

From Theorem 9.6.1, we see that for any $p > 1$ and

$$f(u) = -|u|^{p-1}u,$$

there exists a global weak solution. Particularly, when $p > 1$ is an odd integer, then

$$f(u) = -u^p$$

satisfies the sign condition. Hence, a global weak solution exists. However, the regularity of the weak solution is extremely challenging, since there is no general regularity results for nonlinear wave equations. Even if we know that $u(x,t)$ is uniformly bounded, $u(x,t)$ may not be smooth. However, if the dimension $n \le 3$, then one can show that a bounded weak solution is indeed smooth if $1 \le p \le 5$ (see Chapter 12 in [8]).

9.6.2 The finite-time blowup

Next, we give some sufficient conditions to prove that a solution of a semilinear wave equation will blow up in finite time.

Theorem 9.6.2. *Assume that there exists a constant $\lambda > 2$ such that*

$$zf(z) \ge \lambda F(z), \qquad \forall z \in R^1.$$

Then, the solution $u(x,t)$ of the problem (9.6.1)–(9.6.3) will blow up in finite time if

$$E(0) := \frac{1}{2}\int_\Omega [|\nabla g_1|^2 + |g_2|^2]dx - \int_\Omega F(g_1)dx < 0.$$

Proof. We use the convexity argument. Assume the problem (9.6.1)–(9.6.3) has a global solution for all $t \ge 0$. Recall the conservation of energy, we have

$$E(t) = \frac{1}{2}\int_\Omega [|u_t|^2 + |\nabla u|^2]dx - \int_\Omega F(u)dx = E(0).$$

Define

$$I(t) = \frac{1}{2}\int_\Omega u^2 dx.$$

A direct calculation shows that

$$I''(t) = \int_\Omega [u_t^2 - |\nabla u|^2 + uf(u)]dx.$$

Let $\alpha > 0$ be a parameter. By using the energy identity and choosing $\alpha = \frac{\lambda-2}{4}$ we find that

$$\begin{aligned}
&I''(t)+(2+4\alpha)E(0)\\
&=(2+2\alpha)\int_\Omega u_t^2dx+2\alpha\int_\Omega|\nabla u|^2dx-(2+4\alpha)\int_\Omega F(u)dx+\int_\Omega uf(u)dx\\
&\geq[(2+2\alpha)\int_\Omega u_t^2dx.
\end{aligned}$$

It follows that

$$I''(t)\geq(2+2\alpha)\int_\Omega u_t^2dx-(2+4\alpha)E(0)>0.$$

On the other hand,

$$\begin{aligned}
I'(t)^2&=\left(\int_\Omega uu_tdx\right)^2\\
&\leq\int_\Omega u^2dx\int_\Omega u_t^2dx\\
&\leq I(t)\frac{I''(t)+\lambda}{1+\alpha}.
\end{aligned}$$

Define

$$J(t)=I(t)^{-\alpha},\qquad t\geq 0,$$

$$\begin{aligned}
J''(t)&=\alpha(1+\alpha)I'(t)^2-\alpha I(t)^{-(1+\alpha)}I''(t)\\
&\leq\alpha\lambda I(t)^{-(1+\alpha)}E(0)<0,\qquad t\geq 0.
\end{aligned}$$

It follows that $J(t)$ is a concave function on $[0,\infty)$. Suppose $J'(t)\geq 0$ for all $t\geq 0$. Then, $I(t)$ is monotone increasing on $[0,\infty)$. It follows that

$$J'(t)<J'(0)+\alpha E(0)\int_0^t I(t)^{-(1+\alpha)}dt\leq J'(0)+\alpha E(0)I(0)^{-(1+\alpha)}t<0,$$

if t is sufficiently large. This is a contradiction. On the other hand, if there exists a point $t_0\geq 0$ with $J'(t_0)<0$, then, by Taylor's expansion, we have

$$J(t)\leq J(t_0)+(t-t_0)J'(t_0),$$

which will be negative if t is sufficiently large, a contradiction.

Consequently, $I(t)$ must blow up in finite time. □

For any $p>0$, consider the following semilinear problem:

$$u_{tt}-\Delta u=\lambda u+|u|^{1+p},\qquad (x,t)\in Q,\tag{9.6.6}$$

$$u(x,t)=0,\qquad (x,t)\in\partial\Omega\times(0,\infty),\tag{9.6.7}$$

$$u(x,0) = g(x), u_t(x,0) = g_2(x), \qquad x \in \Omega. \tag{9.6.8}$$

Let λ_1 be the principle eigenvalue of the Laplace operator associated with a Dirichlet boundary condition on Ω and $\phi_1(x)$ be the corresponding positive eigenfunction with

$$\int_\Omega \phi_1(x)dx = 1.$$

Let

$$\alpha := \int_\Omega \phi_1(x)g_1(x)dx, \qquad \beta := \int_\Omega \phi_1(x)g_2(x)dx.$$

Define

$$\lambda^* := \begin{cases} \lambda_1, & \text{if } \alpha + \beta > 0, \\ \lambda_1 + \frac{\beta^2+\alpha\beta+\alpha^2}{\alpha^2}, & \text{otherwise.} \end{cases}$$

Lemma 9.6.1. *Let $b \geq 0$ and $y(t)$ satisfy the following differential inequality:*

$$y''(t) \geq by(t) + |y(t)|^{1+p}, \qquad t \geq 0$$

with initial conditions:

$$y(0) = \alpha > 0, \qquad y'(0) = \beta \geq 0.$$

Then, $y(t)$ must blow up in finite time.

Proof. Suppose $y(t)$ exists on $[0, T]$ for any $T > 0$. We claim

$$y(t) > 0, y'(t) > 0, \qquad t \in (0, T].$$

If $y(t)$ attains a positive local maximum at $t_0 \in (0, T]$, then at $t = t_0$,

$$y''(t_0) \geq by(t_0) + |y(t_0)|^{1+p} > 0.$$

This is a contradiction. It follows that $y'(t) \geq 0$ on $[0, T]$. Suppose there exists a point $t_0 \in (0, T]$ such that $y'(t_0) = 0$. As

$$y''(t_0) \geq |y(t_0)|^{1+p} > 0,$$

we see for a sufficiently small $\varepsilon > 0$,

$$y'(t_0 - \varepsilon) < y'(t_0) = 0,$$

i.e., a contradiction.

Suppose $y(t)$ exists on $[0, \infty)$ and

$$\lim_{t\to\infty} y(t) = c < \infty.$$

Then, there exists a sequence $t_n \to \infty$ such that $y'(t_n) \to 0$. As $b \ge 0$ and $y(t) \ge 0$, hence,

$$\int_0^\infty y(s)^{1+p} ds \le -\beta < \infty,$$

which implies, $c = 0$, a contradiction since $y(t)$ is monotone increasing on $[0, \infty)$ and $y(0) = \alpha > 0$. Consequently, $y(t)$ must blow up in finite time. □

Theorem 9.6.3. *Suppose $\lambda > \lambda^*$ and $\alpha > 0$. Then, the solution of the problem (9.6.6)–(9.6.8) blows up in finite time.*

Proof. The method is similar to the case for a parabolic equation. Suppose the problem (9.6.6)–(9.6.8) has a global solution. Define

$$y(t) := \int_\Omega u(x,t)\phi_1(x)dx,$$
$$y(0) = \alpha > 0, \; y'(0) = \beta \ge 0.$$

Define

$$T^* := \sup\{T : y(t) > 0, \forall t \in [0, T]\}.$$

We claim $T^* = \infty$. Indeed, if $T^* < \infty$,

$$y''(t) = (\lambda - \lambda_1)y + \int_\Omega |u|^{1+p}\phi_1(x)dx \ge (\lambda - \lambda_1)y + y(t)^{1+p}, \; 0 < t < T^*.$$

By Lemma 9.6.1, $y(t)$ is monotone increasing on $[0, T^*)$, a contradiction, since $y(0) = \alpha > 0$. Thus $T^* = \infty$. Now, we see that $y(t)$ satisfies all conditions in Lemma 9.6.1 in $[0, \infty)$. It follows that $y(t)$ must blow up in finite time. □

By using the same argument, we can obtain the blowup result for the Neumann boundary condition.

Corollary 9.6.1. *Let $u(x,t)$ be a solution for the problem (9.6.6),(9.6.8) with the following Neumann boundary condition:*

$$\nabla_\nu u(x,t) = 0, \qquad (x,t) \in \partial Q. \tag{9.6.9}$$

If $\lambda \ge 0$ and

$$\alpha := \int_\Omega g_1(x)dx > 0, \; \beta := \int_\Omega g_2(x)dx \ge 0,$$

then, $u(x,t)$ must blow up in finite time.

Proof. With the Neumann boundary condition (9.5.7), we define

$$y(t) = \int_\Omega u(x,t)dx.$$

Then, $y(t)$ satisfies

$$y''(t) \geq \lambda y(t) + \int_\Omega |u(x,t)|^{p+1} dx,$$
$$y(0) = \alpha > 0, \; y'(0) = \beta \geq 0.$$

Now,

$$|y(t)|^{1+p} \leq \int_\Omega |u|^{1+p} dx |\Omega|^{\frac{p}{p+1}}, \qquad t \geq 0.$$

It follows that

$$y''(t) \geq \lambda y(t) + k_0 |y(t)|^{1+p},$$

where $k_0 > 0$.

Again by Lemma 9.6.1, $y(t)$ blows up in a finite time. □

9.7 Remarks and notes

In this section we studied a general linear system of first-order equations. The characteristic method is introduced to solve the system. One section is devoted to the investigation of Maxwell's system and its global solvability is established. We also studied a class of the conservation law and illustrate how the solution may develop shock waves when the initial condition satisfies certain conditions. These materials are very standard and students with basic ODE knowledge can easily understand them. To show the difference from the linear equations and systems, we studied a class of semilinear wave equations. Some global existence and finite-time blowups are proved by using elementary analysis from ODEs. This part of the material is intended for students as motivation for further research in PDEs.

9.8 Exercises

1. By introducing a suitable new variable, rewrite the wave equation in one-space dimension as a system of first-order PDEs. Does the conclusion hold for the case when the space dimension is higher than one?
2. Find the explicit solution for the following first-order equation

$$u_t + (1 + x^2) u_x = 1, \qquad (x,t) \in R^1 \times R^1_+,$$
$$u(x,0) = 1.$$

3. Find the explicit solution for the following Burgers' equation:

$$u_t + u u_x = 0, \qquad (x,t) \in R^1 \times R^1_+,$$

$$u(x,0) = g(x),$$

where $g(x)$ is defined by

$$g(x) = \begin{cases} 1, & \text{if } x < 0, \\ -1 & \text{if } x \geq 0. \end{cases}$$

4. Let $\mathbf{A}$ and $\mathbf{B}$ be smooth vector fields in a bounded domain $\Omega \subset R^3$ with

$$\nu \times \mathbf{B} = 0, \qquad x \in \partial\Omega.$$

Prove

$$\int_\Omega [\mathbf{A} \cdot (\nabla \times \mathbf{B})]dx = \int_\Omega [(\nabla \times \mathbf{A}) \cdot \mathbf{B}]dx.$$

5. Let $R = [0, L] \times [0, H] \times [0, K]$ be a rectangle in R^3 and ν be the outward unit normal of ∂R. Compute

$$\nu \times \mathbf{E} = 0, (x, y, z) \in \partial R$$

in terms of each component of $\mathbf{E}$.

6. Let R be the rectangle as in Exercise 5. Suppose $\mathbf{H}(x)$ is a C^1-vector field and

$$\nabla \cdot \mathbf{H} = 0, \qquad x \in \bar{R}.$$

What boundary conditions does $\mathbf{H}$ satisfy on $\partial\Omega$?

7. Let $\Omega \subset R^n$ be a bounded domain and $\Omega_0 \subset \Omega$ with $|\Omega_0| > 0$. Suppose $\sigma(x,t) \geq \sigma_0 > 0$ on Ω_0. Prove the solution $(\mathbf{E}, \mathbf{H})$ of (9.4.1)–(9.4.4) decays to 0 as $t \to \infty$.

8. Suppose $f(u) \in C^1(R^1)$ and $u(x,t) \in C^1(R \times R^1_+) \bigcap L^1(R \times R^1_+)$ satisfies

$$u_t + f(u)_x = 0.$$

Prove, for any $v(x,t) \in C^{1,1}(R^1 \times [0, \infty))$ with $v(x,t) \to 0$ if $|x| + |t| \to \infty$,

$$\int_0^\infty \int_{R^1} [uv_t + f(u)v_x]dxdt = \int_{R^1} u(x,0)v(x,0)dx.$$

9. Let $f(u) \in C(R^1)$. Suppose there exists a smooth curve

$$x = s(t), \qquad t \geq 0,$$

that separates $R^1 \times R^1_+$ into two regions. Suppose $u(x,t)$ is a smooth solution of the conservation law

$$u_t + f(u)_x = 0$$

except on the curve $x = s(t)$. Define the jump

$$[[u]] = \lim_{x \to s(t)+} u(x,t) - \lim_{x \to s(t)-} u(x,t), \qquad t \geq 0.$$

Prove

$$s'(t)[[u]] = [[f(u)]], \qquad t \geq 0.$$

(Hint: use the integral identity in Exercise 8.)

10. Let $u(x,t)$ satisfy the Burgers' equation with viscosity $\varepsilon > 0$:

$$u_t + uu_x = \varepsilon u_{xx}, u(x,0) = f(x),$$

where $f(x) \in C_0^1(R^1)$. Prove the limit of $u_\varepsilon(x,t)$ as $\varepsilon \to 0$ exists.

APPENDIX

Some essential results in ordinary differential equations

In this appendix we state some basic results in ordinary differential equations for the readers' convenience. The interested readers may consult standard textbooks such as [27] to find more detailed proofs.

A.1 The solution of a first-order differential equation and stability

A.1.1 Existence and uniqueness for first-order differential equations

Let $a(t)$, $f(t)$ be known functions defined on $[0, \infty)$. For a linear first-order differential equation,

$$y' + a(t)y = f(t), y(0) = y_0.$$

One can find the explicit solution

$$y(t) = y_0 e^{-\int_0^t a(\tau)d\tau} + \int_0^t f(s)e^{\int_t^s a(\tau)d\tau} ds,$$

provided that $a(t)$ and $f(t)$ are integrable over $[0, T]$ for any $T > 0$.

For a nonlinear ODE, we need much stronger conditions in order to have a unique solution. The following theorem is typically proved in standard textbooks.

Theorem A.1.1. *Let $f(t, y)$ be continuous with respect to t in $[0, \infty)$ and locally Lipschitz continuous with respect to y in R^1. Then, for any $(t_0, y_0) \in R^1_+ \times R^1$ the first-order differential equation:*

$$\begin{aligned} \frac{dy}{dt} &= f(t, y), \qquad t \geq t_0, \\ y(t_0) &= y_0 \end{aligned}$$

has a unique solution $y = y(t)$ in $[t_0, t_0 + h]$ for some $h > 0$. Moreover, the solution can be extended to the maximal interval $[t, T^*)$ as long as $y(t)$ is bounded in $[0, T]$

with any $T < T^*$ and

$$\lim_{T \to T^*} y(t) = \infty.$$

The solution $y(t)$ obtained in Theorem A.1.1 is also continuously dependent upon the initial value. However, the solution may show a huge difference when the Lipschitz constant is suitably large.

A.1.2 The stability analysis for an autonomous equation

Recall a solution $y = y(t)$ of an ODE is said to be asymptotically stable if $y(t)$ has a finite limit, say y_0, as $t \to \infty$.

For an autonomous equation

$$y' = f(y),$$

we call $y = y_0$ an equilibrium solution if $f(y_0) = 0$.

Theorem A.1.2. *Suppose that y_0 is an equilibrium solution for the differential equation*

$$y' = f(y),$$

where $f(y)$ is a differentiable function. Then, one of the following cases holds:

(a) *if $f'(y_0) < 0$, the $y = y_0$ is asymptotically stable;*
(b) *if $f'(y_0) > 0$, then $y = y_0$ is unstable;*
(c) *if $f'(x_0) = 0$, no conclusion can be drawn.*

For the case (c), one needs to analyze the solution behavior as t evolves.

A.2 Basic results for second-order differential equations

For a second-order differential equation

$$y'' = f(t, y, y'),$$

with initial conditions:

$$y(t_0) = y_0, \, y'(t_0) = y_1,$$

the sufficient conditions for the general existence and uniqueness are similar to the general system of the first-order ODEs:

$$\mathbf{Y}' = \mathbf{F}(t, \mathbf{Y}).$$

A special linear case of the second-order ODE is the following theorem.

Theorem A.2.1. *Let $p(t)$ and $q(t)$ be continuous in $[0, T]$. Then, the second-order linear equation*

$$y'' + p(t)y' + q(t)y = g(t), \qquad t \geq 0,$$

subject to initial values:

$$y(0) = y_0, y'(0) = y_1,$$

has a unique solution $y(t)$ in $[0, T]$. Moreover, the general solution can be expressed by

$$y(t) = c_1 y_1(t) + c_2 y_2(t) + y_c(t),$$

where $y_1(t)$ and $y_2(t)$ are linearly independent solutions for the homogeneous equation, while $y_c(t)$ is a particular solution for the nonhomogeneous equation.

In particular, for an second-order ODE with constant coefficients

$$ay'' + by' + cy = g(t),$$

one can find the explicit solution representation. The general solution of the above equation depends on the *characteristic equation*

$$ar^2 + br + c = 0.$$

Case 1: $b^2 - 4ac > 0$.

Let r_1 and r_2 be two real roots of the characteristic equation. Then, the general solution of the homogeneous equation is equal to

$$y(t) = c_1 e^{r_1 t} + c_2 e^{r_2 t}.$$

Case 2: $b^2 - 4ac < 0$.

Let $r_1 = d_1 + id_2$ and $r_2 = d_1 - id_2$ be the complex roots of the characteristic equation. Then, the general solution of the homogeneous equation is equal to

$$y(t) = e^{d_1 t} \left[c_1 \cos(d_2 t) + c_2 \sin(d_2 t)\right].$$

Case 3: $b^2 - 4ac = 0$.

Let $r = r_1 = r_2 = -\frac{b}{2a}$ be the roots of the characteristic equation. Then, the general solution of the homogeneous equation is equal to

$$y(t) = c_1 e^{rt} + c_2 t e^{rt}.$$

A.3 Dynamics of the solution of a linear system of ODEs

Let

$$A = \begin{bmatrix} a_{11} & a_{12} \\ a_{21} & a_{22} \end{bmatrix}.$$

Consider an 2×2 linear system of ODE:

$$\mathbf{Y}' = A\mathbf{Y}.$$

The asymptotic stability of the steady-state solution depends on the eigenvalues of the matrix A.

Theorem A.3.1. *The steady-state solution* $\mathbf{Y} = 0$ *must satisfy one of the following cases:*

- **(a)** *a saddle point if A has two distinct eigenvalues with one positive and one negative;*
- **(b)** *a nodal sink if A has two real and negative eigenvalues;*
- **(c)** *a nodal source if A has two real and positive eigenvalues;*
- **(d)** *a spiral sink if A has two complex conjugate eigenvalues with a negative real part;*
- **(e)** *a spiral source if A has two complex conjugate eigenvalues with a positive real part;*
- **(f)** *a center if A has two pure complex conjugate eigenvalues.*

From Theorem A.3.1, we see that the steady-state solution $\mathbf{Y} = 0$ is asymptotically stable for cases (b) and (d) and unstable for the cases (a), (c), (e), and (f). The same idea can be extended into an $n \times n$ system of ODEs.

Theorem A.3.2. *The steady-state solution* $\mathbf{Y} = 0$ *to the linear system*

$$\mathbf{Y}' = A\mathbf{Y}$$

must have one of the following cases:

- **(a)** *asymptotically stable if the real part of all eigenvalues of A is negative;*
- **(b)** *unstable if at least one of the eigenvalues of A is positive;*
- **(c)** *undetermined if some of the eigenvalues are equal to* 0.

In applications, the above results are often used in dealing with the stability analysis for a nonlinear autonomous system:

$$\mathbf{Y}' = \mathbf{F}(\mathbf{Y}) = (f_1(y_1, \cdots, y_n), \cdots, f_n(y_1, \cdots, y_n))^T.$$

Suppose $\mathbf{F}$ is a differentiable vector field. Define the Jacobian matrix of $\mathbf{F}$:

$$A := \left(\frac{\partial f_i}{\partial y_j}\right)_{n \times n}.$$

Theorem A.3.3. *Let* $\mathbf{Y}_0$ *be an equilibrium solution for the nonlinear system*

$$\mathbf{Y}' = \mathbf{F}(\mathbf{Y}).$$

Then, the stability of $\mathbf{Y} = \mathbf{Y}_0$ *is the same as the equilibrium solution for the linear system*

$$\mathbf{Y}' = A\mathbf{Y},$$

where A *is the Jacobian matrix of* $\mathbf{F}$ *at* $\mathbf{Y} = \mathbf{Y}_0$.

In nonlinear dynamics analysis, one of the complications comes from the fact that some eigenvalues of the matrix A are equal to 0. One needs to find more information about the Hessian matrix of $\mathbf{F}$ in order to determine the stability of the equilibrium solution.

APPENDIX

B Sobolev spaces

B.1 Weak derivatives and Sobolev spaces

Let $\Omega \subset R^n$ be a region with a smooth boundary $\partial\Omega$. Let $\alpha = (\alpha_1, \cdots, \alpha_n)$ be a multiindex with $|\alpha| = \sum_{k=1}^{n} \alpha_k$.

B.1.1 Weak derivatives

Definition B.1.1. Let $u(x), v(x) \in L^1_{loc}(\Omega)$. We say that $v(x) = D^\alpha u(x)$ in the weak sense if

$$\int_\Omega u(x) D^\alpha \phi(x) dx = (-1)^{|\alpha|} \int_\Omega v(x)\phi(x) dx$$

for every test function $\phi(x) \in C_0^\infty(\Omega)$

Clearly, if $u(x)$ is $|\alpha|$-th differentiable in the classical sense, then the above integral identity holds. Conversely, it may not be true. All elementary properties such as the product rule and quotient rule for the classical derivative also hold for the weak derivative as long as all the resulting functions belong to $L^1_{loc}(\Omega)$.

As we will see below, a function that has a weak derivative has some regularity. The weak derivative is stronger than the generalized derivative in the sense of distribution. For example, we know that

$$\delta(x) = \frac{dH(x)}{dx}$$

in the sense of distribution. However, $H(x)$ is not differentiable in the weak sense.

B.1.2 Sobolev spaces

Let $k \geq 1$ be an integer and $p \geq 1$.

Definition B.1.2. (Sobolev spaces) Let $k = |\alpha|$ and define the set

$$W^{k,p}(\Omega) = \{\text{all functions with } |\alpha|\text{th-derivative exist and } ||u||_{W^{k,p}(\Omega)} < \infty\},$$

where

$$||u||_{W^{k,p}(\Omega)} = \sum_{|\alpha| \leq k} ||D^\alpha u||_{L^p(\Omega)}.$$

It can be shown that $W^{k,p}(\Omega)$ with the above norm is a Banach space. In particular, when $p = 2$ it is a Hilbert space. We use the notation

$$H^k(\Omega) = W^{k,2}(\Omega).$$

The inner product of $H^k(\Omega)$ is defined by

$$(u, v) = \sum_{|\alpha| \le k} < D^\alpha u, D^\alpha v > .$$

When $p = \infty$, the norm for $W^{k,\infty}(\Omega)$ is defined by

$$||u||_{W^{2,\infty}(\Omega)} = \sum_{|\alpha| \le k} ess.sup_\Omega |D^\alpha(\Omega)|.$$

Definition B.1.3.

$$W_0^{k,p}(\Omega) := \text{ the closure of } C_0^k(\Omega) \text{ under the norm } W^{k,p}(\Omega).$$

In particular, when $p = 2$,

$$H_0^k(\Omega) = W_0^{k,2}(\Omega).$$

A nice result ([23]) is that the norm of $W^{1,p}(\Omega)$ for any $p > 1$ is equivalent to

$$||\nabla u||_{L^p(\Omega)} + ||u||_{L^p(\partial\Omega)}.$$

In particular, for $W_0^{1,2}(\Omega)$, its norm is equivalent to $||\nabla u||_{L^2(\Omega)}$. A more general result for a bounded domain with a Lipschitz boundary is that the norm of $W^{1,p}(\Omega)$ is equivalent to

$$||\nabla u||_{L^p(\Omega)} + p(u),$$

where $p(u)$ is a continuous seminorm for $W^{1,p}(\Omega)$ (see [31]).

In applications, one may need a Sobolev space $H^{-1}(\Omega) := W_0^{1,2}(\Omega)'$, the dual space of $W_0^{1,2}(\Omega)$. It can be characterized by the following representation. For every $f \in H^{-1}(\Omega)$, there exists a set of functions $\{f_i(x)\}_{i=0}^n$ in $L^2(\Omega)$ such that

$$< f, v >= \int_\Omega \left[f_0 v + \sum_{i=1}^n f_i v_{x_i} \right] dx, \qquad \forall v \in H_0^1(\Omega),$$

with the norm

$$||f||_{H^{-1}(\Omega)} := inf \left\{ \sum_{i=0}^n ||f_i||_{L^2(\Omega)} : f_i \text{ satisfies the above identity} \right\}.$$

B.2 Smooth approximation and Sobolev embedding

The next result shows that every function in a Sobolev space can be approximated by a smooth function sequence.

B.2.1 Smooth approximation

Theorem B.2.1. *Let Ω be a bounded domain with C^1-boundary. Suppose $u(x) \in W^{k,p}(\Omega)$. Then, there exists a sequence $u_m(x) \in C^\infty(\Omega)$ such that*

$$||u_m - u||_{W^{k,p}(\Omega)} \to 0, \qquad \text{as } m \to \infty.$$

With the above smooth approximation, we can define the value on a hypersurface $S \subset \Omega$ for a function in a Sobolev space, which is called the trace of the function on S, denoted by $Tr(u)(x)$.

For example,

$$W_0^{k,p}(\Omega) = \{u \in W^{k,p}(\Omega) : Tr(u)(x) = 0 \text{ on } \partial\Omega\}.$$

B.2.2 Sobolev embedding

A function $u(x)$ in a Sobolev space $W^{k,p}(\Omega)$ has some regularity. The following embedding theorem gives a precise regularity.

Theorem B.2.2. *Let Ω be a region in R^n with C^1-boundary. Suppose $u(x) \subset W^{k,p}(\Omega)$.*

(a) If $kp < n$, then $u(x) \in L^q(\Omega)$ and

$$||u||_{L^q(\Omega)} \leq C||u||_{W^{k,p}(\Omega)},$$

where

$$q = \frac{pn}{n - kp}.$$

(b) If $kp > n$, then $u(x) \in C^{k-[\frac{n}{p}]-1,\gamma}(\bar{\Omega})$ with

$$\gamma = \begin{cases} \left[\frac{n}{p}\right] + 1 - \frac{n}{p}, & \text{if } \frac{n}{p} \text{ is not an integer,} \\ \text{any number in } (0,1), & \text{if } \frac{n}{p} \text{ is an integer.} \end{cases}$$

The above embedding theorem shows that a function will be continuous or differentiable if the order of the weak derivative is suitably higher. A more subtle result is the following compactness embedding.

Theorem B.2.3. *Let U be a bounded domain subset in R^n with C^1-boundary $\partial\Omega$. Then, the embedding operator I_e from $W^{k,p}(\Omega)$ into $L^q(\Omega)$ is compact for any $q \in$*

$[1, q^*)$ *where*

$$q^* = \frac{np}{n-p}.$$

B.3 Sobolev spaces with t-variable

B.3.1 The space $L^p((0,t); X)$ and $V_2(Q_T)$

In dealing with time-dependent PDEs, the time derivative needs a special attention. Let X be a Banach space. Define

$$L^p(0,T]; X) = \{u(x,t) : [0,T] \to X, ||u||_X \in L^p(0,T)\}.$$

The norm for $L^p((0,T]; X)$ is denoted by

$$||u||_{L^p(0,T);X)} = \left[\int_0^T ||u||_X dx\right]^{\frac{1}{p}}.$$

One can easily extend $L^p(0,T); X)$ to other Sobolev spaces such as $H^1((0,T); X)$, etc.

For parabolic equations, the weak solution has a better regularity with respect to the t-variable. One needs to introduce a Sobolev space $V_2(Q_T)$ as follows:

$$V_2(Q_T) := \{u(x,t) \in C([0,T]; W^{1,2}(\Omega))\}.$$

The norm for $V_2(Q_T)$ is given by

$$||u||_{V_2(Q_T)} := \max_{0 \le t \le T} ||u||_{L^2(\Omega)} + ||\nabla u||_{L^2(Q_T)}.$$

B.3.2 A compact embedding result

Theorem B.3.1. *(Aubin–Lions lemma) Let X_0, X and X_1 be three Banach spaces with $X_0 \subset X \subset X_1$. Suppose that X_0 is compactly embedded in X and that X is continuously embedded in X_1. For $1 \le p, q \le \infty$. Let*

$$W = \{u \in L^p([0,T]; X_0) \mid \dot{u} \in L^q([0,T]; X_1)\}.$$

(i) *If $p < \infty$, then the embedding of W into $L^p([0,T]; X)$ is compact.*
(ii) *If $p = \infty$ and $q > 1$ then the embedding of W into $C([0,T]; X)$ is compact.*

References

[1] R.A. Adams, Sobolev Space, Academic Press, New York, 1975.

[2] Daniel Arovas, Lecture Notes in Nonlinear Dynamics, University of San Diego, 2014.

[3] Jerrold Bebernes, David Eberly, Mathematical Problems from Combustion Theory, Springer-Verlag New York, Inc., 1989.

[4] John R. Cannon, The One-Dimensional Heat Equations, Encyclopedia of Mathematics and Its Applications, vol. 23, Cambridge Press, Oxford, UK, 1985.

[5] Philippe G. Ciarlet, Linear and Nonlinear Functional Analysis with Applications, SIAM, Philadelphia, PA, 2013.

[6] John B. Conway, A Course in Functional Analysis, 2nd edition, Springer-Verlag, New York, 1990.

[7] Emmanuele DiBenedetto, Partial Differential Equations, Birkhäuser, Boston, MA, 1995.

[8] L.C. Evans, Partial Differential Equations, AMS Graduate Studies in Mathematics, vol. 19, Amer. Math. Soc., Providence, RI, 2010.

[9] A. Friedman, Partial Differential Equations of Parabolic Type, Prentice Hall, NJ, 1964.

[10] A. Friedman, Variational Principle and Applications, John Wiley & Sons, Inc., New York, 1982.

[11] H. Fujita, On the blowing up of solution of the Cauchy problem for $u_t = \Delta u + u^{1=\alpha}$, Journal of the Faculty of Science, University of Tokyo, Section A, Mathematics 16 (1966) 105–113.

[12] D. Gilbarg, N.S. Trudinger, Elliptic Partial Differential Equations, 3rd edition, Springer, New York, 1998.

[13] L. Hörmander, The Analysis of Linear Partial Differential Operators I, Grundlehren der Mathematischen Wissenschaften, vol. 256, Springer-Verlag, Berlin, 1983.

[14] Bei Hu, Blow-up Theories for Semilinear Parabolic Equations, Lecture Notes in Mathematics, vol. 2018, Springer-Verlag, Berlin, 2011.

[15] Bei Hu, Hong-Ming Yin, The profile near blowup time for solution of the heat equation with a nonlinear boundary condition, Transactions of the American Mathematical Society 346 (1994) 117–136.

[16] Erwin Kreyszig, Advanced Engineering Mathematics, 9th edition, John Wiley & Sons, Inc., Hoboken, NJ, 2006.

[17] Lishang Jiang, Mathematical Modeling and Methods of Option Pricing, World Scientific Publishing Co., Singapore, 2005.

[18] Bei Hu, Hong-Ming Yin, On critical exponents for the heat equation with a nonlinear boundary condition, Annales de l'Institut Henri Poincaré, Analyse non Linéaire 13 (1996) 707–732.

[19] Gary M. Lieberman, Second-Order Parabolic Differential Equations, World Scientific, New York, 1996.

[20] O.A. Ladyzenskaja, N.N. Uralceva, Linear and Quasilinear Elliptic Equations, Academic Press, New York, 1968.

[21] O.A. Ladyzenskaja, V.A. Solonikov, N.N. Uralceva, Linear and Quasilinear Equations of Parabolic Type, AMS Translation Series, vol. 23, Amer. Math. Soc., Providence, RI, 1968.

[22] Hans Levy, An example of a smooth linear partial differential equation without solution, Annals of Mathematics 66 (1) (1957) 155–158.

[23] Vladimir G. Mazja, Sobolev Spaces, Spinger-Verlag, Berlin, 1985.

[24] C.V. Pao, Nonlinear Parabolic and Elliptic Equations, World Scientific, New York, 1992.

[25] Joselp M. Power, Lectures Notes in Gas Dynamics, University of Notre Dame, Notre Dame, 2019.

[26] Mark A. Pinsky, Partial Differential Equations and Boundary-Value Problems with Applications, 3rd edition, McGraw-Hill, Inc., New York, 1998.

[27] John Polking, Albert Boggess, David Arnold, Differential Equations with Boundary Value Problems, 2nd edition, Pearson Education, Inc., Upper Saddle River, NJ, 2006.

[28] J. Simon, Compact sets in the space $L^p(O, T; B)$, Annali di Matematica Pura ed Applicata 146 (1986) 65–96.

[29] J. Smoller, Shock Waves and Reaction-Diffusion Equations, Springer-Verlag New York, Inc., 1983.

[30] Walter A. Strauss, Nonlinear Wave Equations, CBMS by AMS, Providence, RI, 1990.

[31] R. Temam, Infinite-Dimensional Dynamical Systems in Mechanics and Physics, Applied Mathematical Sciences, vol. 68, Springer-Verlag, New York, 1988.

[32] G.M. Troianiello, Elliptic Differential Equations and Obstacle Problems, Plenum Press, New York, 1987.

[33] H. Yamabe, On a deformation of Riemannian structures on compact manifolds, Osaka Journal of Mathematics 12 (1960) 21–37.

[34] Hong-Ming Yin, $L^{2,\mu}$-estimates for parabolic equations and applications, Journal of Partial Differential Equations 10 (1) (1997) 31–44.

[35] Hong-Ming Yin, On a reaction-diffusion system modeling infectious diseases without life-time immunity, European Journal of Applied Mathematics 33 (2021) 803–827.

[36] Hong-Ming Yin, Maxwell's Equations and Device Modeling, Lecture notes in Washington State University, 2005. To be published by SIAM Book Series "Studies in Applied and Numerical Mathematics".

Index

H

I

J

K

L

M

Q

R

S

T

U

V

W